Hydroclimatology
Perspectives and Applications

Hydroclimatology provides a systematic structure for analyzing how the climate system causes time and space variations (both global and local) in the hydrologic cycle. Changes in the relationship between the climate system and the hydrologic cycle underlie floods, drought, and possible future influences of global warming on water resources. Land-based data, satellite data, and computer models contribute to our understanding of the complex time and space variations of physical processes shared by the climate system and the hydrologic cycle.

Blending key information from the fields of climatology and hydrology – which are not often found in a single volume – this is an ideal textbook for students in atmospheric science, hydrology, Earth and environmental science, geography, and environmental engineering. It is also a useful reference for academic researchers in these fields.

MARLYN L. SHELTON is Professor Emeritus of Atmospheric Science at the Department of Land, Air and Water Resources, University of California, Davis, USA. He is a member of the American Water Resources Association, the Association of American Geographers, and the Association of Pacific Coast Geographers. He has been elected as a Woodrow Wilson Dissertation Fellow and a Fellow of the Ohio Academy of Science. He has authored and co-authored 4 previous books, published numerous scientific papers, and served as an associate editor for the *Journal of the American Water Resources Association* from 1991 to 1999. He has given interviews to newspapers and television stations on topics such as climate change, El Niño, floods, drought, and air quality, and has presented lectures and workshops for the US Forest Service and the Natural Resources Conservation Service.

Hydroclimatology

Perspectives and
Applications

MARLYN L. SHELTON
University of California, Davis, USA

CAMBRIDGE
UNIVERSITY PRESS

University Printing House, Cambridge CB2 8BS, United Kingdom

One Liberty Plaza, 20th Floor, New York, NY 10006, USA

477 Williamstown Road, Port Melbourne, VIC 3207, Australia

314-321, 3rd Floor, Plot 3, Splendor Forum, Jasola District Centre, New Delhi - 110025, India

79 Anson Road, #06-04/06, Singapore 079906

Cambridge University Press is part of the University of Cambridge.

It furthers the University's mission by disseminating knowledge in the pursuit of education, learning and research at the highest international levels of excellence.

www.cambridge.org
Information on this title: www.cambridge.org/9781108462099

First published 2009
First paperback edition 2018

A catalogue record for this publication is available from the British Library

Library of Congress Cataloging in Publication data
Shelton, Marlyn L.
Hydroclimatology : perspectives and applications / M. L. Shelton.
   p.   cm.
Includes bibliographical references.
ISBN 978-0-521-84888-6
1. Hydrologic cycle.   2. Climatology.   3. Hydrometeorology.   I. Title.
GB848.S54   2009
551.48–dc22

                                                    2008040772

ISBN 978-0-521-84888-6 Hardback
ISBN 978-1-108-46209-9 Paperback

Contents

Preface

Droughts, floods, heatwaves, and other extreme weather events often have disastrous consequences for society and for the infrastructure that provides our goods and services. An increasing global population with an increasing population occupying areas subject to extreme weather events has heightened awareness of the potential impact of climate and weather and extreme events on our daily lives. This new awareness is occurring at a time when a consensus in the scientific community supports the idea of climate change and that at least a part of the change in recent decades is due to human activity. Against this backdrop we have advances in satellite and computer technology that permit us to examine natural processes in ways that were not possible in the recent past. Hydroclimatology is an area that benefits from these advances as it endeavors to improve understanding of the linkages between the climate system and the hydrologic cycle.

A global view provides a sense of the immensity and complexity of the Earth's climate system and the hydrologic cycle. An important suite of climatic processes involves atmospheric moisture, atmospheric energy storage in the form of latent heat, and energy transport by the atmosphere. The heating and cooling of the atmosphere and atmospheric motion define a climatic perspective easily related to the atmospheric branch of the hydrologic cycle that is dominated by moisture transport accomplished by the mobile atmosphere. At regional and local scales, additional processes are introduced into the climatic and hydrologic cycle perspectives as land surface differences exert strong influences on the exchanges of energy and mass between the Earth's surface and the atmosphere. Climate-related fluxes at the Earth's surface are vertically oriented, and hydrologic processes are altered by the character of soil and vegetation. The perspectives of climate and the hydrologic cycle at the Earth's surface have separate sets of variables that complement atmospheric processes but require different observational data.

Hydroclimate incorporates the atmosphere, the oceans, and the land surface and how these realms are coupled by exchanges of energy, mass, and momentum. A comprehensive treatment of the physical processes involved in linking the atmosphere, oceans, and land surface is complicated by our incomplete understanding of many of the natural processes and their variable nature at different time and space scales. Since many earth science sub-disciplines are involved, choices must be made to keep the topic manageable. Consequently, an effort is made in this book to provide a sense of the complexity and interconnectedness of hydroclimatic processes without going into excessive detail in any one area.

This book is intended for students studying atmospheric science and/or climatology and those specializing in hydrology. Chapters 1 and 2 set the conceptual structure of hydroclimatology. Two climate paradigms are introduced to complement the recognized atmospheric and terrestrial branches of the hydrologic cycle and their links with the climate system. Measurement and estimation of hydroclimatic variables are addressed in Chapters 3, 4, and 5. Atmospheric data in Chapter 3 are familiar to atmospheric science students, while data measurements at the Earth's surface covered in Chapter 4 are familiar to hydrology students. Remote sensing in Chapter 5 focuses on satellite and radar data specifically relevant to hydroclimatic analysis. Chapter 6 addresses the runoff process and is intended to provide background in hydrology for atmospheric science students. Hydrology students will benefit most from the spatial and temporal variability of atmospheric phenomena and the interaction of surface and atmospheric events emphasized in Chapters 7 and 8. Floods and drought, Chapters 9 and 10, respectively, provide opportunities for all students to examine the circumstances surrounding the occurrence of extreme weather events and the role played by complex atmospheric circulation features and distant climatic circumstances that influence these events.

The goal of this book is to promote understanding of hydroclimatic diversity, the link between climate and the water resource, and the possible influence of climate change on the future hydroclimate and water resource. Recent hydroclimatic studies utilize contemporary data observation methodologies and improved estimation techniques to achieve expressions of relevant variables. Complex scientific questions arise in efforts to understand the relationships between the climate system and the hydrologic cycle. The impacts of natural climate variability and human-induced change contribute to the complexity. Floods, drought, desertification, agriculture and food production, municipal and industrial water supplies, and water quality are some of the areas requiring carefully formulated plans for sustaining future development. Floods and drought addressed in Chapters 9 and 10 illustrate the character of the complex problems. Faced with such challenges, hydroclimate provides a structure for systematic analysis of atmospheric, hydrologic, and biologic variables related to these areas of human concern.

The analytical perspective employed in this book is based on principles that portray hydroclimate as the relationship between flows or exchanges of energy and moisture between the atmosphere and the Earth's surface. The water balance provides the operational framework for characterizing hydroclimate, the spatial and temporal variations of hydroclimate, and hydroclimate resulting from altered future conditions. Real-world hydrologic events occur within the context of a history of climatic variations in magnitude and frequency. These events have the best chance of being understood when analyzed within the spatial framework of regional and global networks of changing atmospheric circulation patterns and land surfaces processes.

The late Professor Douglas B. Carter shared his vision of climate expression with me, and his vision became the foundation for the dual-climate paradigm developed in this book. The book concepts evolved from 10 years of teaching undergraduate and graduate students in ATM 115 and ATM 215 at the University of California, Davis. I profited greatly from student comments and from discussions with my colleagues at UC Davis for which I am grateful. I am indebted to David Jones who applied his professional skills in extracting data from many digital data archives and converting the data into attractive and informative global and regional maps. The competence, courtesy, and patience of Matt Lloyd and the other staff of the Cambridge University Press were invaluable in the preparation of this book.

This book is dedicated to my family. I am especially indebted to my wife, Sue, whose love, encouragement, understanding, and assistance were constant during the book's lengthy preparation. My son, Kirk, daughter-in-law, Rachel, and my grandchildren, Scott and Emma, are the promise that the continuing search for understanding of God's magnificent world is in good hands.

1

The realm of hydroclimatology

1.1 Water as a unifying concept

Water is an essential resource for humans and for natural ecosystems. Satellite images of the Earth show convincing evidence of an abundance of water on the planet. Unfortunately, only a small percentage of the total water volume is available as freshwater suitable for humans and many natural ecosystems. The relatively small volume of freshwater is further constrained by an uneven distribution over the globe that is paradoxical to the image of Earth as a water planet. Approximately one-third of the world's population lives in countries where the freshwater supply is less than the recommended per capita minimum, and 70 percent of all freshwater withdrawals from lakes, rivers, and groundwater is for crop irrigation to provide food (Entekhabi *et al.*, 1999). Such disparities in water supply and water demand require understanding the underlying physical processes that account for spatial and temporal differences in the occurrence and magnitude of the water supply.

The physical characteristics of water are significant in accounting for how the freshwater supply is sustained. A combination of natural processes collectively recognized as the hydrologic cycle provides the mechanism for the natural redistribution of water among the land, oceans, and atmosphere. Water is the only chemical compound that occurs in natural conditions as a solid, liquid, and gas. The transformation of water from one physical state, or phase, to another is a critical factor in the transportability of water. Water's phase changes and transportability in each of its phases are the foundations of the hydrologic cycle which constantly replenishes and redistributes the relatively small volume of freshwater.

Expanding knowledge of land–atmosphere interactions in the closing decades of the twentieth century heightened awareness of the strong coupling

1

between climate and land surface hydrological processes embracing the hydrologic cycle. The long-term interest in the total hydrologic cycle shared by the disciplines of hydrology and climatology was magnified as the twenty-first century began with the emergence of non-traditional datasets and new investigative techniques applied to an expanding array of water-related problems. Accelerated interest in the hydrologic cycle was especially evident in the field of hydroclimatology that overarches the disciplines of hydrology and climatology.

1.1.1   Hydrology

Modern hydrology is broadly defined as the science that studies the occurrence and movement of water on and under the Earth's surface, water's chemical and physical properties, water's relationship to biotic and abiotic environmental components, and human effects on water (Ward and Robinson, 2000). Hydrology has a predominant land surface orientation and emphasizes processes involved in the land phase of the hydrologic cycle. Hydrology employs the sciences of biology, chemistry, ecology, mathematics, and physics to focus on solving water resource and water management problems concerned with water use, water control, and water quality. A watershed is often the most convenient spatial unit for integrative and synthesizing studies of hydrologic problems, but spatial scale variations are ultimately determined by the nature of the problem being examined.

1.1.2   Climatology

Climatology is an applied science that examines the fluxes of energy, mass, and momentum among the land and ocean surfaces and the atmosphere. These fluxes are integral parts of the climate system modulated by both external and internal factors (Peixoto, 1995). The vertical and horizontal fluxes of energy and mass that are central to climatology are components of physical and dynamical phenomena operating at various scales as an integrated and interactive spatial system that links the land and ocean surfaces with atmospheric circulation. Collectively these components embrace a broad spectrum of thermodynamic and hydrodynamic processes that display identifiable seasonal variations. The atmospheric general circulation is an expression of these seasonal variations, and the general circulation determines the concurrent array of weather patterns (Bryson, 1997). Therefore, the atmospheric phase of the hydrologic cycle is a climate-related phenomenon and one that is expected to display seasonal variability. Climatology requires knowledge of chemistry, mathematics, and physics and a thorough understanding of the atmosphere's physical and chemical interactions with the ocean and land surfaces. Climatologists apply this knowledge to understand atmospheric circulation and dynamics and the atmospheric role in climate variability and climate change. Modern climatology has a strong relationship with computers because simulation is an important aspect of climate science through its serving as the platform for climate experimentation (Petersen, 2000). Climate system models strive to reveal global energy and circulation conditions, flood and drought recurrence, the influence of land surface changes on climate, and climate's role in a variety of social, economic, and environmental problems.

1.1.3   A broadened climate perspective

The traditional concept of climate as the mean atmospheric condition expressed as average temperature, precipitation, and other weather variables and possibly higher moment statistics of these variables for a specified period (Peixoto, 1995), such as a month, a season, or a year, has a limited role in contemporary hydroclimatology. The climate variables are the same as those variables relevant to meteorology, but they are applied at different time and space scales. Climate expressed as the average weather or the average state of the atmosphere is almost synonymous with "statistical meteorology", and it is easy to see why meteorologists have some propriety about the province of climatology (Bryson, 1997). The movement of the atmosphere is the dynamic of interest behind the quantitative distributions used to depict climatic fields, but climate variables owe their importance to forecasting purposes in meteorology. Averaging is what makes the focus climatic. Individual weather events are explained by the incursion of air masses and fronts and by the vertical arrangement of the atmosphere. Consequently, averages and normals make climate appear statistical and useful as a descriptive tool. Probability estimates of extreme events are a logical ancillary feature of averages, and probabilities are widely used in construction and engineering professions.

A definition of climate exclusively concerned with the atmosphere ignores the coupling that exists between the atmosphere and the land and ocean surfaces. Therefore, the traditional view of climate as a static or constant entity does not fulfill the requirements of a contemporary construct of the climate system. The modern climate system (Fig. 1.1) is depicted as five subsystems linked by exchanges of energy, mass, and momentum among the subsystems. The coupling of the subsystems results in a dynamic climate undergoing constant change. This is a more accurate depiction of the actual natural processes involved in the continuous redistribution of energy, moisture, and momentum accomplished by the atmosphere's close interaction with the land, oceans, vegetation, and snow and ice at the Earth's surface. Climate is a direct response

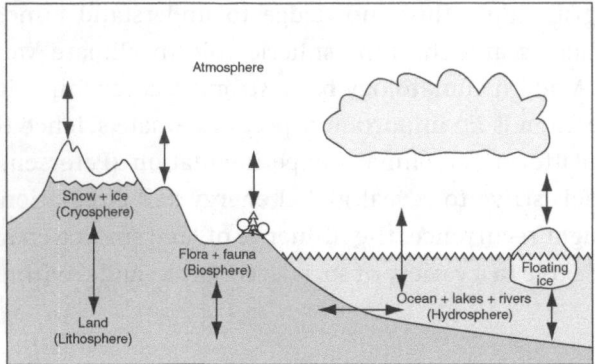

Fig. 1.1. The climate system. Arrows depict generalized fluxes of energy, mass, and momentum linking the five subsystems.

Fig. 1.2. GOES visible spectrum image of Earth on 19 April 2006. (Image courtesy of NOAA and the National Environmental Satellite, Data, and Information Service from their website at http://www.goes.noaa.gov/goesfull.html.)

to vertical and horizontal fluxes resulting from the coupling of the subsystems, and the atmosphere has an important transportation role. This dynamic climate perspective includes the more restrictive traditional concept of climate based on the mean atmospheric condition at the Earth's surface, but it emphasizes climate expressed as a physical system (Peixoto, 1995). Consequently, modern hydroclimatology is best viewed within the context of a thermodynamic and hydrodynamic system driving energy and moisture exchanges at the land and ocean surfaces while being influenced by these same energy and moisture fluxes. The resulting transport of energy and mass is evident in the structure and distribution of clouds revealed by satellite imagery (Fig. 1.2).

1.2   The global hydrologic cycle

The global hydrologic cycle is a logical unifying theme for hydroclimatology. For practical purposes, the global hydrologic cycle is a closed circulation for water's three phases. Within the structure of the general systems perspective commonly employed in the earth sciences, the hydrologic cycle is a subsystem and centerpiece of the global climate system. Consequently, the occurrence and movement of water assumes a primary role in both climatology and hydrology even though some illustrations of the hydrologic cycle lack comprehensive portrayal of the atmospheric transport role. Recognizing the full natural array of phenomena included in the hydrologic cycle is aided by conceptualizing the hydrologic cycle as consisting of two branches. The terrestrial branch encompasses continental processes and the atmospheric branch provides an energy and moisture redistribution mechanism (Fig. 1.3).

The terrestrial branch of the hydrologic cycle consists of the inflow, outflow, and storage of water in its various forms on and in the continents and in the

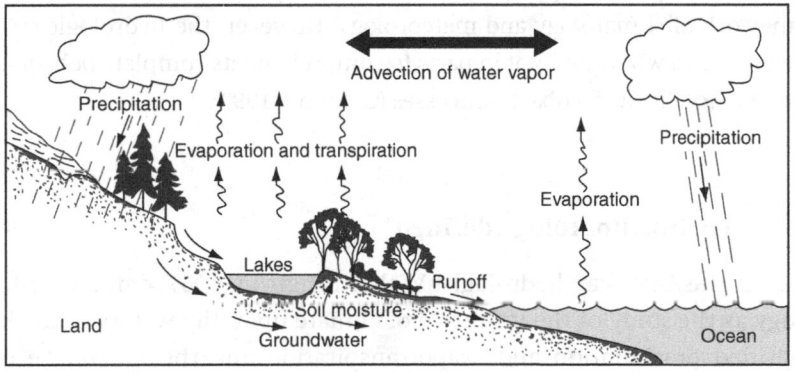

Fig. 1.3. The hydrologic cycle emphasizing both atmospheric and terrestrial branches.

oceans. The primary focus of the terrestrial branch is the natural processes at or near the land surface that ultimately produce surface and subsurface runoff and directly influence cycles of other materials that shape the Earth's surface (Stricker *et al.*, 1993). It is evident that the terrestrial branch is concerned with those processes commonly associated with hydrology.

The atmospheric branch of the hydrologic cycle consists of precipitation, evaporation, and the atmospheric transport of water mainly in the vapor phase. The two branches join at the interface between the atmosphere and the Earth's surface. The outflow of water from the Earth's surface through evaporation and transpiration is the inflow of water for the atmospheric branch. Precipitation, the atmospheric output, is a gain for the terrestrial branch of the hydrologic cycle. The atmosphere's mobility and ability to induce phase changes leading to precipitation establish the atmospheric branch as the forcing for the terrestrial branch of the hydrologic cycle. The dynamics of the hydrologic cycle are regulated by sources and sinks of atmospheric water vapor, by the thermodynamics of phase transitions, and by the dynamics of atmospheric general circulation (Peixoto, 1995).

The atmospheric branch of the hydrologic cycle is coupled with atmospheric general circulation and the transport of water vapor and the liquid and solid water in clouds. Recognizing the transport function of the hydrologic cycle is necessary to comprehensively portray how the hydrologic cycle is sustained. In addition to transporting moisture from the oceans to the continents, the atmosphere has an important role transporting energy vertically and horizontally as well as modulating the radiative forcing at the Earth's surface. The mobility of the atmosphere, and its capacity to force phase transitions of water establish the atmosphere as a forcing function for the terrestrial branch of the hydrologic cycle. The atmospheric branch of the hydrologic cycle is based on the dynamics of the general circulation of the atmosphere and is linked to the terrestrial branch by precipitation and evaporation (Peixoto, 1995). The role of vertical and horizontal transport of energy and mass and the delivery of precipitation places the atmospheric branch within the general framework of climatology and meteorology. However, the hydrologic cycle must be viewed as a whole and not in parts to comprehend its complete behavior and its complex non-linear feedback processes (Chahine, 1992).

1.3   Hydroclimatology defined

The American hydrologist Walter Langbein (1967) defined hydroclimatology as the study of the influence of climate upon the waters of the land. He identified precipitation and evapotranspiration and the imbalance of these climatic elements as the focus of hydroclimate. However, subsequent advances in understanding natural processes complemented by development of contemporary measurement techniques, data acquisition, and analytical tools suggest this perspective is too restrictive for modern science. Modern hydroclimatology requires a more holistic view that emphasizes a process orientation and a role in a variety of environmental systems ranging from water quantity and quality to stream habitats. Although the significance of the imbalance between precipitation and evapotranspiration remains valid, additional conceptual elements are needed to account for an expanding range of concerns.

The perspective adopted in this book is that hydroclimatology is an approach to studying moisture in its three phases in the atmosphere and on the Earth's surface. This realm is the intersection of climatology and hydrology, and it includes energy and moisture exchanges between the atmosphere and the Earth's surface and energy and moisture transport by the atmosphere. Emerging from this array of moisture fluxes and storages is a conceptual framework defining the occurrence of hydrologic events within their climatological context. The climatological context is a specific array or pattern of atmospheric pressure and circulation identifiable with a particular hydrologic event for a given location. For example, extreme events such as floods and drought tend to have well-defined atmospheric features related to the land surface event. This approach is consistent with the hydroclimatology perspective suggested by Kilmartin (1980) and Hirschboeck (1988). Such a robust concept of hydroclimatology includes hydrometeorology and knowledge of evaporation, runoff, interception, groundwater recharge and other surface or near-surface water relations (Mather, 1991). It does not attempt to define hydroclimatology as dealing with problems of the borderline between climatology and hydrology or the width of that borderline, which varies according to the view of the individual investigator. Perhaps most importantly, this approach to hydroclimatology does not attempt to set the breadth of the field but recognizes the opportunities for continual expansion with advances in understanding natural processes.

Hydroclimatology viewed as hydrologic events driven by climatically related energy and moisture fluxes and storages requires distinguishing between climatology and meteorology and hydroclimatology and hydrometeorology. Meteorology is the study of the weather or the day-to-day state of the atmosphere emphasizing variations in temperature, precipitation, pressure, wind, cloudiness, and humidity for a specific location. Meteorology employs physics and mathematics to explain short-term atmospheric motion and related phenomena. Hydrometeorology is the application of meteorology to problems involving the hydrologic cycle, the water balance, and the rainfall statistics of storms. In practice, hydrometeorology is concerned with measurement and analysis of precipitation data involving extrapolation of point data to spatial units, determination of rainfall probabilities, computing the frequency of intense storms, evaluating flood hazards, and design of local hydraulic structures. The boundaries of hydrometeorology are not distinct and the problems explored often overlap those of climatologists, hydrologists, cloud physicists, and weather forecasters.

One of the strengths of the hydroclimate concept is that it is robust. It is applicable to the study of a broad range of natural processes. It is equally useful for examining human modification of the climate system or the hydrologic cycle. Hydroclimate emphasizes study of the precipitation–evapotranspiration difference and the consequences of the imbalance. Precipitation and evapotranspiration are due to different meteorological, physical, and biological causes. For any given location they are not often the same in either amount or distribution through the year. The character of this imbalance is the basis for defining the hydroclimatic significance of an event or set of conditions. Hydroclimatic significance provides the structure for examining a series of important questions. What is the nature of the imbalance? What accounts for the imbalance? What are the ramifications of the imbalance in terms of how water is processed at the land–atmosphere interface?

1.4   Emergence of the hydrologic cycle

An early record of the importance of water for human life can be found in Genesis, the first book of the Bible. In this account of creation, light is provided on day one by the Sun, Moon, and stars. Separation of waters below the sky from waters above it occurs on day two, and day three begins with the separation of land and oceans.

Contemporary thought recognizes that energy from the Sun warms the Earth and water dominates the distribution of heat over the planet (Langenberg, 2002), and the related energy and moisture transfers constitute the global hydrologic cycle. Consequently, the light and water present at the beginning of time represent the ingredients needed for the hydrologic cycle.

The full context of contemporary hydroclimatology emerges from the historical pursuit of knowledge to understand the Earth's atmosphere and the hydrologic cycle. The early work was motivated by individual interests and curiosity, but over time the accumulation of information formed a coherent body of knowledge. False starts and imprecise ideas often related to mythology occurred, but these were identified and corrected or abandoned. An overview of the development of climatology and hydrology indicates the similarities and differences in how these two fields developed, the role of the hydrologic cycle in their development, and the ultimate emergence of hydroclimatology out of the two disciplines.

1.4.1   Speculation period

The rise of climatology as a science is closely related to developments in meteorology and to the human capacity to obtain more and improved atmospheric observations and measurements. The earliest evidence of human interest in the atmosphere was a concern for phenomena recognized in today's world as belonging to the field of meteorology. Climate is a more abstract concept than weather, and in these early days people did not travel extensively and were less likely to observe climatic differences between places (Linacre, 1992). However, interest in climate evolved as understanding of atmospheric processes improved, and a close coupling of climatology and meteorology characterizes much of their early history. Around 3000 BC, Mesopotamian astronomers and mathematicians studied clouds and thunder and were the first to identify winds according to the direction from which they blow. At about this same time, Egyptian astronomers and mathematicians recognized that the seasonal position of the Sun in the sky is a basic factor underlying climate differences. Climate was mentioned in the writings of the Xia dynasty in China (2100–1600 BC), and weather details were recorded in China as early as 1500 BC.

The earliest written record recognizing the global hydrologic cycle is attributed to the author of the book of Ecclesiastes around 1000 BC (Nace, 1974). Pre-800 BC texts in India may be the earliest indication of human understanding of the atmospheric branch of the hydrologic cycle (Ward and Robinson, 2000). However, the human necessity for water required numerous responses that predate writings about the hydrologic cycle. As early as 3000 BC, occupants of the Indus Valley in India constructed water supply, irrigation, and drainage systems, and Egyptians constructed a rock-fill dam between 2950 and 2750 BC. A variety of water facilities were constructed in Assyria, Babylonia, Israel, Greece, Rome, and China before the Christian era.

1.4.2   Greek and Roman era

Early Greek philosophers were interested in weather phenomena, the atmosphere, and the hydrologic cycle. Herodotus (440 BC) compared the climate of places, and Hippocrates in 400 BC wrote about weather and health and the dangers of drinking polluted water. Aristotle wrote a comprehensive meteorological treatise in 334 BC that served as the basis for weather theory for the next 2000 years. Erastosthenes described climate in terms of the Sun's position in the sky in 200 BC. In general, early Greek philosophers embraced the basic idea of the hydrologic cycle and proposed a variety of explanations for the origins of rivers and springs. Some proposals portrayed reasonable constructs, but underground mechanisms were imaginary.

The sustained influence of Aristotle's meteorological treatise was partially due to Roman philosophers devoting little interest in the atmosphere. Roman philosophers were more concerned with hydrology and benefited from practical knowledge gained from construction of great hydraulic works. They expanded on the explanations of rivers and springs proposed by the Greeks, and Vitruvius, a Roman architect and engineer, in 100 BC conceived that groundwater is derived from rain and snow infiltrating from the surface. Many consider this theory to be the forerunner of modern hydrologic cycle concepts. Although Ptolemy (AD 130), a Greek astronomer living in Alexandria, created a map that divided the known world of the second century into seven roughly determined climatic zones, Roman philosophers, at the fall of the Roman Empire in AD 476, had contributed little toward understanding the atmosphere.

1.4.3   Middle Ages

With a few exceptions, advances in understanding the atmosphere and the hydrologic cycle by Western scholars languished between AD 400 and 1500 during the period known as the Middle Ages. The attention these topics received during this period came largely from other world regions. In the late tenth century, the Persian scholar Karaji described the basic principles of hydrology (Pazwash and Mavrigian, 1981). Norse poems of the ninth to twelfth centuries contained descriptions of the hydrologic cycle indicating recognition of the roles of ocean evaporation, condensation, cloud formation, and precipitation on the land (Ward and Robinson, 2000). Except for the introduction of the wind vane in AD 850, few advances in understanding the atmosphere were achieved, but Islamic scholars during the ninth to twelfth centuries translated and expanded on the work of the Greeks and Romans. By the middle of the fifteenth century, extended sea voyages opened new trading areas and the broadened knowledge of ocean winds acquired as a result of these voyages contributed to formulation of theories regarding global wind patterns.

1.4.4   Observation period

Early Greek and Roman theories of the hydrologic cycle remained dominant until the sixteenth century when Leonardo da Vinci in Italy and Palissy in France used field measurements to assert that the water in rivers comes from precipitation (Biswas, 1970). The observation-based approach to the hydrologic cycle was advanced in the late seventeenth century when Perrault and Mariotte in France and Halley in England provided a quantitative basis for the basic principles of the hydrologic cycle by showing that precipitation supplied the water in rivers and streams and moisture circulated between the land, oceans, and atmosphere. Measurements enabled scientists to draw correct conclusions on the observed hydrologic phenomena and the mass balance concept was established.

Scientific analysis of the atmosphere progressed in the sixteenth and seventeenth centuries with the development of weather instruments that provided data from which laws applicable to the atmosphere could be derived. The hygrometer and anemometer were designed by da Vinci in 1500, Galileo invented the thermometer in 1593, Torricelli introduced the barometer in 1643, and Pascal demonstrated the reduction in atmospheric pressure with increasing elevation up a mountain in 1647. Boyle discovered the fundamental relationship between pressure and volume in a gas in 1662, and Townley introduced the first European rain gauge in England around 1676. Halley recognized the relationship between the general atmospheric circulation and the Sun's heat over the Earth, and he constructed a comprehensive map of global winds in 1683 and a map of the trade winds and monsoon circulations in 1686.

1.4.5   Modernization era

Weather instruments were improved and standardized in the eighteenth century, and network measurements of precipitation began before 1800 in Europe. Ideas essential to understanding the atmosphere slowly evolved through the eighteenth century as contributions increased from a growing number of scholars and intellectual activity shifted westward out of Greece and Italy. Notable advances in understanding the atmosphere during this period included Hadley's treatise on tropical circulation published in 1735, Black's introduction of latent heat in 1760, and Erasmus Darwin's explanation of cloud formation by adiabatic expansion in 1788.

In hydrology, the eighteenth century was marked by advances in mathematical applications to fluid mechanics and hydraulics by European scientists that formed the foundation for understanding hydraulic principles. This experimental work formed the foundation for the modern science of hydrology. LeClerc described the Earth's hydrologic cycle in 1744, and the use of "hydrology" in approximately its current meaning began in 1750. Improvements in methods to quantify streamflow during this period complemented advances in measuring temperature and precipitation.

Around 1800, John Dalton established the nature of evaporation and the present concepts of the global hydrologic cycle. Mulvaney in 1851 described the time of concentration concept that is the basis for the rational method of runoff computation. In 1856, Darcy developed the law of flow through porous media that represented one of the final obstacles to understanding groundwater flow and the hydrologic cycle. The last half of the nineteenth century saw the beginning of the association of hydrology and civil engineering as scientists examined the relations between rainfall amounts and streamflow rates and estimated flood flows in designing bridges and other structures. Saint-Venant derived the equations of one-dimensional surface water flow in 1871, and Manning developed an equation for open channel velocity in 1891. These nineteenth century activities led to significant advancements in understanding the basic relations between precipitation and evaporation over an area and the runoff from that area (Mather, 1991). The fundamental principles and concepts produced by this early work emphasize water on the land. This perspective is now recognized as the terrestrial branch of the hydrologic cycle (Stricker *et al.*, 1993).

In the nineteenth century, insights into the atmospheric branch of the hydrologic cycle were provided by improved understanding of the general circulation of the atmosphere. Redfield explained the circular nature of storms in 1831, and Gaspard Gustave de Coriolis formulated the "Coriolis force" in 1835. Ferrel proposed his three-cell model of atmospheric motion in 1856, and Coffin prepared world wind charts in 1875. The systematic use of balloons to monitor the free atmosphere began in 1892. Weather instrument networks expanded beyond Europe early in the nineteenth century. Instrument networks were well established in India by 1820, and the Smithsonian Institute began gathering climatological data nationwide in the United States in 1847.

1.4.6   Twentieth century and beyond

In the first half of the twentieth century, more details of the character of the atmosphere were discovered. Systematic data collection in the atmosphere using aircraft began in 1925, and radiosondes were successfully deployed in 1928. The nature of the jet stream was first investigated in 1940. Although climatology became more quantitative in the early twentieth century, there was a decline in interest as climatology was perceived as being primarily descriptive, merely statistical meteorology, or only concerned with climate classification (Linacre, 1992). In contrast, hydrology gained formal recognition as a science in the first half of the twentieth century. Early in the period hydrology was characterized by empiricism and qualitative description, but this was replaced by more theoretical and quantitative approaches by the mid twentieth century. The first watershed-scale measurement of land use change effects on streamflow in the United States was undertaken at Wagon Wheel Gap, Colorado, in 1911 using two small watersheds. Other significant advances were the work of Sherman (1932) who introduced the unit hydrograph, and Horton's (1940) ideas on infiltration, soil moisture accounting, runoff, and that hydrologic processes operate at a variety of scales. Recognition of water quality as an aspect of hydrology developed during this period along with the formation of government agencies and international organizations concerned with various aspects of hydrology.

In the second half of the twentieth century, climatology joined other sciences undergoing rapid changes in response to new technology, data sources, and analytical tools. An especially important development for climatology was the increased availability of atmospheric data from ground sources and satellites and the development of coupled large-scale atmospheric and biosphere models. The development of radar, rockets, and satellites by the mid twentieth century supported expanded exploration of the atmosphere. The introduction of high-speed computers in the 1950s supported data storage and analysis and the solution of mathematical equations describing atmospheric behavior. The first meteorological satellite was launched by the United States on 1 April 1960, and the first geostationary satellite was launched in 1966. These developments stimulated a rapid increase in understanding meteorological processes and in focusing climatology on the solution of practical problems. Archived satellite imagery and data became sufficient to support studies of a variety of current and future climate questions. Advances in computer technology made possible the processing of equations of motion, thermodynamics, and conservation of mass for multiple-level atmospheric general circulation models and coupled atmosphere–ocean–biosphere models. However, model improvements continue to be sought for better physical representations of clouds, soil moisture, snowcover, and sea ice (Burroughs, 2001).

Detailed field studies designed to understand routes followed by water in reaching the stream channel expanded rapidly in hydrology during the last half of the twentieth century. Often these field studies incorporated subjecting rational hydrologic principles to mathematical analysis to gain understanding of natural processes. The temporal and spatial variability of natural processes presents a great challenge to achieving understanding of how water arrives at the stream channel.

The emergence of international and interdisciplinary programs late in the twentieth century was significant in forging the contemporary image of hydro-climatology and developing a more sophisticated understanding of the hydro-logic cycle. The Global Atmospheric Research Program (GARP) was conceived in the 1960s with over 140 nations participating. The GARP Atlantic Tropical Experiment (GATE) initiated in 1974 focused on understanding the relationship between tropical cloud clusters and the atmospheric general circulation. The Global Energy and Water Cycle Experiment (GEWEX) initiated in 1988 by the World Climate Research Program strives to improve understanding and prediction of precipitation and evaporation processes at the global scale by promoting improved understanding of the hydrologic cycle's role in the climate system. GEWEX goals are pursued through a series of related programs including the GEWEX Continental-Scale International Project (GCIP), the GEWEX Americas

Prediction Project (GAPP), the Climate Variability and Predictability (CLIVAR) Study, the Climate and Cryosphere (CliC) Project, and the GEWEX Cloud System Study (GCSS). The goal of GCSS, organized in the early 1990s, is to improve cloud parameterizations as a specific component of the hydrologic cycle for climate and numerical weather prediction models. The Coordinated Enhanced Observing Period (CEOP) is designed to collect a comprehensive dataset from observation, satellite, and model sources covering all aspects of the hydrologic cycle (Randall *et al.*, 2003). The Global Precipitation Climatology Project (GPCP) is the GEWEX program devoted to producing community analysis of global precipitation (Adler *et al.*, 2003). These programs involve scientists and engineers from numerous disciplines representing several nations, and their goal is to develop understanding of the hydrologic cycle that can be transferred to other regions of the world. Major GEWEX accomplishments include the creation of multiyear global datasets of clouds, precipitation, water vapor, surface radiation, and aerosols.

1.5   Two climates for two hydrologic cycles

Contemporary hydroclimatology recognizes that the atmosphere has a central role in delivering moisture to a specific location. In addition, the atmosphere provides the climatic framework for the Earth's surface energy and moisture fluxes and balances that drive the hydrologic cycle. Understanding the role of these components is required for a comprehensive view of hydroclimatology that extends beyond the traditional context of climatology and beyond the boundaries of hydrometeorology. A dynamic atmosphere within the context of the climate system provides a transport function for energy and mass identified as the atmospheric branch of the hydrologic cycle. This atmospheric transport component serves as the driving force for the global hydrologic cycle and makes possible closure of the terrestrial branch of the hydrologic cycle. Consequently, it is convenient to identify two hydrologic cycles in terms of an atmospheric branch and a terrestrial branch. Obviously, both branches are necessary to form the global hydrologic cycle. Recognizing the coupled roles of the atmospheric branch and the terrestrial branch of the hydrologic cycle provides a rational construct for the total hydrologic cycle.

The differences between the atmospheric branch and the terrestrial branch of the hydrologic cycle suggest that climate expressed as fluxes of energy and mass may be different for these two entities. The atmospheric branch of the hydrologic cycle is most closely related to a dynamic and mobile atmosphere characterized by both vertical and horizontal energy and mass transfers. The terrestrial branch of the hydrologic cycle has an Earth surface orientation and energy and mass fluxes are dominantly vertical. The energy and mass fluxes and the transport role embraced by the atmospheric and terrestrial branches of the hydrologic cycle do not lend themselves readily to traditional climatology designations because they overarch approach and scale concepts commonly employed as criteria in climatological designations. Furthermore, just as there is a rationale for distinguishing between the atmospheric and terrestrial branches of the hydrologic cycle a rationale exists for distinguishing between the climate that is dominantly atmospheric and the climate that is dominantly associated with the Earth's surface. To facilitate ease of reference these two climatic constructs are identified as "climate of the first kind" and "climate of the second kind", respectively.

The atmospheric transport role is the basis for equating "climate of the first kind" with the atmospheric branch of the hydrologic cycle. Atmospheric moisture transport is a central component of the hydrologic cycle in the same way that latent heat provided by phase changes of water is a significant contributor to the atmospheric redistribution of energy and moisture. A "climate of the second kind" is aligned with the disposition of precipitation as outputs of evapotranspiration, streamflow, and storage of soil moisture and groundwater at the Earth's surface. This is the realm of the terrestrial branch of the hydrologic cycle. The energy and mass fluxes related to "climate of the first kind" and "climate of the second kind" drive other natural systems and reflect the intermingling of climate and hydrology which emphasizes the coupling of the two branches of the hydrologic cycle.

1.5.1   Climate of the first kind

The atmosphere is responsible for moving energy and moisture globally with regular schedules and patterns. Storm delivery occurs as a specific component of the dynamic atmosphere. The persistence of the atmospheric time and space patterns gives rise to an expected state frequently expressed in terms of averages or probabilities commonly equated with climate. These statistics reflect the popular interest in periods longer than those characteristic of weather events. The widespread use of the climatic normal is an expression of long-term conditions of temperature and precipitation related to the general circulation of the atmosphere. The climatic normal is the average for a recent 30-year period, for example 1971–2000. The period used in this computation is updated every decade as more recent data become available, but as Bryson (1997) asserts there are no true climatic normals. An atmospheric focus is the basis for climate of the first kind that is fundamentally a circulation-based paradigm.

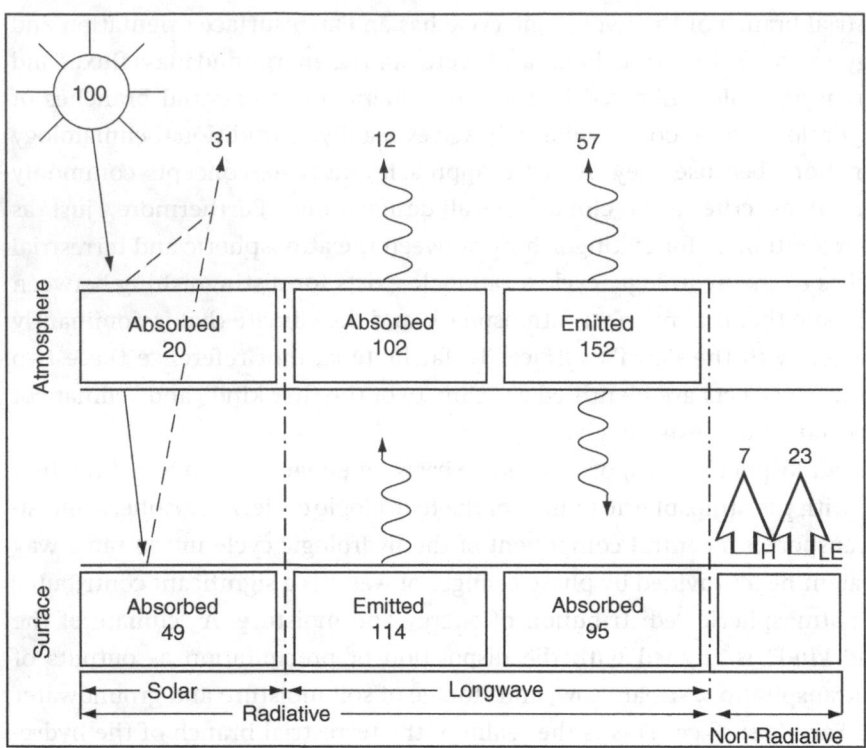

Fig. 1.4. The Earth and atmosphere radiation balance and related non-radiative energy flows. Units are percentage of global mean solar radiation (100 units = 342 W m^{-2}).

Climate in the atmosphere is characterized by the transmission, absorption, and reflection of solar and terrestrial radiation (Fig. 1.4). Solar energy is the main input, but clouds and atmospheric gases reflect more sunshine away to space than they absorb. Note the units in Figure 1.4 are relative quantities based on 100 units of solar radiation at the top of the atmosphere. Most of the absorbed input of solar radiation takes place near the bottom of the atmosphere. Terrestrial radiation is generated continually by all surfaces that receive radiant energy whereas insolation is periodic. The greenhouse effect of the atmosphere selectively absorbs nearly all terrestrial radiation and radiates a larger quantity of atmospheric radiation back to Earth. The total flow of terrestrial radiation actually exceeds the total for insolation because the terrestrial flux is sustained by both insolation and atmospheric longwave radiation. These processes are examined in more detail in Chapter 2.

The Earth and the atmosphere are heated unevenly in both space and time by the fluxes of insolation and longwave radiant energy. Tropical locations receive more radiation than they emit (Fig. 1.5), and the atmosphere becomes warm and

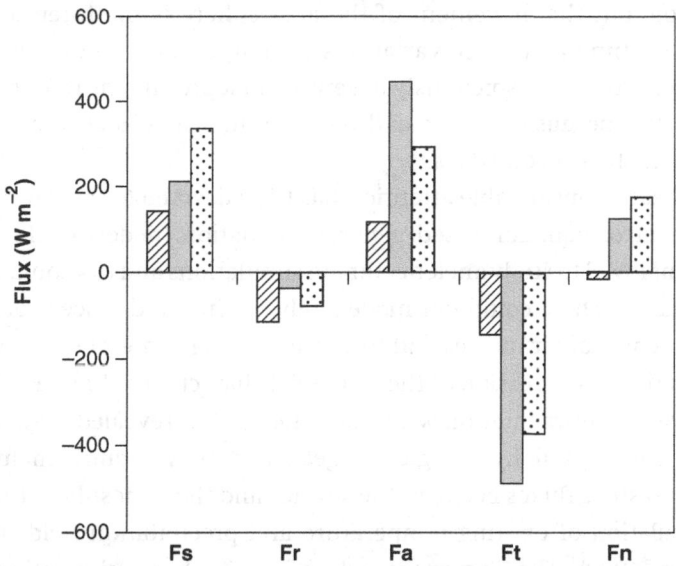

Fig. 1.5. Daily averaged radiation fluxes for the South Pole (light stippled column), for a northern Australia (11.5° S) tropical grassland (gray column), and for a mid-latitude (44° N) meadow (heavy stippled column). Fs is incoming solar radiation, Fr is reflected solar radiation (albedo), Fa is incoming atmospheric longwave radiation, Ft is outgoing terrestrial longwave radiation, and Fn is net radiation. (Compiled from data in Gay, 1979; Linacre and Geerts, 1997; and Beringer and Tapper, 2002.)

moist in response to the storage of heat and vapor fluxes from the surface. Polar latitudes receive smaller quantities of radiant energy than tropical locations, and they radiate less intensely. Nevertheless, these regions emit more radiant energy than they receive, and the atmosphere is relatively cold and dry. Furthermore, Figure 1.5 shows that the long hours of daylight during the high-sun season at mid-latitude locations can produce high solar radiation values and larger net radiation than at lower latitudes during specific periods. The global imbalance of energy and moisture in the atmosphere creates atmospheric pressure gradients that initiate global-scale atmospheric motion. The global atmospheric circulation triggered in this way provides the systematic framework for global air transport more extensive and persistent than features responding to local or regional influences. These topics are addressed in Chapter 7.

The fundamental control of climate of the first kind is atmospheric circulation with energy and moisture fluxes imposed within the circulation pattern. Other influences commonly cited, such as latitude, elevation, and distance from water, are associated features, or false controls, rather than causal factors. The associated features help to explain the fluxes of energy and moisture responsible for many of the global atmospheric pressure features that initiate air circulation, but the movement of the atmosphere is the foremost factor responsible for time and space variations in temperature and precipitation. Evapotranspiration is conspicuously absent from a prominent role in climate of the first kind because it is a small quantity unless it is accumulated for periods greater than several days.

In recent years, considerable attention has been devoted to the development and refinement of equation-based general circulation models (GCMs). These have been employed to study ancient climate simulations and to simulate future climate scenarios. These computer models rely on time and space averaging of data to an extent that air masses and fronts are averaged out of existence. This may be the ultimate expression of the statistical character of climate of the first kind. However, experimentation with these models has revealed they are quite sensitive to continental hydrologic budgets, and they require inclusion of energy and moisture fluxes between the surface and the atmosphere to achieve realistic simulation of existing temperature and precipitation fields (Szilagyi and Parlange, 1999). The important role of atmospheric circulation is not diminished, rather these models suggest a concept of climate is needed that extends beyond the atmosphere to incorporate the role of the Earth's surface processes and the atmospheric redistribution of heat and moisture transferred into the lower atmosphere from the surface.

1.5.2   Climate of the second kind

The interface between the atmosphere and the various material systems that lie beneath it is the focus for climate of the second kind. The climate of the interface is primarily forced by vertical endowments, storages, and return fluxes of energy and moisture. This distinctive perspective of climate at the interface has emerged largely from investigations into energy budget and hydrologic cycle questions addressing the coupling of climate and hydrology. The interface is conceptually a zone extending from a few meters below the Earth's surface and into the atmosphere to the top of the atmospheric boundary layer that has a typical depth of about one kilometer (Andrews, 2000). This zone is inhabited extensively by vegetation that extends upward into the atmosphere and downward into the soil creating a transition zone for fluxes of energy and mass between the land and the atmosphere. Strong vertical discontinuities of physical properties characterize this transition zone. These sharp contrasts are responsible for the strong vertical divergences in radiative fluxes of energy and the conversion of energy between radiation, sensible heat, and latent heat. As a consequence, there are large vertical gradients of sensible heat and latent heat and considerable differences in phase changes of water at the interface. Also, a strong vertical flux of water exists. Consequently, climate of the second kind includes surface, plant, soil, and gravity water. Groundwater in dead storage is outside the conceptual boundaries of the interface.

Climate of the second kind provides a means of estimating the outputs of evapotranspiration and streamflow and storages of soil moisture and groundwater. Quantification of these fluxes and storages is a fundamental requisite for developing budgets of inputs, storages, and outputs for consecutive periods of days, weeks, months, or seasons. The budgets deal mainly with exchanges in the vertical across the interface, and they employ accounting procedures based on the principles of the conservation of energy and mass. The climatic activity at the interface is principally vertical, and there is little that is atmospheric about climate of the second kind. Energy received at the interface as radiant energy is greatly affected by the atmosphere, but a radiation budget would exist in the absence of the atmosphere. The Moon is a convenient illustration of this reality.

Hydroclimatology is a dynamic budget of water fluxes vertically together with the energy endowment that drives evapotranspiration or the return flow of moisture to the atmosphere. Precipitation and evapotranspiration are the main contestants for dominance of the flow of water at the Earth's surface. The terrestrial branch of the hydrologic cycle is a manifestation of precipitation gains versus evapotranspiration losses at the surface. While evapotranspiration is tiny on a daily basis and is difficult to measure, it accumulates with time to attain an important stature in problems that deal with a period of a week, month, season, or year.

Water resources are seasonally replenished and discharged by climate of the second kind. Even the behavior of lakes and oceans is dependent upon this conception of climate's seasonal timing. The ecology of natural vegetation, agriculture, and forestry is timed by both the availability of soil moisture and the energy to extract it for plant processes as predominantly evapotranspiration. Plants keep only a tiny portion of the moisture extracted by roots. The great bulk of the moisture is passed into the atmosphere. Irrigation of agricultural crops is an attempt to overcome the water deficiency resulting from climate of the second kind.

Controls for climate of the second kind are the forcing functions that drive natural systems at the Earth–atmosphere interface. Fundamentally, these forcing functions are energy and moisture availability and the coincidence of the energy demand for moisture and the moisture supply. Moisture storages in the soil and as groundwater are important related processes that can be considered as secondary controls. While these storages are influenced by physical characteristics of the soil and strata, they respond to the supply and demand of energy and moisture at a specific site.

Evapotranspiration ascends to a prominent role in climate of the second kind and is the primary control for this climate concept. Evapotranspiration is driven by radiant energy, but the presence of moisture is required for transformation of

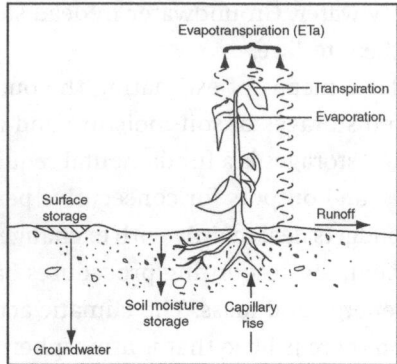

Fig. 1.6. Sketch of precipitation partitioning at the Earth's surface and plant utilization of soil water.

the radiant energy endowment at the Earth–atmosphere interface. For short intervals, evapotranspiration is a small quantity and is difficult to measure convincingly. As an alternative, numerous empirical techniques are available to estimate evapotranspiration or the climatic demand for water. Recognizing that the climatic moisture demand may exceed available moisture, the idea of potential evapotranspiration (ETp) was introduced almost simultaneously by Thornthwaite *et al.* (1945) and Penman (1948) to identify the maximum possible moisture loss limited only by the energy endowment. Numerous methodologies have appeared subsequently to quantify energy-driven evapotranspiration. In contrast, actual evapotranspiration (ETa) is the transfer of moisture from the surface to the atmosphere in response to both the energy demand and the available moisture supply.

While ETa is an upward directed flow of moisture away from the Earth's surface, the source of this moisture is precipitation that is a downward vector toward the surface in the hydrologic cycle. Precipitation is capricious and spatially variable, but the energy demand for moisture is more regular in both space and time. Moisture storage in the upper soil layers plays an important role in supplying moisture between precipitation events (Fig. 1.6). Soil moisture storage occurs from the Earth's surface downward for a few meters in the zone where water and air share the void spaces among solid soil particles. The moisture balance of the soil layer and the average soil moisture content are critical to land area climate (Hartmann, 1994). The soil layer and associated vegetation determine whether precipitation is quickly evaporated, absorbed by the soil, or becomes streamflow.

Climate of the second kind is clearly a manifestation of the hydrologic cycle and the important role of climate as the forcing function underlying the fluxes and storages of moisture expressed as precipitation, evapotranspiration, soil moisture, groundwater, and stream discharge. With the development of expanded data networks and satellite technology for remote data assessment, the water balance provides an analytical framework for examining the spatial and temporal characteristics of wetting and drying periods, and annual variations in the water resources (Hartmann, 1994).

1.6   Hydroclimatic data

Present hydroclimate observations and archived data in the United States use the English system of units. Almost all other nations and the scientific literature use Système Internationale (SI) units for these variables. Unit conversions from one system to another are widely available and are not presented here. Only SI units are used in this book.

Symbols for representing hydroclimatic data are another issue. The quantities to be represented far exceed the capacity of the Greek and Roman alphabets. Also, some symbols commonly used in climatology are used for a different purpose in hydrology. One example is the Greek letter delta (Δ). The symbol Δ in climatology represents the slope of the saturation vapor pressure curve at the mean wet-bulb temperature of the air. In hydrology, the symbol Δ is used to indicate "a change in" a quantity. The symbols used in this book are based on common usage in both disciplines. Consequently, there are a few instances where a symbol has more than one meaning. However, the context in which the symbol appears clearly identifies the intended nature of the symbol. This approach is preferred to introducing a symbol set unique to this book.

Climate data are viewed traditionally as point data. Instruments for collecting climate data are designed to measure precipitation, temperature, radiation, wind speed, evaporation, pressure, and humidity for a specific location. The point sensors are located on the ground, a tower, or attached to balloons and are intended to record data for a given location. Streamflow is a volume measurement integrating processes operating over an area. Consequently, observed climatic and hydrologic data are not immediately comparable due to the contrasting nature of point and areal data. Computational methods available for estimating areal data from point data and remote sensing technology provide improved spatial data that alleviate point observational biases in estimating spatial variability.

1.7   Data quality

No meteorological instrument precisely follows the quantity it is measuring. Routinely deployed meteorological instruments must be robust to withstand exposure to all weather events, and the necessary robustness is achieved at the expense of instrument sensitivity (Linacre, 1992). It is common to identify first- and second-order instrument responses which refer to the differential equations used to approximate how the instrument responds to the variable it is designed to sense. The World Meteorological Organization (WMO, 1996) describes instrument measurement procedures for meteorological variables.

The utility of hydroclimatic data is a function of several factors related to data collection and recording. The primary consideration is the nature of the observation provided by an instrument and its influence on data reliability. An instrument properly installed, carefully maintained, and conscientiously observed is expected to provide reliable data. Unfortunately, errors are likely to occur even when instruments are observed with the greatest care, and inconsistencies are introduced by changes in observation and reporting times. Precipitation data can be particularly misleading due to the tendency to continue the use of a station name although the equipment is relocated. An especially subtle problem occurs when the site remains the same but the measuring device is changed, generally under the guise of modernization. Changes in the environment surrounding the observation station influence data reliability by altering the exposure of instruments. Other sources of data contamination are variations in the height of measurements or ground cover (Changnon and Kunkel, 2006). Stream gauge data suffer many similar types of reliability problems, but the most significant is streamflow regulation and/or land-use changes that alter runoff.

Archived data are commonly accepted as reliable information without regard for hidden inconsistencies. However, archiving is not a guarantee that the data are reliable. Electronic data transfer from observation sites directly to an archive has reduced data errors, but errors are still present in stored records. Use of archived hydroclimatic records requires awareness of quality control protocols.

A substantial difficulty in all hydroclimatic analysis is the problem of insufficient and incomplete data that hinder proper evaluation of the information content of basic data. Insufficient data is often a time-series issue in that the length of record is inadequate to reveal the range of important hydroclimatic behaviors. Spatial data are insufficient when their resolution is unable to portray representative variability. Incomplete data result from an inconsistency in spatial or temporal data collection.

Insufficient data present a challenge more difficult to overcome than incomplete data. Insufficient temporal data may be valid, but are short-term. To be of greatest hydroclimatic utility, long-duration data are desirable. However, the discussion of scientific instruments in Sections 1.4.4 and 1.4.5 highlights the irregular spatial and temporal patterns of instrument development and dispersal.

Hydroclimatic records in many regions extend back only a few decades, but hydroclimate variations on time scales of decades to centuries are particularly important. Some European locations have records covering 300 years or more, but there are few records for pre-1850 outside Europe and Eastern North America (Parker *et al.*, 2000). The instrument record of climate data for North America contains few stations established before 1800 and the western United States has a limited number of stations with reliable records prior to 1900 (Jones and Bradley, 1995). Concern for data sufficiency increases as the climate becomes more arid due to greater precipitation variability and the scattered location of stations. Arid zone hydroclimatology requires careful analysis on a case-by-case basis because few generalizations are valid (Stockton and Meko, 1990). When the written and instrumental data are of short duration, statistical and analytical techniques can be employed to extend the record to supplement direct measurements as a means of achieving comprehensive portrayal of hydroclimate.

Incomplete data issues arise when, with very few exceptions, the available hydroclimatic information has not been consistently measured in space or time. Values may be missing due to an instrument malfunction or by a failure of the observer to record the value. Various techniques are employed to estimate missing time-series data. The specific technique depends on the nature of the hydroclimatic element and the temporal nature of the missing value. Linacre (1992) summarizes commonly employed methods for estimating missing hydroclimatic data and the errors of estimation associated with these procedures. With careful attention to the specific requirements of missing data estimation, the incomplete data problem can be overcome and the time-series can be made complete.

Review questions

1.1 What role did the hydrologic cycle play in the historical development of the disciplines of climatology and hydrology?

1.2 What are the components of the atmospheric branch of the hydrologic cycle?

1.3 What are the components of the terrestrial branch of the hydrologic cycle?

1.4 How are the concepts of "climate of the first kind" and "climate of the second kind" related to the hydrologic cycle?

1.5 What is meant by the climatological context?

2

The climate system and the hydrologic cycle

2.1   Climate and water

Climate variability directly impacts the water resource and has stimulated expanding interest in the scientific study of the global hydrologic cycle. Water is a fundamental aspect of most people's lives in humid regions where its supply is taken for granted, but water is highly valued in supply-limited arid regions. The prospect of climate change and an altered water supply imposes a broadened dimension for everyone to understand the relationships among climate, the hydrologic cycle, and the water resource. In addition, water's role in geochemical and biological processes involving climate and landscape change is receiving increased attention. Although the hydrologic cycle displays large annual, seasonal, and regional variations, large-scale modifications of the Earth's surface due to human activity affect the water balance of the continents and of the atmosphere and oceans. Numerical simulations suggest that water vapor in the atmosphere significantly amplifies enhancement of the greenhouse effect with ramifications for altering the climate system (Burroughs, 2001). Against this backdrop, understanding the relationship between the climate system and the hydrologic cycle assumes a high priority.

The conceptual development of hydroclimatology recognizes that climate is the driving force for the hydrologic cycle. Interactions of precipitation, evapotranspiration, interception, percolation, soil moisture storage, groundwater recharge, other surface and near-surface water relations, and streamflow are all hydroclimatology components. Consequently, hydroclimatology provides a perspective for analyzing a broad range of water resource problems involving current and future states of the climatic and hydrologic systems. The precise manner for viewing these fluxes is influenced by the size of the areal unit considered and this emphasizes the necessity for considering the concept of scale.

2.2   Scale considerations

Land–atmosphere interactions cannot escape the difficulties arising from the many different space and time scales that are either inherent in the hydroclimatic system or imposed by methods of examination. Spatial scales result from a variety of factors. The physical characteristics of the interface components, such as vegetation height or topographic features, are an evident basis for defining a fixed spatial scale. Spatial scales are also defined by external forcing processes represented by precipitation cells or cloud-affected radiation fields and by dynamic processes that create boundary layers and other features of various dimensions. The measurement process and even computational factors can establish spatial scales (Clark, 1985).

Interactions among climate, biosphere, and hydrosphere are scale dependent owing to the combination of non-linear dynamics with spatial and temporal variability that arises from threshold conditions influencing some fluxes (Dooge, 1995). Scale is important because understanding scale interactions allows the application of information known for one scale in the analysis of processes to be used at another scale (Wood, 1995). This is a necessary step when the emphasis is on process and physical understanding. This level of understanding is required if we are to move between scales with ease. Recognizing that the hydrologic cycle has two major branches is important in moving between scales.

Water is one of the crucial links between the various components of the climate system. With the close link between climate and the hydrologic cycle, much of the spatial and temporal variability in the hydrologic cycle is introduced through the atmospheric branch that is a shared element of the climate system and the hydrologic cycle. At the global scale, this influence is especially evident in the rising and sinking motions associated with the tropical Hadley cell circulation. It is also evident in the relative frequency and seasonality of mid-latitude cyclonic storms. It is perhaps less obvious how an approach through climatology will assist in the solution of water problems, but a key element in the solution of the water balance is scale. At the local spatial scale and instantaneous time scale, many physical characteristics and fluxes are considered as constants and treated deterministically. At the watershed scale, variability is substantial and characteristics and fluxes are treated as probability-distributed (Niemann and Eltahir, 2004). Since the underlying premise of hydroclimatology is that climate drives the hydrologic cycle, climate as the forcing provides precisely the foundation required to begin an assessment of water issues at scales ranging from a single plant or a small plot to a watershed or a region. The details of how these applications are achieved are addressed in the following chapters.

2.3   Dynamic climate

Global climate has changed through time and continues to change and to display variability (Burroughs, 2001; Lockwood, 2001). Both climate change and variability can be a response to internal or external forcing of the climate system. Passive forcings involve modulation of faster responding components by slower response time components. Active forcings result from variations and instabilities of the climate system dynamics and by coupled interactions between climate system components. Passive forcings are known as stochastic forcings due to their random evolution while active forcings are called dynamic forcings because a strong dynamical response is required in the slower component (Bigg *et al.*, 2003). It is now becoming evident that one component of climate change and variability is related to natural processes and another component is attributed to human activities (IPCC, 2001). Aside from concerns for the causal factors, the important point is that an abundance of evidence shows that climate is dynamic and not static. This has significant applied ramifications because it warns us that past experience provides a limited view of potential future conditions.

Climate change and variability in both space and time have significant practical importance related to climate forcing of the hydrologic cycle. Energy and moisture are the active factors of climate whose coincidence is managed through the hydrologic cycle. The partitioning of precipitation among competing environmentally determined options within the hydrologic cycle includes runoff. Runoff is the water that appears in rivers, lakes, and streams and along with groundwater represents the water supply available for human use. Recurring droughts and declining water supplies are powerful reminders of human dependence on climate as the source of an essential natural resource.

While drought emphasizes a short-term climate variation, the potential for long-term climate change represents an additional need for comprehensive knowledge of the climate system and hydroclimatology. Many scientists expect that human alteration of atmospheric trace gases will produce global warming by the middle of the twenty-first century that will cause climates to be different from those of the present (Burroughs, 2001; IPCC, 2001). The present climate serves as a benchmark for identifying and assessing the magnitude of climate change. State-of-the-art general circulation models do not provide precise indications of regional climate change expected to accompany global warming, and this contributes additional confusion to the already complex issue of understanding hydroclimatology. The large-scale averaging in GCMs has the effect of smoothing regional-scale processes of greatest concern for understanding hydroclimatic problems. However, several techniques are available for subgrid-scale parameterizations that permit the GCM areal averages to be downscaled (Shelton, 2001). Improvements in representing land-surface hydrological processes in coupled atmosphere–land–ocean GCMs provide another avenue for achieving better representation of regional hydroclimatic processes.

Intuitively, we expect a warmer Earth to be wetter as a result of accelerated evaporation serving as a forcing to increase precipitation within the hydrologic cycle. However, regional climate differences can be expected to produce broad geographical variations in the temperature and precipitation response. An additional complication suggested by model simulations is that the relationship between warmer climates and the intensity of the hydrologic cycle may result from the sensitivity of sea surface temperatures to a warmer climate. Consequently, warmer climates can be associated with either an increase or a decrease in the intensity of the hydrologic cycle (Yang et al., 2003), and there may be latitudinal and seasonal differences related to vegetation's role in recycling water (Dirmeyer and Brubaker, 2006). The present climate must be understood before the ramifications of climate change can be fully appreciated.

2.4   The climate system

The climate system provides a paradigm that combines climate of the first kind and climate of the second kind in a single conceptual structure. The atmosphere and the surface media are equal components linked by exchanges of energy and moisture. In this way, the climate system is portrayed as the atmosphere, hydrosphere, lithosphere, cryosphere, and biosphere and the related flows of energy and moisture between the spheres (see Fig. 1.1). The climate system is regarded as continuously evolving with parts of the system leading and other parts lagging in time. The highly non-linear interactions between the subsystems tend to occur on many time and space scales. The subsystems are not always in equilibrium with each other and may not be in internal equilibrium (Lockwood, 2001).

2.4.1   Subsystems

The atmosphere is a mixture of different gases and aerosols, but small concentrations of radiatively active gases play a major role in determining the amount of energy transmitted by the atmosphere and stored in the atmosphere. Larger concentrations of non-radiatively active gases are important contributors to atmospheric pressure and in initiating atmospheric motion. Atmospheric motion is basically a problem of convection under the influence of rotation. What emerges when we disregard the irregular details of the flow is a pronounced tendency for atmospheric motion to be organized on a global scale into a scheme recognized as global atmospheric circulation. An imposed change on the atmosphere results in a much shorter response time than that of any other climate system component. Response or relaxation time is the time it takes for a system to re-equilibrate to a new state after a small perturbation has been applied to its boundary conditions or forcing. The atmospheric response time is on the order of days to weeks and is due to its relatively large compressibility and low specific heat and density which make the atmosphere fluid and unstable (Peixoto and Oort, 1992).

Oceans, seas, rivers, lakes, and subsurface water constitute the hydrosphere. However, over one-half of the solar radiation reaching the Earth's surface is absorbed by the oceans and this gives them a dominant climate role except in local settings near other surface water features. The oceans provide the atmospheric surface temperature boundary condition for over 70% of the globe, and they provide 85% of the water vapor flux into the atmosphere, which provides energy to the atmosphere (Bigg *et al.*, 2003). The oceans display a much slower circulation than the atmosphere and greater thermal inertia than the lithosphere. The total mass of the oceans is about 280 times that of the atmosphere, but the heat capacity of the oceans is nearly 1200 times larger. The greater density of water than air leads to greater mechanical inertia and slower oceanic circulation, but the ocean circulation is effective in redistributing heat and freshwater globally. Higher specific heat of water than land accounts for relatively smaller surface temperature changes of water compared to land. The response time of oceans varies within a range that extends from weeks to months in the upper mixed layer, which has a thickness of about 100 m, to centuries or millennia in the deep ocean. The dynamic nature of the ocean surface properties allows great scope for feedbacks between the ocean and atmosphere involving energy, chemical, and gaseous components (Bigg *et al.*, 2003).

The cryosphere includes ice sheets, glaciers, and sea ice that represent the largest freshwater reservoir on Earth, but the freshwater is largely held in long-term storage. The cryosphere's importance to the climate system is mainly due to its high reflectivity for solar radiation and its low thermal conductivity, even though seasonal changes in continental snowcover and sea ice produce large variations in intra-annual energy balances. In addition, the cold surfaces of the cryosphere have a stabilizing influence on the atmospheric general circulation because the cold surfaces at the poles tend to be related to high-pressure regions. Extended ice fields in Antarctica and Greenland and other continental and alpine glaciers have slow response times often measured in hundreds or thousands of years (Peixoto and Oort, 1992).

The lithosphere is the rigid outer layer of the Earth that includes the crust and the uppermost part of the upper mantle. The crust has continental pieces that form land surfaces and topography and ocean basin pieces that form marine topography. The continental surfaces are the most significant from the perspective of the climate system, and the lithosphere is regarded as an almost permanent feature of the climate system. The exception is a shallow upper active layer where a small heat capacity means that land plays a small role in heat storage. Nevertheless, there is strong interaction between the lithosphere and the atmosphere through the transfer of mass, angular momentum, sensible heat, and kinetic heat, and soil moisture has a strong influence on albedo, evaporation, soil thermal conductivity, and local variations in the surface energy balance (Peixoto and Oort, 1992).

Terrestrial and marine flora and fauna occupy the biosphere. The presence or absence of terrestrial vegetation or structural or functional vegetation changes can influence surface albedo, evaporation, soil moisture, runoff, and the carbon dioxide balance in the atmosphere, oceans, and on the land. Human interaction with the climate system through agriculture and urbanization are additional factors (Peixoto and Oort, 1992). Marine biota have an especially important role in the ocean biological pump and the rate of carbon dioxide cycling, and changes in the production of dimethyl sulfide which contributes to sulfate aerosol production in the atmosphere.

2.4.2   Water in the climate system

Water is an active ingredient in the global climate system and vast quantities of water are continuously on the move in the climate system. The physical properties of water ensure that it plays a major role. Its high thermal capacity provides a mechanism for moderating mid-latitude winter temperatures. The variation of saturation vapor pressure with temperature is the factor that causes oceanic surface temperatures at low latitudes to be limited by evaporation to values near 29 °C, thereby limiting tropical marine air temperatures to about the same value. The substantial amount of energy involved in phase changes governs the passage of solar energy through the atmosphere. The infrared radiative characteristics of water vapor permit it to act as the principal agent of energy loss from the atmosphere through infrared radiation to space. Through its role in energy exchange processes and related dynamical processes, water is a major contributor to governing the temperature of the free atmosphere.

Water continually moves between the subsystems. This movement provides both moisture relocation and vertical and horizontal energy transfers. However, the quantity of water is finite for practical purposes. It changes state or phase and moves from one storage or subsystem to another. The residence time in each of the storages is vastly different. Average moisture residence in the atmosphere is three days, but the cryosphere may store moisture for tens to hundreds of thousands of years. An average water molecule must wait a very long time in the ocean, in an ice sheet, or in a deep aquifer between brief excursions into the atmosphere (Hartmann, 1994).

The atmospheric emphasis of climate of the first kind is evident in the climate system structure. The atmospheric envelope of gases is the most variable and most rapidly responding component of the climate system. Climate of the second kind embraces the combined influences of the other four spheres.

2.5   The atmospheric subsystem

The atmosphere is a thermo-hydrodynamical system characterized by three attributes that are key determinants of the Earth's climate. Atmospheric composition in terms of the presence and volume of separate gases is a critical beginning point for understanding the climate system. The atmosphere's thermodynamic state as specified by pressure, temperature, and specific humidity is an extension of atmospheric composition. The atmosphere's three-dimensional motion field results from the combined influences of atmospheric composition and its thermodynamic state. Each of these characteristics is important in defining the Earth's climate, but atmospheric composition is the fundamental building block for the other characteristics.

Earth's atmosphere is a relatively thin gaseous envelope distributed almost uniformly over the surface. The dry atmosphere is composed mostly of molecular nitrogen and oxygen and trace amounts of numerous other gases. Listings of the gases and their total mass are available in introductory level meteorology and climatology books. In the vertical direction, 50% of the mass of the atmosphere is found below 5.5 km and more than 99% is found below an altitude of 30 km. Up to the mesopause at an altitude of about 78 km, atmospheric composition is practically uniform for concentrations of nitrogen, oxygen, other inert gases, and carbon dioxide. In contrast, water vapor is concentrated predominantly in the lower troposphere, and ozone is concentrated in the middle stratosphere.

Atmospheric composition is a primary determinant of the Earth's climatic response to radiant energy. Molecular nitrogen and oxygen, the most abundant atmospheric gases, are not radiatively active due to their diatomic structure and the absence of dipole movement even when vibrating (Hartmann, 1994). In contrast, the atmospheric gases important for the absorption and emission of radiant energy make up less than 1% of the atmospheric mass. The radiatively active trace gases include water vapor, carbon dioxide, ozone, methane, and nitrous oxide. However, the trace gases display a selective reaction to radiant energy attributable to the configuration of the gas molecules and the frequencies of the radiant energy photons (Hartmann, 1994). The atmospheric response to solar and terrestrial radiation is discussed in Sections 2.8 and 2.9.

2.6   Feedbacks

Exchanges of energy and moisture among the atmosphere and the other components of the climate system form a complex network linking the five spheres. These internal linkages represent ways in which the climate system components interact with each other, and they aid our understanding of how a perturbation or climatic forcing in one part of the system may produce effects in other parts of the system (Burroughs, 2001). The changes or responses in the subsystems made possible by the internal linkages represent processes known as feedback mechanisms. Feedbacks link the subsystems into a dynamic, self-regulating system. Feedback mechanisms act as internal controls of the system and result from a coupling or mutual adjustment among two or more subsystems (Peixoto and Oort, 1992). As the effect is transferred from one subsystem to another it is modified in character or in scale. The strength of a feedback can be quantified. Positive feedbacks amplify the magnitude of the response or act as a secondary forcing mechanism in the same direction as the initial forcing factor. Negative feedbacks dampen or reduce exchanges between or among subsystems by acting as a secondary forcing mechanism in the opposite direction to that of the initial forcing factor. The climatic forcings that trigger feedback processes are categorized as external and internal perturbations to the climate system. External forcings are natural events, and internal perturbations are attributed to natural causes and human-related actions.

Understanding the feedback process is aided by a feedback loop model commonly used in electrical engineering. The electrical feedback loop is based on the idea that a portion of the output signal of a system is used to modify the input signal. The relationship is expressed as

$$V_2 = \frac{G}{1-f} V_s \qquad\qquad (2.1)$$

where V_2 is the final signal output, G is the signal gain or amplification of the system, f is the system feedback, and V_s is the initial input signal. For a single feedback case, negative feedback is related to $f < 0$ and positive feedback results from $0 < f < 1$ (Peixoto and Oort, 1992).

Applying feedback concepts to the climate system requires expressing the terms in Equation 2.1 as climate variables. To accomplish this, regard ΔT_{final} as the signal output V_2 and $\Delta T_{forcing}$ as the signal input V_s. The term $G/(1 - f)$ is the gain or amplification with feedback and is basically a transfer function represented as G_F (Peixoto and Oort, 1992). The climatic feedback relationship is then expressed as

$$\Delta T_{final} = G_F(\Delta T_{forcing}) \qquad\qquad (2.2)$$

where ΔT_{final} is the overall change in temperature between the initial and final equilibrium states, $\Delta T_{\text{forcing}}$ is the temperature change due to the perturbation, and G_F expresses the feedback process. With only one feedback mechanism operating, solution of Equation 2.2 is simple when the magnitudes of G_F and $\Delta T_{\text{forcing}}$ are known. However, multiple feedbacks introduce significant complexity because feedbacks are neither additive nor multiplicative and may be either positive or negative. Moreover, feedbacks do not occur uniformly across the globe, and global averages may poorly represent local or regional conditions.

Numerous feedback processes exist within the climate system and multiple feedbacks impose complexity that is often poorly understood. Some of the widely discussed feedbacks are water vapor–greenhouse, cloud–radiation, ocean–circulation, snow/ice–albedo, vegetation–albedo, and temperature–thermal emission. However, many are complex processes that contain both negative and positive feedback elements. The cloud–radiation and ocean–circulation relationships are examples of influences of both positive and negative feedbacks. The snow/ice–albedo relationship is a positive feedback case, and the temperature–thermal emission relationship illustrates a negative feedback when viewed in simplest terms.

2.7   The hydrologic cycle

The movement of water among the subsystems of the climate system constitutes the hydrologic cycle. While the global water amount remains effectively constant on time scales of thousands of years, the water changes state between solid, liquid, and gaseous forms as it travels through the hydrologic system (Hartmann, 1994). Obviously, the atmospheric and terrestrial branches of the hydrologic cycle have to be considered as a whole in order to be meaningful. Until relatively recently, the hydrologic cycle was studied in a piecemeal way considering only the earth-bound branch. This was due largely to the fact that atmospheric data were relatively sparse. Even the land data on evaporation, evapotranspiration, and precipitation have a high degree of uncertainty, and precipitation data for the oceans remain scarce. Improved knowledge of atmospheric water vapor has been a major contributor to characterizing the hydrologic cycle for various time and space scales.

The atmospheric branch of the hydrologic cycle is fundamentally concerned with moisture transport as shown by the depiction of the hydrologic cycle in Figure 1.3. The terrestrial branch is clearly unable to accomplish the water vapor transport required to sustain terrestrial precipitation. The importance of atmospheric moisture transport is illustrated by the maldistribution in space and time of evapotranspiration that puts moisture back into the atmosphere to be

Table 2.1. *Earth's global water volume by storage category*

Category	Volume (km^3)	Percentage of total
Oceans	1 350 000 000	97.40
Polar ice caps, glaciers	27 500 000	1.984
Groundwater	8 200 000	0.592
Soil moisture	70 000	0.005
Lakes	205 000	0.015
Rivers	1700	0.001
Atmosphere	1300	0.001
Total water	1 385 978 000	100.0
Fresh water	35 978 000	2.60

Adapted from Baumgartner and Reichel, 1975, and Speidel and Agnew, 1988.

redistributed. In this way, the atmosphere is the central component of both the climate system and the hydrologic cycle.

The global water inventory in Table 2.1 is the traditional starting point for examining the hydrologic cycle. The dominance of the oceans in storing water contrasts markedly with the relatively small quantity present in the atmosphere. Since the atmospheric water vapor volume at any moment if condensed would produce a layer of water 25 mm deep spread evenly over the Earth's surface, it is evident that the atmospheric reservoir must be replenished constantly. This notion is supported by theoretical and observational evidence that the annual amount of water that moves through the atmosphere is about 950 mm (Adler *et al.*, 2003). This amount enters the atmosphere by evapotranspiration and returns to the surface as precipitation.

The emphasis on water volume in Table 2.1 often overshadows the contribution water cycling makes to climate in terms of energy exchanges in the atmosphere and at the land surface. Water and ice clouds account for about half of the Earth's reflectivity of solar radiation (IPCC, 2001). This reflected insolation is not available to participate in other energy exchange processes (see Fig. 1.4). However, water vapor is the most important natural greenhouse gas accounting for about two-thirds of the atmosphere's natural greenhouse effect (Kump, 2002). In this way, water vapor is an important participant in the absorption and emission of atmospheric thermal radiation. Evaporation of water from the Earth's surface accounts for most of the surface cooling that balances heating by absorption of solar and thermal radiation (Rosen, 1999). This energy from the surface is transported into the atmosphere as latent heat. Condensation of water vapor from the surface releases this energy to warm the atmosphere and drive atmospheric circulation.

2.8   The radiation balance

The major forcing for the climate system is energy received from the Sun. The Earth intercepts a fraction of the radiant energy emitted by the Sun. A surface perpendicular to the solar ray at the upper edge of the Earth's atmosphere and at the mean Earth–Sun distance receives $1367 \, \mathrm{W \, m^{-2}}$ or the solar constant (S_c), but its true value is uncertain by $4 \, \mathrm{W \, m^{-2}}$ (Hoyt and Schatten, 1997). The solar radiation spread evenly over the spherical surface of the entire globe equates to $342 \, \mathrm{W \, m^{-2}}$. This relationship exists because the shadow area of a spherical planet is a circle with the same radius as the sphere but an area one-fourth that of the sphere. The radiant energy delivered to the upper edge of the Earth's atmosphere is available to be transmitted, reflected, or absorbed by the atmosphere and reflected or absorbed at the Earth's surface.

The Earth's radiation balance is an accounting of the disposition of the intercepted solar energy for a specified period. In its simplest form, the radiation balance tells us that the amount of solar radiation absorbed by the atmosphere and the Earth's surface is equal to the amount of thermal radiation emitted by the Earth's surface and atmosphere back to space for periods greater than one year (Burroughs, 2001). By assuming that both the Sun and Earth behave as blackbody radiant energy absorbers and emitters (see Section 2.8.2), the simplest expression of the radiation balance takes the form

$$\text{Solar radiation absorbed} = \text{Earth radiation emitted} \tag{2.3}$$

which complies with the conservation of energy expressed in the First Law of Thermodynamics for a closed system. Quantitative expression of the individual solar and thermal radiant energy fluxes involved in Equation 2.3 requires familiarity with the physical laws that determine the properties of solar radiation and thermal radiation. An overview is presented in the following sections to facilitate development of the basic principles required for expanding Equation 2.3 to include the relevant radiant energy fluxes. Expanded treatments of these topics are found in Peixoto and Oort (1992), Guyot (1998), and Andrews (2000).

2.8.1   Electromagnetic radiation

Any object with a temperature above absolute zero ($-273.15 \, °\mathrm{C}$ or $0 \, \mathrm{K}$) emits radiant energy. Radiant energy in the form of electromagnetic energy travels in a wave form and has properties determined by the temperature of the emitting object. The wave is a series of adjacent crests and troughs. The energy can be characterized by the speed, wavelength, or the frequency of the waves, but the wavelength is used most often in climatology. Wavelength is the distance between two adjacent crests. However, the speed of an electromagnetic wave is a constant, and the speed must equal the product of the wavelength and the frequency. The frequency is the number of waves passing a given point per unit time. Thus, longer wavelengths of electromagnetic energy have lower frequencies and shorter wavelengths have higher frequencies. Electromagnetic energy requires no intervening medium for transmission.

Electromagnetic radiation can be thought of as either a wave or as a stream of particles that represents movement of energy through space. A photon is a single particle or a discrete unit of electromagnetic energy. It represents the smallest amount of energy transported by an electromagnetic wave of a specific frequency. Wave theory is commonly employed when scattering of radiation by particles and surfaces is the primary concern (Hartmann, 1994). Thinking of radiant energy as discrete parcels or photons is the common convention when radiation absorption and emission are of primary interest.

Differences in the energy levels of electromagnetic radiation are related to photon energy, E_{ph}, which is proportional to its frequency. The relationship is expressed as

$$E_{ph} = h\nu = \frac{hc^*}{\lambda} \tag{2.4}$$

where h is Planck's constant (6.63×10^{-34} J s), ν is the wave frequency in cycles per second, c^* is the speed of light (3.00×10^8 m s^{-1}), and λ is the wavelength in micrometers, μm ($1 \, \mu$m $= 10^{-6}$ m). The difference in photon energy is important when electromagnetic energy interacts with matter. Equation 2.4 tells us that high-frequency and short-wavelength photons have high energy. Low-frequency and long-wavelength photons have low energy. The significance of these differences is seen when the selective response of atmospheric gases to radiation is addressed in Section 2.9.

2.8.2 Blackbody radiation

Understanding absorption and emission of radiant energy is aided by using the concept of blackbody radiation. A blackbody is a hypothetical object that emits or absorbs electromagnetic radiation with 100% efficiency. An object that emits radiation that is uniquely related to the temperature of the object is a blackbody radiator. The energy flux is known as blackbody radiation since it corresponds to emission from a surface with unit emissivity at all wavelengths. Emissivity is the ratio of the actual energy emission of an object to the blackbody emission per unit area at the same temperature. The dependence of blackbody emission on temperature is expressed by the Stefan–Boltzmann law as

$$F^* = \varepsilon \sigma T^4 \tag{2.5}$$

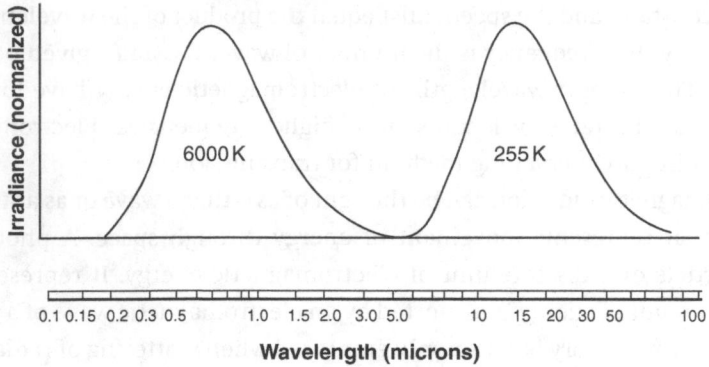

Fig. 2.1. Normalized blackbody emission spectra as a function of wavelength for the Sun (6000 K) and Earth (255 K). (From Peixoto and Oort, 1992, Figure 6.2. Used with kind permission of Springer Science and Business Media.)

where F^* is the blackbody radiative energy emission in $W\,m^{-2}$, ε is the emissivity for a blackbody and is dimensionless, σ is the Stefan–Boltzmann constant ($5.67 \times 10^{-8}\,W\,m^{-2}\,K^{-4}$), and T is temperature in K. The Stefan–Boltzmann law is an integral of Planck's law over all frequencies for the entire wavelength domain. Using the average flux density of the Sun's photosphere ($6.4 \times 10^{7}\,W\,m^{-2}$) and rearranging Equation 2.5, the effective emission temperature of the Sun's photosphere is estimated as 6000 K. Neither the Sun nor Earth is a perfect blackbody, but they conform approximately to Planck's law.

A second characteristic of blackbody radiant energy that is a corollary of Planck's law is that the wavelength most strongly emitted by an object is inversely proportional to its absolute temperature. This relationship is known as Wien's displacement law and is expressed as

$$\lambda_{\max} = \frac{k}{T} \tag{2.6}$$

where $\lambda_{\max}$ is the wavelength of maximum energy emission in μm, k is a constant (2898 μm K), and T is the absolute temperature in K. An important characteristic evident from Wien's law is the hotter an object, the smaller the wavelength of maximum emission. Conversely, the cooler an object, the longer the wavelength of maximum emission. In Wien's law we find the basis for distinguishing between solar and planetary radiation in the electromagnetic spectrum. Solar radiation emitted from the photosphere at a temperature of about 6000 K has a peak emission at 0.5 μm in the visible portion of the spectrum. Earth's radiation with a temperature of 255 K peaks at 10 μm in the far-infrared portion of the spectrum. Ideal Planck curves (Fig. 2.1) for the Sun and Earth integrated over the Earth's surface and all solid angles show the solar and terrestrial fluxes are equal when normalized according to their maximum values. These conditions are typical of the mid-latitudes, a solar elevation of 40°, and diffuse terrestrial radiation (Goody and Yung, 1989). The emission peaks and their related curves establish a physical basis for distinguishing between solar radiation as shortwave radiation and terrestrial radiation as longwave radiation.

2.8.3   Solar radiation

Solar radiation originates from a distant source and can be treated as parallel unidirectional radiation. The solar radiation arriving at the upper edge of the Earth's atmosphere extends over a broad range of wavelengths from very short gamma rays, up to visible wavelengths, and on to the near infrared (Fig. 2.1). Solar radiation is commonly called shortwave radiation because it occurs at wavelengths between 0.1 and 4.0 μm. About 99% of the incident solar radiation occurs in the spectral interval from 0.15 to 4.0 μm. The shortest wavelengths in the ultraviolet portion of the spectrum at 0.1 to 0.4 μm account for about 9% of the total solar radiation received at the upper edge of the atmosphere. Visible light occurs at wavelengths between 0.4 and 0.7 μm and accounts for 45% of the total solar radiation received. Wavelengths of 0.7 to 4.0 μm occupy the near-infrared portion of the solar spectrum and account for 46% of the incident solar radiation.

Much of the Sun's ultraviolet radiation is absorbed by ozone and oxygen in the upper atmosphere as solar radiation penetrates the Earth's atmosphere (Fig. 2.2). Water vapor and carbon dioxide in the lower atmosphere absorb much of the infrared radiation beyond 1.2 to 2 μm. This accounts for the solar radiation reaching the Earth's surface being reduced to a range from 0.3 to 2 μm (Burroughs, 2001). Section 2.9 treats the selective reduction of solar radiation in the atmosphere.

It is important to recognize that about 30% of the incident solar radiation at the upper edge of the Earth's atmosphere is not available to be absorbed. This energy is reflected back to space by the atmosphere and the Earth's surface and constitutes the Earth's albedo (α_p) or reflectivity (see Fig. 1.4). The left side of Equation 2.3 can then be stated as

$$\text{Solar radiation absorbed} = S_c(1 - \alpha_p)\pi r_p^2 \tag{2.7}$$

where S_c is the solar constant of 1367 W m^{-2}, α_p is the planetary albedo of 0.30, and r_p is the Earth's radius of 6370 km. The globally averaged absorbed solar radiation at the top of the Earth's atmosphere is about 235 W m^{-2}. This is the amount of radiant energy that must be returned to space by terrestrial and atmospheric thermal radiation for the system to remain in thermal equilibrium.

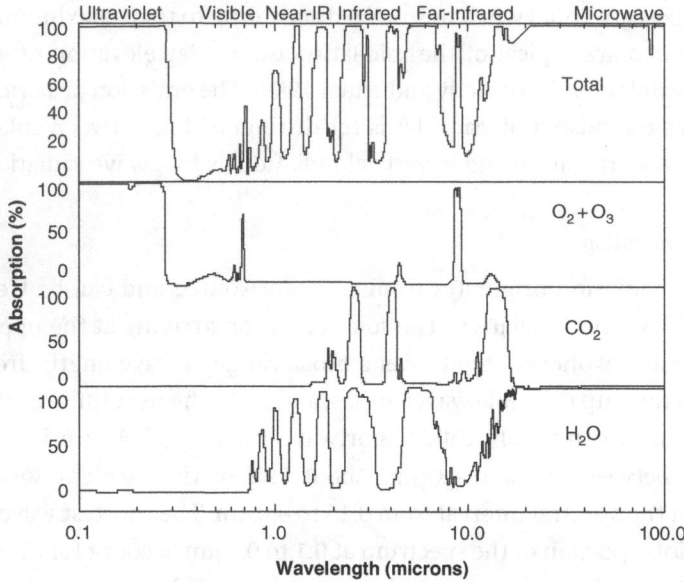

Fig. 2.2. Absorption spectra for the total atmosphere and for selected atmospheric gases between the top of the atmosphere and the Earth's surface. (From Peixoto and Oort, 1992, Figure 6.2. Used with kind permission of Springer Science and Business Media.)

2.8.4   Terrestrial radiation

The Earth's radiation is complicated because the land, oceans, and atmosphere are all emitters of thermal radiation. In general, most land surfaces and the oceans can be regarded as blackbody radiators. They radiate across a continuous spectral interval approaching a constant emissivity. The radiative properties of the atmosphere are dominated by a small number of trace gases that constitute a relatively small proportion of the total atmospheric volume. Water vapor, carbon dioxide, and ozone are the primary trace gases of concern for their radiatively reactive properties. These gases absorb specific wavelengths of radiant energy, and they radiate only at specific wavelengths that give them variable emissivities.

As a result of the multiple emitters, terrestrial radiation at any given point is proportional to the fourth power of the surface temperature and the characteristics of the intervening atmosphere. The eventual emitted radiation is different from the radiation emitted by the surface because the surface radiation is absorbed and re-emitted by the atmospheric trace gases whose temperatures are different from the underlying land or ocean surface temperature (Burroughs, 2001). The emitted thermal radiation is called terrestrial radiation even though it comes from both the surface and the atmosphere. Another distinctive characteristic of terrestrial radiation is that emission occurs during both the day and night.

A first approximation expressing emitted terrestrial radiation is achieved by symbolic representation of the right side of Equation 2.3. By assuming the

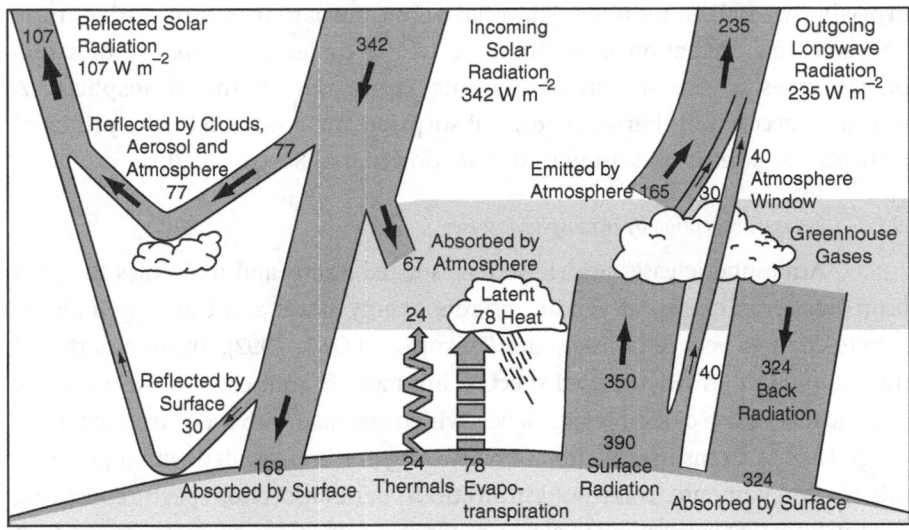

Fig. 2.3. Earth–atmosphere annual global mean radiation balance. Units in $W\,m^{-2}$. (From Kiehl and Trenberth, 1997, Fig. 7. Used with permission of the American Meteorological Society.)

terrestrial emission is like that of a blackbody and that the emission occurs over the surface area of a sphere, the terrestrial radiation emission is

$$\text{Earth radiation emitted} = \sigma T_e^4 4\pi r_p^2 \tag{2.8}$$

where σ is the Stefan–Boltzmann constant defined for Equation 2.5, T_e is the Earth's blackbody emission temperature in K, and r_p is defined for Equation 2.7. The net thermal radiation lost to space from the surface and the atmosphere depicted by solving Equation 2.8 approximates the 235 $W\,m^{-2}$ of outgoing longwave radiation shown in Figure 2.3.

Terrestrial radiation ranges across the spectral range of 4 to 200 μm with a peak emission at about 10 μm (see Fig. 2.1). Nearly 99% of the terrestrial infrared radiation occurs in the spectral range from 4 to 80 μm. However, a majority of terrestrial radiation is concentrated between 8 and 30 μm.

2.9   Selective atmospheric response to solar radiation

Radiant energy arriving from the Sun can be traced through all of the energy transfers and transformations that embrace hydroclimatology. As the photons penetrate the Earth's atmosphere, they are transmitted through the gaseous envelope, or they are reflected or absorbed by gases, particulates, clouds, and water droplets. The reflected photons constitute a major component of the Earth-atmosphere albedo. The photons absorbed by atmospheric gases are surprisingly small in number. The relatively transparent nature of the atmosphere to solar radiation is attributable to the molecular response of atmospheric gases to the stream of photons impinging on the atmosphere. A simplified account of thermal energy absorption and emission by the principal atmospheric gases is presented in this section and in Section 2.10.

2.9.1   *Energy absorption by atmospheric gases*

Atmospheric gases are viewed as isolated atoms and molecules that can absorb and emit energy at certain discrete energy states and can only undergo discrete changes between these states (Peixoto and Oort, 1992). Atoms absorb and emit energy in narrowly defined spectral lines and cannot contain extra energy. They must be at one of their energy levels which resonate with their structure. Each energy level of atoms making up atmospheric gases is associated with a precisely defined energy amount. This results in an atmospheric absorption spectra consisting of many lines that correspond to electronic energy transitions characteristic of each particular atomic species (Peixoto and Oort, 1992). Energy absorption by a molecular gas occurs in bands consisting of a large number of closely spaced spectral lines.

For a photon to be absorbed, its energy must be transferred to the substance absorbing it either in the form of increased internal energy or as heat. The energy of the photon is absorbed if it corresponds to the difference between the energy of two allowable states of the molecule. A range of energies corresponds to each mode of energy storage in a molecule. An atmospheric gas molecule can store energy in rotational, translational, vibrational, or electronic forms.

2.9.2   *Rotational energy transitions*

Rotational energy is related to the spinning of the atoms in a molecule around a common axis perpendicular to the line joining the atoms. A convenient model for diatomic molecules with an electric dipole moment is a spring connecting two balls that are different. Changes in the rate of rotation of molecules require the smallest energy differences and correspond to the energy of photons with wavelengths shorter than about 1 cm (Hartmann, 1994). However, molecules that are symmetric about their center of mass have no dipole moment and resist rotational transitions. Such atmospheric gas molecules include H_2, O_2, and N_2, and those with more complex structures like CO_2, and CH_4. A CO_2 molecule has a symmetric linear arrangement (Fig. 2.4), and the CH_4 molecule has a spherically symmetric arrangement.

2.9.3   *Translational energy transitions*

Translational energy is kinetic energy associated with the movement of molecules from one location to another in the unconfined atmosphere.

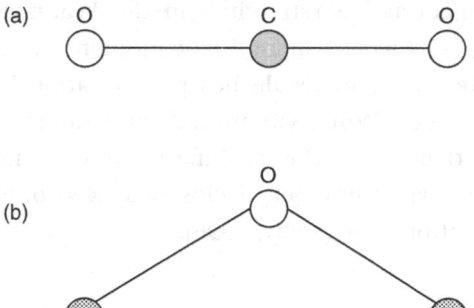

Fig. 2.4. Diagram showing molecular structure as the basis for a molecule's allowable energy transition. (a) A symmetric linear molecule, CO_2, that lacks pure rotational transitions. (b) A bent triatomic molecule, H_2O, that has pure rotation bands and vibration-rotation bands due to its permanent dipole moment.

Rotational and vibrational are other kinetic energy forms, but they are quantized while translational energy is not quantized. Translational energy does not have stationary states in an unconfined spatial domain, but it is a component of the equilibrium energy state. The collision of atmospheric molecules can supply or remove energy to interactions between photons and gaseous molecules. This kinetic or translational energy of a molecule is generally small at terrestrial temperatures, but it is larger than rotational energies and smaller than the energy necessary for vibrational transitions.

2.9.4   Vibrational energy transitions

Vibrational energy is related to rapid variations of the interatomic distances within molecules. The forces of atomic attraction and repulsion are in balance for stable molecules at their appropriate interatomic distance. The absorption of a photon's energy causes an excited state of the binding forces between atoms. For a diatomic molecule, this can be thought of as functioning like a spring between two masses in which the spring expands and contracts. Vibrational energy levels of common atmospheric gases are separated by energies much higher than rotational levels. Therefore, vibrational transitions involve the absorption of shorter wavelength photons than for rotational transitions.

A molecule in an excited vibrational state will have rotational energy and engage in an energy transition that alters both the vibrational and rotational energy content of the molecule. Oscillations around this stable point can involve pure rotational transitions or vibrational-rotational transitions depending upon the molecular structure of the gas. A linear triatomic molecule, like CO_2, responds like a diatomic molecule except its structure permits the molecule to have one bending mode and two stretching modes. Water vapor is a bent triatomic molecule that gives it a permanent dipole moment (see Fig. 2.4). This structure means the water vapor molecule has pure rotation bands in addition to vibration-rotation bands. While vibrational transitions require a photon with a wavelength less than 20 μm, the combination of vibrational and rotational transitions supports a large number of closely spaced photon frequencies representing an absorption band (Salby, 1996).

2.9.5   Electronic energy transitions

Electronic transitions occur when the outer electrons of an atom or molecule are promoted from their ground state to an excited state sufficient to break their molecular or electronic bonds. High-frequency, shortwave solar radiation has more energy per photon than low-frequency radiation and is capable of supporting the high energy requirements of electronic transitions. These transitions occur very fast, and the energies of the electronic states are quantized. Electronic transitions associated with photodissociation or electron excitation correspond to the largest energy differences and are supported by wavelengths shorter than 1 μm (Hartmann, 1994).

In a molecule, the electronic transition can be accompanied by rotational and vibrational transitions. Superposition of rotational and vibrational transitions on the electronic transition results in a combination of overlapping spectral lines appearing as a continuous spectral band (Salby, 1996).

2.9.6   Absorption lines and bands

Most of the photons with wavelengths shorter than 0.2 μm are absorbed in the upper atmosphere through electronic transitions involving photodissociation and ionization of nitrogen and oxygen (see Fig. 2.2). Photons with wavelengths between 0.2 and 0.3 μm are absorbed by electronic transitions with ozone in the stratosphere (Hartmann, 1994). Sunburn caused by radiation at 0.295 to 0.33 μm with a peak at 0.3075 μm demonstrates that some solar radiation at these wavelengths does penetrate to the Earth's surface. Spectral bands are particularly important to electronic energy transitions in the ultraviolet and visible wavelengths but are relatively unimportant for the far-infrared wavelengths (Peixoto and Oort, 1992).

The visible component of solar radiation occurs at wavelengths of 0.4 to 0.7 μm. Photons at these wavelengths are too energetic to be absorbed by vibrational-rotational transitions related to most of the gases in the atmosphere and not energetic enough to support electronic transitions associated with photodissociation. In the absence of transitions corresponding to the photons' energy, the photons have a good chance of passing through the atmosphere without absorption. The result is that the atmosphere is transparent to solar radiation in the visible wavelengths.

Photons of near-infrared solar radiation at wavelengths of 0.7 to 4.0 μm are slightly less energetic than visible solar radiation wavelengths, and they are weakly absorbed by vibrational-rotational transitions involving CO_2, O_3, O_2, and H_2O. These energy transitions account for most of the relatively small magnitude of solar radiation absorption by molecules in the Earth's troposphere (Hartmann, 1994). Consequently, the atmosphere is relatively transparent to solar radiation.

2.10   Terrestrial radiation and the greenhouse effect

About one-half of the solar radiation reaching the upper edge of the Earth's atmosphere is ultimately absorbed by the Earth's surface (see Fig. 2.3). The longwave infrared radiation emitted by the Earth's surface has an important role in the climate system because it provides the largest quantitative energy input to the atmosphere. Atmospheric gases that are transparent to shortwave solar radiation are largely opaque to longwave terrestrial radiation that occurs at wavelengths greater than 4.0 μm. The longwave thermal radiation is absorbed by atmospheric gases and reradiated back to the Earth's surface. The opacity of the atmosphere to longwave radiation and the atmospheric emission of thermal radiation back to the Earth's surface are the fundamental basis for the energy exchanges related to the phenomenon known as the greenhouse effect (Fig. 2.5). The radiant energy flux from the atmosphere to the Earth's surface reveals that the surface receives the greatest energy input from the atmosphere just as the atmosphere receives the greatest energy input from the Earth's surface.

The greater absorption and emission of thermal infrared photons by atmospheric gases is associated with the low photon energies of infrared radiation emitted by the Earth's surface and the corresponding low vibrational and rotational energy transitions of many polyatomic atmospheric gases. Diatomic molecules N_2 and O_2 have no dipole moment, and they lack vibration-rotation transitions at the small photon energies corresponding to terrestrial radiation (Peixoto and Oort, 1992). Therefore, these gases do not interact with electromagnetic radiation in the longwave spectrum.

Water vapor and carbon dioxide are trace gases present in significant amounts in the atmosphere. These polyatomic atmospheric gases display combinations of vibrational and rotational transitions that allow the molecules to absorb and emit photons at a large number of closely spaced frequencies. The water vapor molecule has a permanent dipole moment that supports pure rotation transitions beginning at about 25 μm and extending to longer wavelengths with greater and greater absorption. In addition, the bent triatomic water vapor molecule has a

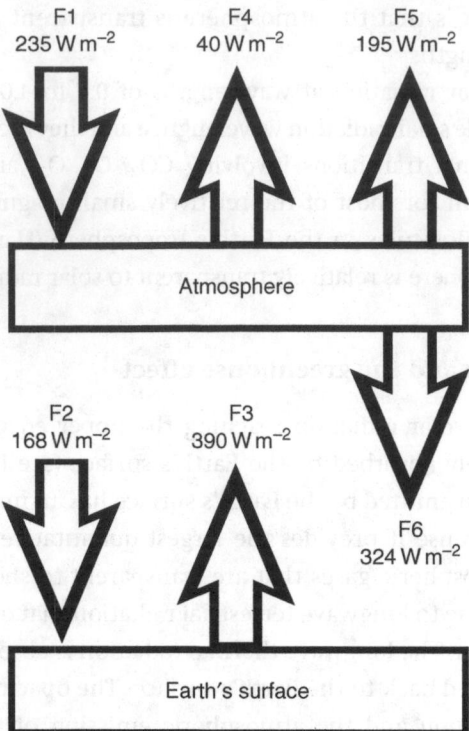

Fig. 2.5. A simple model of the greenhouse effect showing principal solar and thermal radiation fluxes. F1 is incoming solar radiation less albedo, F2 is solar radiation absorbed by the Earth's surface, F3 is terrestrial longwave radiation emitted by the Earth's surface, F4 is terrestrial longwave radiation passing through the atmosphere, F5 is atmospheric longwave radiation emitted to space, and F6 is atmospheric longwave radiation emitted back to the Earth's surface.

vibration-rotation band for bending at 6.3 μm that strongly absorbs terrestrial radiation at wavelengths greater than 12 μm. Carbon dioxide is a symmetric, linear molecule that can develop temporary rotational transitions to accompany vibrational transitions. It has a strong bending mode that supports a strong absorption region at photon wavelengths near 15 μm. This absorption band has a considerable effect even at low carbon dioxide concentrations. Still, carbon dioxide is second to water vapor in terms of importance in atmospheric radiative transfers. In this way, minor trace gas concentrations of these and other polyatomic molecules determine the infrared transmissivity of the atmosphere (Hartmann, 1994).

The practical climatic significance of the selective absorption and emission of radiant energy by atmospheric trace gases is that the atmosphere is warmed from the bottom. The absorption of radiant energy at the bottom of the atmosphere and the output of longwave radiation from the top of the atmosphere establishes an unstable environment that promotes constant turbulence within the troposphere or the lower layer of the atmosphere. The emerging pattern of global climate is an expression of the dynamic nature of the exchanges of energy and moisture sustained by the physical processes ultimately driven by the energy fluxes. Elucidating the underlying processes is aided by conceptualizing the patterns of several components dynamically linked within the climate system.

2.11   Global radiation balance

The radiation balance is a global phenomenon. The flows of radiant energy into and out of the upper edge of the Earth's atmosphere are approximately equal over a time scale of several years. This equilibrium state is the expected behavior for the climate system and a stable climate. The fluxes of energy may not balance over shorter periods, or a perturbation in the system may cause the climate system to seek a new equilibrium state. Whatever the condition of the global climate system, individual sites or regions typically experience an imbalance in radiation exchanges. Recognition of the global balance is an important initial step in understanding the global energy system and its related regional components.

The coupling of blackbody absorption and emission is fundamental to the Earth–atmosphere system radiation balance. The conservation of energy requires that the solar radiation absorbed must be balanced by the planetary radiation emitted. Although the Earth is not a perfect blackbody, it conforms closely with Equation 2.5. Therefore, the relationship between absorbed and emitted radiation for the Earth–atmosphere system is relatively straightforward. The planetary radiation balance is expressed as

$$S_c(1 - \alpha_p)\pi r_p^2 = \sigma T_e^4 4\pi r_p^2 \tag{2.9}$$

where all the terms are defined for Equations 2.7 and 2.8. Using an albedo of 0.30 gives an emission temperature, or an effective radiation temperature, for the Earth without an atmosphere of 255 K. This temperature represents the thermal state when the energy received by the Earth from the Sun balances the energy lost by the Earth back into space. However, the global observed mean surface temperature is 288 K, and a temperature of 255 K is characteristic for the atmosphere at an altitude of 5.5 km. These observations indicate additional factors must be considered in deriving the radiation balance for the Earth's surface, but the radiation balance for the top of the Earth's atmosphere is purely radiative.

Radiant energy absorbed and emitted at the top of the atmosphere varies geographically and seasonally. Zonal means of both fluxes based on latitude

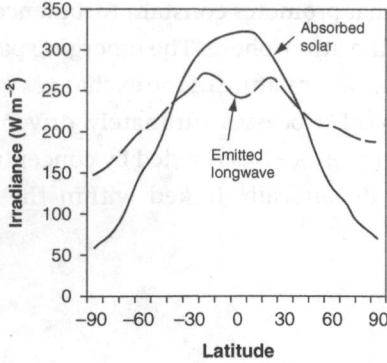

Fig. 2.6. Annual mean absorbed solar radiation and emitted longwave radiation zonally averaged. (Data courtesy of the NASA Langley Research Center Atmospheric Science Data Center from their website at http://eosweb.larc.nasa.gov/.)

circles and calculated for a year or longer reveal the strong influence of the latitudinal gradient of insolation (Fig. 2.6). The curve for absorbed solar radiation is offset slightly to the Southern Hemisphere due to the eccentricity of the Earth's orbit around the Sun. It should be remembered that the absorbed solar radiation is the energy available for driving the Earth's atmospheric circulation (Peixoto and Oort, 1992). The emitted longwave radiation curve has a high plateau between about 30° N and 30° S with a slight dip over the equator. Emitted longwave radiation declines less abruptly at higher latitudes than absorbed solar radiation creating a region of energy surplus between about 40° N and 40° S. Energy deficit regions occur poleward of these latitudes in both hemispheres. To sustain an equilibrium between the radiative processes acting to warm the low latitudes and to cool the high latitudes, poleward heat transport by the atmosphere and the oceans is needed to offset the radiative differences (Trenberth and Solomon, 1994).

Sections 2.9 and 2.10 address the atmospheric response to solar and thermal radiant energy flows. Radiant energy absorption by the atmosphere means the atmosphere is a source of thermal radiant energy that augments solar heating of the Earth's surface. Consequently, the radiation balance for the atmosphere is not the same as the radiation balance for the Earth's surface.

Figure 2.3 quantifies the combined influences of atmospheric absorption of solar and thermal radiation summarized in Figure 1.4. The radiation balance at the top of the atmosphere is the same as that expressed in Equation 2.7, and the emission temperature is 255 K. The Earth's surface temperature must be warmer because it receives radiant energy from the Sun and radiant energy transmitted back to the surface from the atmosphere. These fluxes between the

Earth's surface and the atmosphere define the greenhouse effect resulting from the atmospheric response to radiant energy.

Assuming the Earth–atmosphere system is in thermal equilibrium, the radiation balance at the Earth's surface is expressed as

$$\frac{S_c}{4}(1 - \alpha_p)\left(\frac{1 + \tau_s}{1 + \tau_t}\right) = \sigma T_g^4 \tag{2.10}$$

where τ_s is transmitted solar radiation, τ_t is transmitted thermal radiation, T_g is the emission temperature for the Earth's surface in K, and all other variables are defined for Equations 2.5 and 2.7. Values in Figure 2.3 support an approximate value for τ_s of 0.78 indicating strong transmission and weak absorption of solar radiation by the atmosphere. The value for τ_t is 0.17 indicating weak transmission and strong absorption of thermal radiation by the atmosphere (Andrews, 2000). Solving Equation 2.10 using these values results in a surface temperature of 283 K which is close to the observed mean value of 288 K. It is important to note that Figure 2.3 includes two non-radiative processes that contribute to heating the atmosphere. These energy transfers are expected to account for some of the difference between the observed temperature and the calculated temperature from Equation 2.10.

2.12   Surface radiation balance

At the Earth's surface, the concept of net radiation is a useful paradigm. In its simplest form, net radiation is expressed as

$$R_n = K{\downarrow} - K{\uparrow} + L{\downarrow} - L{\uparrow} \tag{2.11}$$

where R_n is net all-wave radiation, $K{\downarrow}$ is incoming shortwave solar radiation, $K{\uparrow}$ is the outgoing shortwave radiation or the albedo of the surface, $L{\downarrow}$ is the long-wave emission from the atmosphere directed downward to the Earth, and $L{\uparrow}$ is the longwave emission from the Earth's surface. On an annual basis, Figure 2.3 shows that the Earth's surface is a net absorber of radiant energy and net radiation for the Earth's surface is $102\,\mathrm{W\,m^{-2}}$. This condition is commonly referred to as an energy surplus and is contrasted with an energy deficit in which net radiation is a negative value. Specific locations commonly experience a period of net radiation surplus during the day and a period of net radiation deficit at night. Hourly net radiation data for Davis, California, (39° N) for two days (Fig. 2.7) show seasonal changes in the magnitude of the net radiation surplus and in the period of the net radiation deficit. Locations at high latitudes have greater contrasting seasonal summations of energy surplus and energy deficit, and these seasonal contrasts diminish at lower latitudes.

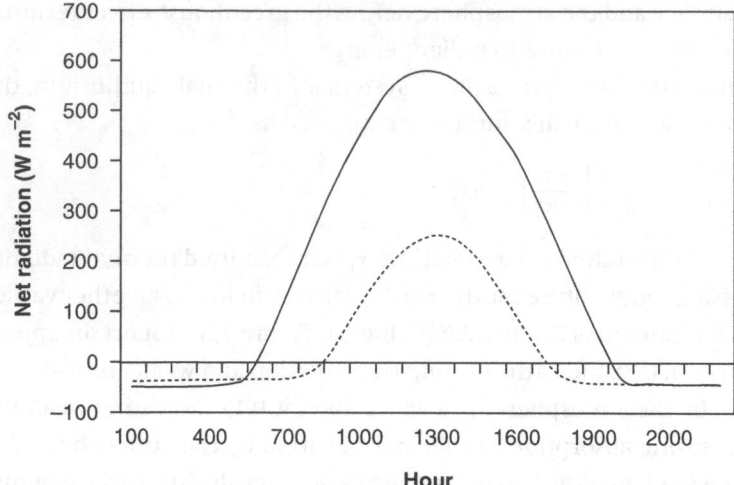

Fig. 2.7. Net radiation at Davis, California (39° N). Dotted line is 15 January 2006, and solid line is 15 July 2006. (Data courtesy of the California Department of Water Resources from their website at http://wwwcimis.water.ca.gov/.)

Conduction does not appear in Figure 2.3 because the average annual flux into and out of the soil is zero. Also, photosynthesis is not included because it is a relatively small quantity of about 1% globally of solar radiation, and this is significantly smaller than the radiation or other non-radiative fluxes. The heat flow associated with Earth's radioactive heating is disregarded due to its small magnitude of 0.0002% of the radiant energy received from the Sun. The fact that the total radiation at the surface and at the top of the atmosphere must balance is a considerable constraint that lends confidence to the assigned values (Kiehl and Trenberth, 1997).

2.13   Planetary energy balance

At the Earth's surface, positive quantities of radiant energy are available to be transformed into non-radiative energy forms. These non-radiative energy transfers are important components in helping to maintain the energy balance of the Earth–atmosphere system as shown in Figure 2.3. The largest portion of radiant energy absorbed at the Earth's surface is used to evaporate water. The radiant energy is transformed into latent heat, which is attached to the water molecules and travels with them until they return to the liquid or solid state in the atmosphere. This represents a significant transfer of energy from the Earth's surface to the atmosphere. The second largest allocation of net radiant energy is to the atmosphere in the form of sensible heat. Air is a poor conductor of heat, and atmospheric warming by conduction from the Earth's surface is limited to a depth of a few centimeters. However, warming the bottom of the atmosphere imparts a bounce to the air that supports transferring the heated air by convection. The smallest quantity of net radiant energy goes to underlying soil layers or is used to melt snow and ice (Peixoto and Oort, 1992). For a land area, the partitioning of net radiation, R_n, among alternative energy fluxes is expressed as

$$R_n = LE + H + G \tag{2.12}$$

where LE is the latent heat flux into the atmosphere resulting from evapotranspiration at the surface and condensation within the atmosphere, H is the sensible heat flux to the atmosphere by conduction and convection resulting from the temperature difference between the surface and the overlying air, and G is the soil heat flux into the ground by conduction.

The energy balance for water is more complex because the depth where the energy storage flux is considered negligible can be as much as several kilometers for areas where deep water is formed. In addition, horizontal energy flux may be a greater consideration in water due to the greater mobility of water compared to a land surface (Rosen, 1999). Even for land areas, Equation 2.12 does not include terms such as latent heat of fusion, oxidation, and heat transfer by precipitation that can be important locally or for limited time intervals. Another significant characteristic of the energy balance is that it is negative at night and during winter at high latitudes. During these periods, thermal radiation dominates the radiation balance and non-radiative fluxes provide energy to the surface to compensate for thermal radiation losses.

2.14   The water balance

The water balance is a conceptual structure supporting a quantitative assessment of moisture supply and demand relationships at the land–atmosphere interface on a daily, weekly, monthly, or annual basis. Precipitation provides the moisture supply, and the moisture demand is represented by evapotranspiration. The water balance traces the partitioning of water among the various pathways inherent to the hydrologic cycle (Arnold et al., 1999). This concept has the most immediate application for land areas because the difference between precipitation and evapotranspiration defines the amount of water that is available to become runoff and is available for human use. Its utility is expanded by its transferability to the atmosphere and as the basis for a global water balance (Oki, 1999). In addition, this accounting of moisture supply and demand can be applied at different time and space scales as a way of characterizing climate in terms of the coincidence of precipitation and the energy demand for moisture expressed as evapotranspiration.

The relationship between precipitation and evapotranspiration is the fundamental climatic influence driving multiple processes at the land–atmosphere interface. In most instances, the adequacy of available moisture to satisfy the energy-driven moisture demand is a critical determinant. However, while precipitation is a routinely measured climate variable there are problems and limitations associated with measuring or estimating evapotranspiration. A number of techniques have been developed to quantify the maximum evapotranspiration flux and the quantity that actually occurs. Both of these evapotranspiration characterizations play a role in the water balance and through their relationship emphasize the role of soil moisture storage (Eagleson, 1978). Techniques for measuring and estimating evapotranspiration are presented in Chapter 4.

2.14.1   Earth's surface water balance

The water balance for the Earth's surface is an instructive tool for considering the link between the climate system and the hydrologic cycle. The water balance is a robust concept that can be applied at various time and space scales. At the global scale, the water balance connotes the long-term accounting of all water in its various forms and states. An annual accounting at the global scale is often used in the same sense as the hydrologic cycle. At smaller time and space scales, a strict water balance does not exist in that moisture inflows and outflows are not equal. In this case, the term water budget is often used to describe the moisture accounting for a selected period. Today, the terms water balance and water budget are often used interchangeably with the context defining the time and space scales.

The surface water balance is an accounting of what happens to precipitation within the Earth–atmosphere interface. It is expressed as

$$\frac{\partial St_S}{\partial t} = -H_D \cdot R_S - H_D \cdot R_U - (ET - P) \tag{2.13}$$

where St_S is water storage at and below the surface, t is time, H_D is horizontal divergence, R_S is surface runoff, R_U is subsurface runoff, ET is evapotranspiration, P is precipitation, and all units are in cm or mm (Oki, 1999). The water balance equation and the energy balance equation are the two fundamental equations of hydrology (Eagleson, 1994) and are foundational equations in climatology. Other processes, such as interception, infiltration, and deep percolation, are subsumed under these five variables, but the influence of other processes is largely incorporated in the quantitative expression of these five variables except at the shortest time intervals. In fact, the long-term average hydrologic balance commonly assumes the storage term is small and can be ignored. This simplification is based on the assumption that lake, reservoir, and groundwater levels remain approximately the same when averaged for a decade or more. Therefore, the long-term surface water balance is presented often in the abbreviated form

$$R_S = P - ET \tag{2.14}$$

where the variables are defined for Equation 2.13. This expression emphasizes streamflow as the residual of precipitation depleted by evapotranspiration.

2.14.2   *Atmospheric water balance*

The atmospheric water balance is expressed in a fashion similar to the surface water balance. Conceptually, it is important to recognize that the atmospheric balance is the complement of the surface balance (Oki, 1999). Precipitation minus evapotranspiration is the net moisture flux from the atmosphere to the surface, and it is given the opposite sign in the atmospheric water balance to indicate a net flux away from the surface. The atmospheric water balance is expressed as

$$\frac{\partial W}{\partial t} = -H_D \cdot R_A + (ET - P) \tag{2.15}$$

where W is precipitable water, R_A is the horizontal export of water by atmospheric motion primarily in the form of water vapor, and the other terms are defined above. Precipitable water is the column storage of water vapor assuming the atmospheric content of water in solid and liquid phases is small. Therefore, precipitable water is the amount of liquid water that would result if all the water vapor in a unit column of atmosphere were condensed (Peixoto and Oort, 1992).

Expressing the atmospheric water balance in the form of Equation 2.15 establishes that R_A is analogous to R_S in the surface water balance shown in Equation 2.13. The availability of high-resolution atmospheric data provides a basis for estimating the atmospheric water balance. Water vapor flux convergence/divergence is calculated for various scales, and integrating the horizontal transport of water vapor with respect to pressure quantifies R_A or aerial runoff (Peixoto and Oort, 1992).

2.14.3   *Global water balance*

Combining the equations for the surface and the atmosphere produces the water balance for the Earth–atmosphere system in which the exchanges of water across the surface and for the atmosphere cancel each other. The term $ET - P$ common to Equations 2.13 and 2.15 establishes the connection between the terrestrial and the atmospheric branches of the hydrologic cycle. The combined equation reduces to

$$St_S + W = -R_S - R_A \tag{2.16}$$

where the variables are defined previously. For periods of a year or more, the changes in surface water storage and precipitable water in the atmosphere are assumed to be zero. The net horizontal transport by the atmosphere and surface runoff are equal and opposite in sign. This means that water delivered to the continents by atmospheric transport (R_A) must equal the runoff conveyed by rivers (R_S). The consequence of this reasoning is that when the total global

Table 2.2. *Water balance of the continents and oceans*

	Precipitation		Evaporation		Runoff [a]	
	(km³)	(mm)	(km³)	(mm)	(km³)	(mm)
Continents	111 100	746	71 400	480	39 700	266
Africa	20 743	696	17 334	582	3 409	114
Antarctica	2 376	169	389	28	1 987	141
Asia	30 724	696	18 519	420	12 205	276
Australia	7 144	803	4 750	534	2 394	269
Europe	6 587	657	3 761	375	2 826	282
North America	15 561	645	9 721	403	5 840	242
South America	27 965	1564	16 926	946	11 039	618
Oceans	385 000	1066	424 700	1176	−39 700	−110
Arctic Ocean	826	97	452	53	374	44
Atlantic Ocean	74 626	761	111 085	1133	−36 459	−372
Indian Ocean	81 024	1043	100 508	1294	−19 483	−251
Pacific Ocean	228 523	1292	212 655	1202	15 868	90
Global	496 100	973	496 100	973	0	0

[a] Runoff for oceans represents advected moisture into the ocean basin (positive value) or advected moisture out of the ocean basin (negative value).
Adapted from Baumgartner and Reichel, 1975, and Speidel and Agnew, 1988.

hydroclimatic system is considered over a long period, mean precipitation is counterbalanced by mean evapotranspiration. This is the basis for the important hydroclimatic premise that global precipitation and global evapotranspiration are equal. Consequently, if global evapotranspiration increases in response to an increased energy input, then global precipitation will increase or the converse can occur. Global models indicate that a 1% change in global temperature will produce a 3% change in global evaporation. By extension, a 1% change in global temperature will produce a 3% change in global precipitation.

This reasoning leads to the water balance for the continents and the oceans shown in Table 2.2. However, precipitation and evapotranspiration need not be equal for the continents because pieces or segments of the global system are being considered. At this scale, the maldistribution of water for the continents is evident. South America has the greatest precipitation, evapotranspiration, and runoff. Antarctica has the smallest precipitation and evapotranspiration. Precipitation for Asia and Africa is the same, but Africa has the smallest runoff emphasizing the importance of the energy and moisture relationship. North America is closest to the "land" average based on the runoff to precipitation ratio. The runoff to precipitation ratio expresses the fraction of precipitation that is available for human use. Oceanic data indicate the Atlantic and Indian oceans are major sources of atmospheric moisture while the Pacific Ocean is nearly in balance. The highest oceanic evaporation is 30% greater than the largest land value for South America, but Atlantic Ocean precipitation is half the value for South America.

The same conceptual structure is applicable to the classic hydrologic equation for land areas smaller than continents. The hydrologic equation is an expression of the conservation of mass

$$P = ET + \Delta St + qr + qs + qd \tag{2.17}$$

where ΔSt is change in storage in lakes, rivers and streams, soil, and groundwater; qr is the net runoff over the soil surface, qs is the net lateral subsurface flow, and qd is the deep subsurface drainage. Runoff or streamflow is the sum of qr, qs, and qd. Therefore, the hydrologic equation can be simplified to

$$P = ET + \Delta St + R_S \tag{2.18}$$

This form emphasizes the collective presence of runoff as the water observed in a stream channel without regard for the path it followed in arriving there. This is similar to the water balance expressed as Equation 2.14, but the variables are expressed in a different order.

The hydrologic equation is useful for doing water inventories or water balances for land areas of various sizes and for different time increments. It is important to recognize that the term "water balance" has several quite different meanings depending on the time and space scales being considered. On the macroscale, water balance can be used in the same sense as the hydrologic cycle. On the mesoscale, it is common to consider the water balance of a major drainage basin. On the microscale, one might investigate the water balance of a field or individual tree.

Whatever the focus, the water balance is an accounting of what happens to precipitation for some selected period. Seasonal water balances for land areas poleward of 50° N were calculated by Dirmeyer and Brubaker (2006) for the period 1979–2003 to study changes in moisture recycling. The long-term water balance for the continental United States based on observational data for 1337 watersheds is reported by Sankarasubramanian and Vogel (2003). The monthly water balance for Red Bluff, California (Fig. 2.8), illustrates the changing seasonal relationship between precipitation and evapotranspiration in a setting with contrasting energy and moisture maxima. At the regional and local scale, the water balance provides an analytical framework for examining the spatial and temporal characteristics of wetting and drying periods, and annual variations in the water balance are a useful basis for distinguishing climate differences (Hartmann, 1994). Though simple in

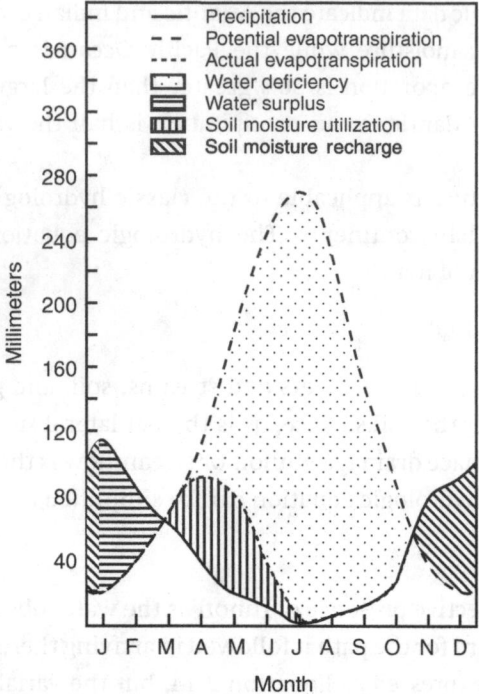

Fig. 2.8. Water balance for Red Bluff, California (40° N). Terms are defined in the text.

concept, it is often a complex undertaking to try to evaluate the water balance. Few factors that must be quantified for the water balance calculation are measured except for experimental conditions at a small number of locations.

Review questions

2.1 How does the climate system embrace both climate of the first kind and climate of the second kind?

2.2 What is the role of feedbacks in the climate system?

2.3 How does the concept of blackbody radiation contribute to understanding of absorption and emission of radiant energy by the atmosphere?

2.4 How can atmospheric carbon dioxide be a good absorber of terrestrial radiation and a weak absorber of solar radiation?

2.5 How can we demonstrate the greenhouse effect is real?

2.6 How much would the emission temperature of Earth change if the planetary albedo was 15%?

2.7 How are the radiation balance and the energy balance equations related?

2.8 How are the water balance and the hydrologic equation related?

3

Measuring hydroclimate atmospheric components

3.1   An atmospheric focus

The relationship between climate of the first kind and the atmospheric branch of the hydrologic cycle presented in Chapter 1 emphasizes atmospheric transport and mobility as central characteristics of these concepts. The atmosphere's radiative properties are important in determining energy exchanges within the atmosphere and between the Earth's surface and the atmosphere. Resulting radiation variations account for the gradient of radiant energy that provides the potential energy for horizontal and vertical atmospheric motion. Vast quantities of water are continuously in motion in the climate system conditioned by water vapor sources and sinks in the atmosphere (Peixoto, 1995). The dynamics of global atmospheric circulation revealed by its state variables are a logical starting point for observing the transport of energy and moisture inherent to climate of the first kind and the atmospheric branch of the hydrologic cycle.

It is convenient to portray the general condition of the atmosphere as a thermodynamic-hydrodynamic system. It can be characterized by its composition, its thermodynamic state as specified by the three thermodynamic variables, pressure, temperature, and humidity, and its three-dimensional motion field. Precipitation, evaporation, and runoff are fluxes and not state variables even though they are intimately connected with the state of the atmosphere (Peixoto and Oort, 1992). These flux variables that couple the atmospheric and terrestrial phases of climate and the hydrologic cycle are discussed in Chapter 4 in the context of climate of the second kind and the terrestrial branch of the hydrologic cycle.

Comprehensive understanding of hydroclimatic processes requires reliance on both observational data and theory. Using observations and balance equations makes it possible to investigate the mechanisms by which processes, such as transport of energy and water vapor, occur in the atmosphere. Observational data are a fundamental requirement for achieving insight into atmospheric mechanisms and processes. Most of our present knowledge of the physical and dynamical structure of the atmosphere is based on *in situ* observations. However, satellites and other technologies are providing improved techniques for observing atmospheric processes and are supplementing information from conventional observation systems. The sources of data important to hydroclimatology that characterize the atmosphere are grouped as surface data, upper air data, and satellite data to facilitate discussion of their roles. Surface and upper air data are discussed here, and satellite data are discussed in Chapter 5.

3.2   Surface data

Land-based surface data that are particularly important for climate of the first kind and the atmospheric branch of the hydrologic cycle are traditionally viewed as point data. These *in situ* measurements of atmospheric state variables form the framework for atmospheric transfer processes defined by fluid mechanics and thermodynamics. The observed variables contributing to understanding the atmospheric state are radiation, temperature, pressure, and humidity. The WMO network of land-based stations has grown from a few stations in Europe and North America in 1875 to a global network of about 7000 stations that currently report on a daily basis. However, radiation data other than solar irradiance are reported by few of these stations. The WMO's Global Climate Observing System (GCOS) Surface Network (GSN) of 1000 stations was established to ensure an evenly distributed network of homogeneous land surface data for monitoring global climate and to promote availability of the data (Peterson *et al.*, 1997; Parker *et al.*, 2000).

The majority of surface data are acquired by observation of direct measurement instruments. These instruments are read *in situ* in most cases, but remote instrument reading by telephonic or radio transmission is employed by an increasing number of weather and climate observation stations. However, the data are *in situ* measurements because the instruments are located at a specific site whether they are read directly at that site or the data are transmitted to another location for archiving.

A comprehensive treatment of all instruments used to collect hydroclimatic surface data pertaining to climate of the first kind and the atmospheric phase of the hydrologic cycle is beyond the scope of this book. Selected instruments for *in situ* measurement of atmospheric state variables are presented to illustrate the nature of the data collection process for each variable. Instruments commonly found at weather or climate stations are emphasized. Sources devoted to instrument descriptions, instrument mechanics, siting factors, and operation of observation stations should be consulted for detailed information on these topics (e.g. WMO, 1996; DeFelice, 1998; Guyot, 1998).

3.3   Radiation

The driving force for the Earth's hydroclimate is the supply of radiant energy from the Sun. Climate of the first kind and the atmospheric branch of the hydrologic cycle are intimately coupled with the character of solar radiation, how the Earth's atmosphere and surface respond to the incoming solar radiation, and how the Earth and atmosphere radiate energy to space. Measuring these energy fluxes is an important step in defining their time and space variations and understanding how they drive atmospheric circulation and energy, mass, and momentum transport.

Instruments for measuring radiant energy fluxes are deployed at a minority of regularly operated climate stations. The absence of radiation instruments is especially evident in developing countries due to the high maintenance requirements and maintenance expense of these instruments. Many radiation instruments require daily inspection and some require daily calibration. The demands of monitoring these instruments are major obstacles to the establishment of dense instrument networks providing routine observations. In recognition of the sparse data problem, the Baseline Surface Radiation Network (BSRN) proposed by the World Climate Research Program became operational in 1992 and now has about 20 fully operational stations. One of the purposes of the BSRN is to monitor long-term changes in radiation fluxes at the Earth's surface in a variety of climatic zones (Ohmura *et al.*, 1998).

Solar radiation is measured with a variety of instruments that can be broadly grouped as thermal or quantum sensors. These instruments are designed to measure the intensity of radiant energy over broad or narrow spectral bands. Thermal sensors absorb solar radiation and convert it into thermal energy in a form that can be measured. Thermal radiation quantities are expressed in terms of irradiance, which is a measure of the rate of energy received per unit area and has units of $W\,m^{-2}$. Quantum sensors utilize absorbed energy quanta to liberate electrons and produce an electric current (Guyot, 1998). A quantum sensor is sensitive to a specific solar spectral domain and has a high sensitivity and rapid response compared to a thermal sensor. Quantum sensors are used primarily to measure the photosynthetic light spectrum and are not deployed for routine meteorological or climatological observations.

Thermal solar radiation instruments are classified by the type of quantity measured. Pyrheliometers measure direct solar radiation. Pyranometers measure whole hemisphere solar radiation, reflected solar radiation, and diffuse solar radiation when screened from direct solar radiation by an equatorial ring. Pyrgeometers measure incoming longwave atmospheric radiation when facing upward and outgoing longwave terrestrial radiation when inverted. Pyrradiometers, or radiometers, measure all-wave radiation arriving on one plane.

3.3.1   Pyrheliometer

Direct beam solar radiation with wavelengths of 0.3 to 4.0 μm is measured with a pyrheliometer oriented toward the Sun so the receptor surfaces are perpendicular to the incident solar beam. The instrument is attached to a mounting that permits it to follow the Sun for continuous recordings. A pyrheliometer has an aperture with an acceptance angle of 2.5° to 5° to limit its view to the solar disk and a narrow sky annulus (WMO, 1996).

The general design of pyrheliometers is a metal tube with a small opening at one end. The opening is covered with a quartz window to allow passage of the whole solar spectrum while protecting the sensor from wind and contaminants. A rotating wheel with different filters can be added to measure specific spectral bands. A blackened disk containing thermopiles is mounted at the closed end of the tube. A thermopile is an electrical loop of alternating lengths of wire where each length reaches from a surface warmed by radiation to a cool surface (Linacre, 1992). The solar radiation passing through the opening is directed to the rear disk containing the thermopiles. The incident solar radiation produces heating of the surface that is sensed by the thermopile and transformed into an electrical voltage interpretable as a radiation flux in $W\,m^{-2}$. Comparing the measured direct beam irradiance with global and diffuse irradiances requires obtaining the horizontal component of the direct solar irradiance. This is achieved by multiplying the direct solar irradiance by the cosine of the Sun's zenith angle. Pyrheliometers are the most accurate of all radiation instruments, and they are commonly used as calibration standards for working instruments (Bradley, 2003). However, they are usually found only at research stations or laboratories because of their need to track the Sun.

3.3.2   Pyranometer

A pyranometer is an instrument for measuring global solar radiation, or both direct and diffuse solar radiation, onto a plane surface (Beaubien et al.,

Fig. 3.1. Pyranometer providing continuous data. (Photo by author.)

1998). A horizontally mounted pyranometer detects solar radiation originating from all parts of the sky (Fig. 3.1). An inverted pyranometer measures reflected solar radiation. An upward facing pyranometer fitted with a ring device known as an occulting band shades the sensor from direct beam radiation so it measures only diffuse solar irradiance. Pyranometers are exposed continually in all weather conditions and must have a robust design that resists the corrosive effects of humid air (WMO, 1996).

The individual designs of pyranometers vary, but they normally use thermo-electric, photoelectric, pyroelectric or bimetallic elements as sensors (WMO, 1996). One of the most popular instruments is based on thermopiles that measure the thermal difference between a black surface and a white surface with both surfaces exposed to the sun in the same plane. Thermopile designs using alternating horizontal strips, alternating pie-shaped disks, or concentric circle arrays of the thermocouples are widely used in Australia, Europe, and North America. The principle is that radiant heating of the blackened surface of a thermopile relative to the reflectively white cool surface generates an electro-motive force proportional to the energy flux received. An alternating surface design has the advantage that the environmental exposure of the surfaces is virtually identical, and changes in ambient air temperature have the same influence on both surfaces (Beaubien *et al.*, 1998). The thermopiles are protected from the wind and from longwave radiation by a glass hemispherical dome. The spectral range of the pyranometer of 0.3 to 3.0 μm is limited by the transmission of the glass dome. Pyranometer sensitivity may change with time and exposure to radiation due to deterioration of the black paint. At least daily inspection of pyranometers and recorders is desirable, and regular calibration of pyranometers is especially important for achieving reliable measurements (WMO, 1996).

3.3.3   Pyrgeometer

Measuring longwave radiation is a problematic pursuit and has often been abandoned in favor of inferring longwave radiation from measurements of shortwave and all-wave radiation. However, longwave radiation is a significant component of the surface energy budget, and there is a growing demand for good quality longwave radiation measurements (Philiponia et al., 1998). Improvements in materials used to block solar radiation have advanced the capabilities of instrument designs for longwave radiation measurements (WMO, 1996).

Measurements of longwave radiation are achieved using a pyrgeometer. This instrument resembles a pyranometer, but it has a polyethylene or silicon dome with good transmittance for the spectral domain of thermal infrared radiation in the range of 3.0 to 100 μm. Also, the pyrgeometer thermopile has one end connected to a blackened disk serving as the receiving surface and the other end is in thermal contact with the instrument casing. A circuit combining the output of a thermistor embedded in the instrument body and the thermopile gives a voltage proportional to the longwave flux. Consequently, the pyrgeometer works by determining the thermal balance of the instrument itself (Bradley, 2003). The reliability of a pyrgeometer is closely related to the transmittance of the dome, which must be cleaned daily, and polyethylene domes must be replaced frequently. Other factors contributing to instrument uncertainty are a lack of an absolute calibration standard, variation in calibration techniques, and uncertainties in the internal thermal balance of the instrument (Burns et al., 2003).

3.3.4   Pyrradiometer

The surface energy budget is driven by net radiation, but net radiation remains among the most difficult atmospheric parameters to measure (Brotzge and Duchon, 2000). Each of the terms on the right side of Equation 2.9 must be measured. This can be accomplished with a pair of pyranometers and a pair of pyrgeometers with one instrument from each pair facing upward and one downward. The alternative is to use a pyrradiometer or net radiometer that produces a signal due to differential radiation absorption (Fig. 3.2). A large number of pyrradiometers are commercially available to measure the sum of shortwave and longwave radiative exchanges, but most instruments are for research applications and few weather or climate stations measure net radiation on a routine basis.

The pyrradiometer consists of two horizontal blackened plates placed back to back and separated by a thermal insulator to sense the downward and upward directed radiation fluxes. Each blackened disk has an internal thermopile, and

Fig. 3.2. Pyrradiometer (net radiometer) installed over a plowed field. (Photo by author.)

the temperature difference between the two sensor surfaces is proportional to net radiation (R_n). That is

$$R_n = C(T_u - T_d) \tag{3.1}$$

where C is a parameter expressing the rate at which the sensor is warmed or cooled by conduction and convection, T_u is the temperature of the upper surface in °C, and T_d is the temperature of the lower surface in °C. The instrument is mounted with the sensors horizontal so they measure the temperature differential produced between the plates by downwelling and upwelling radiative energy (Bradley, 2003).

Most pyrradiometers use polyethylene hemispherical domes to eliminate natural ventilation and reduce thermal convection from the instrument. Polyethylene is used for the domes because it is transparent to both shortwave and longwave radiant energy between 0.3 and 100 μm. However, the polyethylene domes degrade after a few months of exposure, requiring replacement of the domes and recalibration of the pyrradiometer (Brotzge and Duchon, 2000). The instrument is placed 1 to 2 m above the surface being examined to avoid shading the surface or averaging too large an area.

3.4   Temperature

Temperature plays a central role in radiative and dynamical atmospheric processes inherent in climate of the first kind and the atmospheric branch of the hydrologic cycle. Unfortunately, air temperature is one of the most spatially variable of all climatic elements, and the difference between air temperature and surface temperature can be substantial. Surface temperature is related to radiant energy fluxes through the Stefan–Boltzmann relationship (Equation 2.5) and net radiation (Equations 2.9 and 2.10). Because surface temperature fluctuates in response to net radiation changes, air temperature measurements are usually taken 1 to 2 m above ground level. In addition, a shelter structure is provided to protect instruments from radiation and precipitation (WMO, 1996).

Temperature at a standard weather station is traditionally measured with a mercury thermometer. An experienced observer can assess the temperature reading within 0.1 °C even though the calibration markings are 0.5 °C apart (Linacre, 1992). However, this degree of resolution is likely unjustified due to other temperature measurement errors related to the response time of the instrument, faulty calibration, exposure of the thermometer, or the manner of observation.

Maximum and minimum temperatures are recorded by thermometers with specialized designs. The maximum temperature thermometer is a mercury thermometer with a constriction in the mercury column above the bulb. An increase in temperature forces the mercury to expand through the constriction. Cooling temperatures result in contraction of the mercury. The column breaks at the constriction and the separated column registers the highest temperature reached since the thermometer was last reset. The maximum thermometer is mounted at an angle of about two degrees from the horizontal with the bulb slightly lower to avoid gravity helping to return mercury through the constriction (WMO, 1996).

Minimum temperature thermometers commonly contain alcohol and have an index of dark glass or metal embedded within the liquid column. The surface tension of the alcohol's meniscus pushes the index toward the bulb when temperatures cool. Warmer temperatures cause expansion of the alcohol, but it passes by the stationary index into the upper part of the thermometer. The upper end of the index shows the lowest temperature until the thermometer is reset. The minimum thermometer is mounted almost horizontally, like the maximum thermometer, to prevent the index from being moved by gravity.

Electrical temperature sensing devices are employed at some climate stations. These instruments are thermocouple devices and resistance element sensors that include resistance thermometers, thermistors, and diodes. Temperature is determined by electronic devices as a function of measured changes in electric signals produced by the instruments. Thermocouples self-generate an electric current as a function of temperature while the other electronic thermometers require application of an external signal to detect a change in a physical property (DeFelice, 1998).

A thermocouple is a loop of two different wires connected to a pair of metals with different thermal responses. When one junction has a different temperature than the other, an electromotive force is produced in the circuit and current flows producing a measurable voltage. The magnitude of the force is a function of the temperature difference between the two junctions. Thermocouples are accurate and reliable, work over a wide temperature range, and are compatible with most measuring and recording systems (DeFelice, 1998).

Resistance thermometers are made of metals and alloys whose resistance increases with temperature. These devices are fragile and are commonly used as parts of other measuring devices (DeFelice, 1998). Thermistors are ceramic semiconductors whose resistance decreases with increasing temperature. These sensors are the most widely used of the electrical temperature devices (Guyot, 1998). Diodes are another negative resistance element, but their use is limited by the difficulty in finding two diodes with the exact same features. Nevertheless, they are inexpensive and are used in analog thermometric devices (DeFelice, 1998). The use of sonic thermometers for operational monitoring purposes is still the exception, but Peters *et al.* (1998) demonstrate they can provide routine observational data.

3.5   Atmospheric pressure

Atmospheric transport that is central to climate of the first kind and the atmospheric branch of the hydrologic cycle is accomplished by the response of the Earth's atmosphere to pressure distributions. The dynamically changing distribution of atmospheric pressure is in response to the principles of energy, mass, and momentum conservation underlying the vertical and horizontal pressure gradients that set the atmosphere in motion. Measuring atmospheric pressure at the surface is essential for defining horizontal pressure patterns and related atmospheric circulation features accounting for atmospheric moisture transport.

Atmospheric pressure is the force exerted by the atmosphere on a unit area of the underlying surface. The force associated with atmospheric pressure is due to the mass of the air molecules, the kinetic activity of the gas molecules, and the pull of gravity on the gas molecules. Since weight is the gravitational force acting on a unit mass, atmospheric pressure at a given location on the Earth's surface is the weight per unit area of air above that site. Atmospheric pressure is measured in units of pascals (Pa), but the accepted practice for meteorological applications is to use the unit hectopascal (hPa) which is equal to 100 Pa (WMO, 1996). Standard sea-level pressure is 1013.25 hPa.

3.5.1   The gas law

Atmospheric pressure is one of the three variables that define the thermodynamic state of the atmosphere through the equation of state for a mixture of gases or the ideal gas law. Thermodynamics plays a central role in quantitative understanding of atmospheric phenomena and especially processes related to the general circulation of the atmosphere. The ideal gas law is a combination of Boyle's law and Charles' law expressing the relationship of pressure, volume, and temperature of a gas. The Earth's atmosphere is practically an ideal gas, and the ideal gas law is a basic starting point for understanding atmospheric dynamics. The ideal gas law for dry air is expressed as

$$P = \rho R T \tag{3.2}$$

where P is pressure of dry air in Pa, ρ is the density of dry air ($1.29\,\mathrm{kg\,m^{-3}}$), R is the gas constant for dry air ($287\,\mathrm{J\,kg^{-1}\,K^{-1}}$), and T is ambient air temperature in K.

Equation 3.2 shows that atmospheric pressure can vary in response to changes in either atmospheric density or temperature. Consequently, atmospheric pressure variations must be expected to occur both vertically and horizontally as changes in density and temperature are realized in both dimensions. The attraction of gravity compresses the atmosphere so that the maximum air density is at the Earth's surface. In addition, air temperature decreases vertically in the lower atmosphere with increasing distance above the Earth's surface, which is the major source for heating the atmosphere.

The vertical forces of gravity and atmospheric pressure acting on the atmosphere are almost always in balance. The result is atmospheric pressure decreases exponentially with height, and atmospheric pressure at any selected height is equal to the weight of the atmosphere above that height. The hydrostatic equation relates changes in pressure to changes in height in the atmosphere in full derivative form as

$$\frac{dp}{dz} = -g\rho \tag{3.3}$$

where p is pressure (hPa), z is geometric height (m), g is acceleration due to gravity ($9.81\,\mathrm{m\,s^{-2}}$), and ρ is defined above (Andrews, 2000). The negative sign indicates that pressure decreases with height.

The atmospheric pressure profile (Fig. 3.3) shows that pressure is about 700 hPa at 3000 m, 500 hPa at 5500 m, and 300 hPa at 10 000 m. These pressure values indicate that an increasingly small proportion of the atmosphere is contained in the atmospheric column above a specified level above the surface. Near the surface a change in elevation of 100 m is associated with a pressure

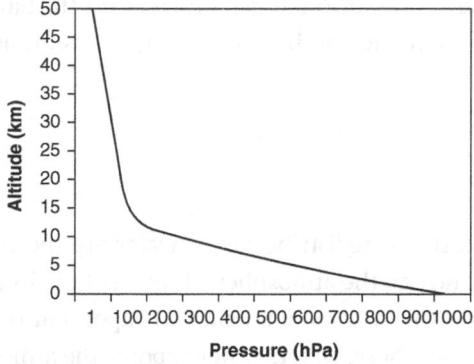

Fig. 3.3. Atmospheric pressure profile from the Earth's surface to the upper atmosphere.

decrease of 10 hPa. Horizontal pressure differences are commonly less than 10 hPa over a distance of 1000 km.

3.5.2   The barometer

A barometer is the instrument used to measure atmospheric pressure. The most common mercurial barometer is a cistern-type instrument that has a glass column closed at one end with the other end submerged in a mercury-filled reservoir. Mercury in the tube adjusts until the weight of the mercury column balances the atmospheric force exerted on the reservoir. The mercurial barometer is very accurate, but it is cumbersome and requires special care in handling and reading the values. The glass tube is attached to a wooden mounting for stability. The mercurial barometer requires adjustments for expansion and contraction of mercury with temperature changes and variations in gravity and latitude (Guyot, 1998).

Aneroid barometers are less accurate but more portable and more easily automated than mercurial barometers. The aneroid barometer is a fully or partially evacuated small, flexible metal box called an aneroid cell. The cell expands and contracts in response to small changes in external atmospheric pressure (DeFelice, 1998). A series of mechanical levers and linkages are attached to a pointer that is calibrated to indicate pressure. The cell can move plates of a capacitor in an electric circuit for automated transmission.

Piezoresistive barometers are increasingly common for both surface and upper air applications. The piezoresistive sensor uses very thin silicon membranes chemically etched on a small wafer. Two electrical resistance elements are deposited along the grain of the crystal, and two other elements are aligned across the grain. This configuration provides opposite resistance changes as the membrane stretches in response to pressure changes, and it alters the balance of an active Wheatstone bridge. Balancing the bridge accomplishes an accurate pressure determination (WMO, 1996).

3.6   Humidity

The atmospheric transport and redistribution of water are the common elements of climate of the first kind and the atmospheric branch of the hydrologic cycle. Water occurs in the atmosphere as a solid, liquid, and vapor, but the vapor phase is the dominant form of atmospheric water. Water vapor is the atmospheric constituent with the greatest effect on atmospheric dynamics and thermodynamics and the atmospheric radiation balance. However, the atmospheric water vapor concentration varies enormously in the troposphere on small spatial and temporal scales. This important variable is a challenge to measure, but knowledge of the distribution of atmospheric water vapor is of fundamental importance to weather and climate, atmospheric radiation studies, and the global hydrological cycle.

Direct measurement of the water vapor content of the atmosphere requires instrument arrays found in laboratories rather than at ground-based observation sites. For routine observations, determining the water vapor content of the atmosphere is achieved using indirect measurement techniques that are complemented by expressing the water vapor content of the atmosphere as humidity.

3.6.1   Humidity expressions

There are several ways to express humidity, and each is controlled to some extent by air temperature. Absolute humidity is the mass of water vapor per unit volume of air and is expressed in units of grams of water vapor per cubic meter of air. Absolute humidity is a measure of the actual amount of water vapor in the air, but it is sensitive to changes in both air temperature and atmospheric pressure. Specific humidity is the mass of water vapor in the air per unit mass of air. Specific humidity is not sensitive to volume changes in the air, and it is expressed in units of grams of water vapor per kilogram of air. Relative humidity is a more easily achieved measure of water vapor in the air, but it is a measure of water vapor relative to ambient air temperature. It expresses the actual amount of water vapor in the air compared to the total amount of water vapor that can exist in the air at its current temperature. Relative humidity (RH) is determined using water vapor partial pressure and saturation vapor pressure and is expressed as a percentage as shown by

$$RH = \frac{e_a}{e_s} \times 100 \qquad\qquad (3.4)$$

where e_a is the actual vapor pressure in hPa and e_s is the saturation vapor pressure at the observed temperature in hPa. Other water vapor indicators include dew point temperature, saturation deficit, and the mixing ratio. The relationship among these indicators is found in standard textbooks.

3.6.2   Humidity measurement

Humidity is measured with devices known generically as hygrometers or as a hygrograph if fitted with a recording device. Hygrometers employ several different measuring methods and instrument designs that are based on the hygroscopic quality of material, on the longitudinal change of human or animal hairs, on cooling caused by evaporation of water, or on the change of electrical resistance (WMO, 1996). The earliest hygrometers used human hair that stretches as it absorbs moisture. The bundle of human hair is linked mechanically to a pointer that is calibrated to record relative humidity.

The instrument commonly found in a weather station is based on temperature differences and is called a psychrometer. A standard mercury or electrical thermometer indicating ambient air temperature is paired with a second mercury or electrical thermometer whose bulb is covered with a wick or sleeve saturated with distilled water. The thermometers are positioned in a ventilated space. Water is evaporated from the wick into the air until equilibrium between the wick and the air is reached. The temperature of the wet-bulb thermometer is lowered by an amount determined by the evaporative cooling. The equilibrium temperature indicated by the wet-bulb thermometer is known as the wet-bulb temperature. The difference between the dry-bulb and wet-bulb temperatures is the wet-bulb depression (DeFelice, 1998). For the wet-bulb thermometer at equilibrium

$$e_a = e_s - AP(T - T_w) \tag{3.5}$$

where e_a is the actual vapor pressure in hPa, e_s is the saturation vapor pressure at T_w in hPa, A is a constant of proportionality with a value of $6.6 \times 10^{-4}(1 + 1.15 \times 10^{-3} T_w)$, P is air pressure in hPa, T is the dry-bulb temperature in °C, and T_w is the wet-bulb temperature in °C. An alternative to Equation 3.5 for determining e_a is to use a psychrometric chart that permits T, T_w, e_a, the dew point temperature, relative humidity, and the mixing ratio to be determined when any two are known (Linacre and Geerts, 1997).

When T_w is known, e_s can be determined from standard tables or graphs based on the Clausius–Clapeyron relationship. The form of the Clausius–Clapeyron equation familiar in atmospheric science and applied to water vapor in the presence of air is

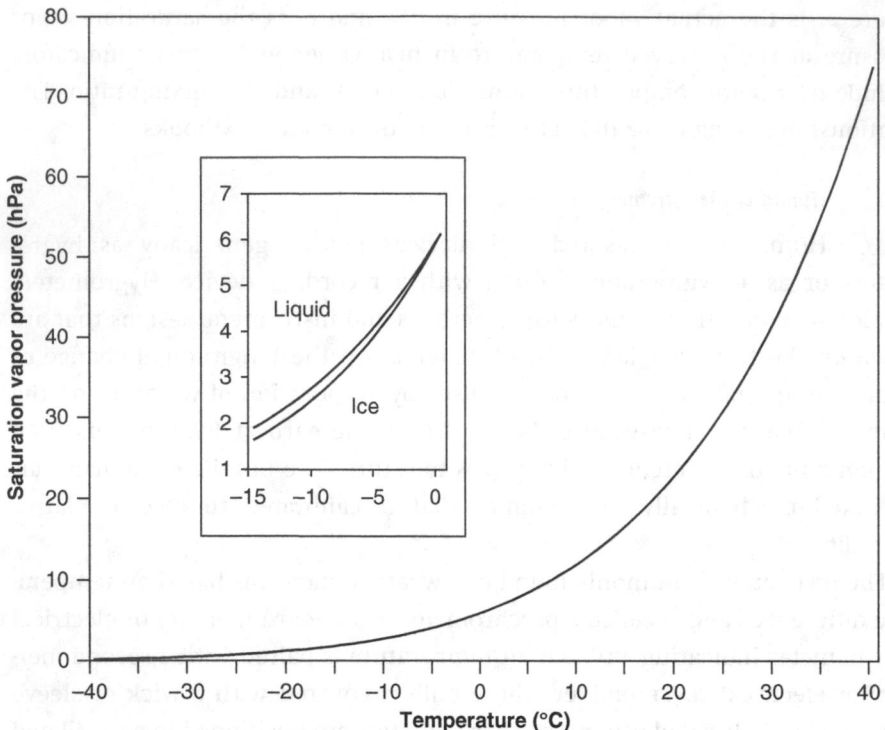

Fig. 3.4. Variation of saturation vapor pressure with air temperature.

$$\frac{de_s}{dT} = \frac{L_v e_s}{R_v T^2} \tag{3.6}$$

where L_v is the latent heat of vaporization ($2.453 \times 10^6\,\mathrm{J\,kg^{-1}}$), R_v is the gas constant for water vapor ($4.615 \times 10^2\,\mathrm{W\,kg^{-1}\,K^{-1}}$), and e_s and T are defined previously (Andrews, 2000). Figure 3.4 shows the relationship between temperature and saturation vapor pressure defined by Equation 3.6. The slope of the curve changes at an exponential rate, and the curve decreases at a different rate at temperatures below $0\,°\mathrm{C}$ depending on whether the medium is liquid or ice. These characteristics are important for determining saturation of the atmosphere and precipitation formation and are described in Chapter 4.

3.7   Radiosonde upper-air measurements

Upper-air climatology is less well known than surface conditions, but upper-air observations provide direct measurements of atmospheric state variables above the Earth's surface. These measurements are an important complementary data source for understanding atmospheric circulation and its transport of energy and mass. The upper-air data provide a vertical dimension that reveals the atmospheric response to time and space variations required to maintain energy, mass, and momentum balances and the circulation patterns shaped by these adjustments. The circulation patterns are major factors in atmospheric moisture transport that is fundamental to climate of the first kind and the atmospheric branch of the hydrologic cycle.

Measuring vertical profiles of atmospheric state variables provides understanding of how the atmosphere responds to and contributes to conditions observed at the Earth's surface. The collection of data in the free atmosphere relies heavily on radiosondes. A radiosonde is a small, lightweight package of battery-powered, expendable meteorological instruments attached to a helium-filled free-flying balloon. The balloon-borne instruments simultaneously measure and transmit meteorological data while ascending through the atmosphere at intervals that vary from 1 to 6 s depending on the type and manufacturer of the radiosonde (Dabberdt et al., 2003). When wind information is processed by tracking the balloon's movement the instrument package is known as a rawinsonde.

Radiosondes provide vertical profiles of atmospheric temperature, pressure, humidity, and wind from the Earth's surface up to a height of 20 to 30 km where atmospheric pressure is 10 hPa (Fig. 3.5). The data are transmitted in a predetermined sequence to a ground receiving station where the data are processed at a fixed time interval. Radiosonde temperature data are supplemented by aircraft observations, particularly over the oceans where radiosonde launches are limited to remote islands or ships (Dabberdt et al., 2003). Combining radiosonde and aircraft data requires consideration that the aircraft may disturb the temperature, pressure, and moisture fields it is measuring to a greater extent than a balloon.

The global radiosonde network includes about 900 upper-air stations and approximately two-thirds of these stations make observations twice daily. Global radiosonde launchings at 0000 and 1200 Coordinated Universal Time (UTC) began in 1957. WMO member countries form part of the Global Observing System of the World Weather Watch program and share their sounding data with other members. The synoptic radiosonde/rawinsonde observing programs are designed to meet real-time operational needs for weather forecasting and analysis requiring simultaneously acquired observations at a large number of locations and with a high vertical resolution. The radiosonde network is predominantly land-based and favors the middle latitudes of the Northern Hemisphere. Unfortunately, there has been a deterioration of the radiosonde network due to the loss in September 1997 of the Omega radionavigational system used to track the sondes and the closure of radiosonde stations in some countries to reduce operating costs. The eight-station Omega system was a global network, but it became too expensive to operate with the success of Global Positioning System (GPS) technology.

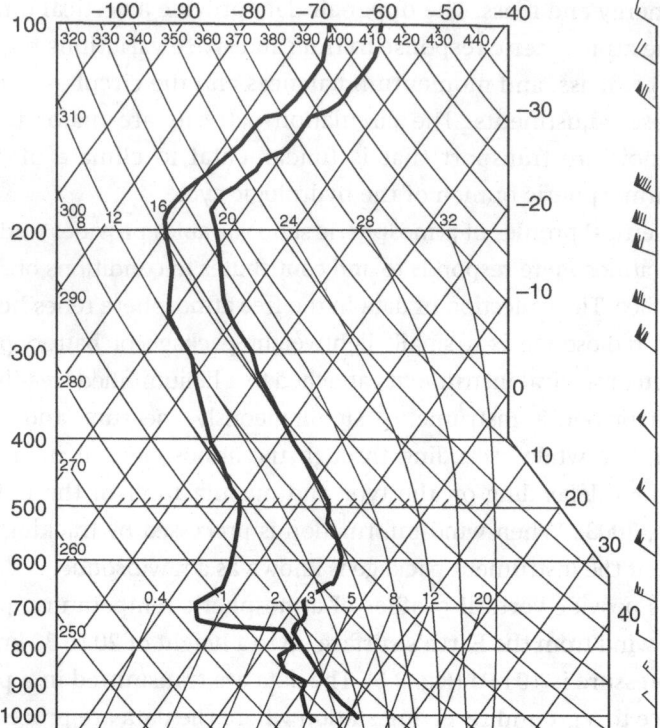

Fig. 3.5. Radiosonde-based atmospheric sounding for 0000 UTC 20 November 2006 at Oakland, California (38° N, 122° W). The bold line plotted to the right is atmospheric temperature, and the bold line plotted to the left is dew point temperature. Pressure coordinates (hPa) are used for the vertical axis. Wind direction and speed are shown along the right margin with the barb pointing into the direction from which the wind blows and the length of the flags indicating wind speed. (Data courtesy of the NOAA Forecast System Laboratory from their website at http://raob.fsl.noaa.gov/.)

3.7.1   Radiosonde instruments

There is no worldwide standard radiosonde, and radiosonde observations are made with a greater variety of sensors and devices than surface observations (Elliott *et al.*, 2002). The most widely used radiosondes are produced by China, Finland, Japan, Russia, and the United States (Luers and Eskridge, 1998). In general, the instruments in these radiosondes have different error and response characteristics and undergo modifications and improvements during the life-time of a particular model.

Temperature sensors in contemporary instrument packages are thermistors, capacitive sensors, resistance wires, bimetallic elements, or thermocouples (WMO, 1996). Radiosondes widely used in the United Kingdom and the United States measure temperature with a temperature-sensitive capacitor consisting of two electrodes separated by a ceramic dielectric (Hudson *et al.*, 2004). Atmospheric temperature measurement by every radiosonde instrument is affected by heating from sources other than the air itself that make the sensor temperature different from the ambient air temperature. A temperature correction is required to determine the actual air temperature. This correction is small at altitudes below 15 km but can be substantial at altitudes between 20 and 30 km. A major difference among all radiosondes is the nature of the temperature correction applied to the temperature sensor value (Eskridge *et al.*, 2003).

The most widely used radiosondes determine atmospheric pressure changes using the change in measured capacitance that occurs between two capacitance plates (Dabberdt *et al.*, 2003). Other radiosondes use a classical aneroid-type sensor that is two electrodes separated by a distance that varies depending on the volume of a partially evacuated and expandable cell. A piezoresistance element is sometimes used for pressure measurements, and a water hypsometer has been used successfully by the Swiss Meteorological Institute (Richner *et al.*, 1996).

Most radiosonde instrument packages have an electric resistive hygrometer to determine humidity. This device uses a thin piece of lithium, a carbon hygristor, or other semiconductor material and measures the resistance which is affected by humidity. An alternative design is two electrodes separated by a thin polymer film. The capacitance varies depending on the amount of water absorbed by the polymer film and on the film temperature. Relative humidity is calculated from the capacitance and temperature data (Hudson *et al.*, 2004).

A radio transmitter sends the sensor data to a ground-based receiving station. The balloon ascends at a rate of 5 to 8 m s^{-1} until it eventually bursts due to the low pressure and low temperature. This rate of ascent allows the measurements to 30 km to be completed in about 90 minutes. The instrument package may be equipped with a parachute for its descent back to the surface, but only 25% of routinely released radiosondes are ever recovered (DeFelice, 1998).

Wind is not measured directly by the radiosonde, but wind directions and velocities are determined from the balloon drift. Determining the balloon drift requires the ability to track the balloon after it is launched or to locate the balloon during its flight. Balloon tracking uses an optical system, a radio signal, or radar. The optical system uses a theodolite to determine the balloon's azimuth and elevation. The radio signal uses a radio theodolite to determine the same information from the transmitter on the balloon. Both of these methods require knowledge of the balloon's height, which is determined from the pressure measurement, and both require clear conditions between the balloon and the ground station. Equipping the balloon-borne package with a radar reflection shield permits the balloon's position to be tracked by radar. Tracking changes in the balloon's position by three or more ground-based stations provides a basis for calculating wind directions and velocities. All three of the tracking methods have reduced accuracy as the distance to the balloon increases and its elevation angle decreases (Dabberdt *et al.*, 2003).

Locating the balloon during its flight is achieved using one of the navigation systems developed for other purposes. The LORAN-C is a radionavigation system developed by the United States, Canada, and Russia for maritime purposes. A transponder on the balloon receives the navigation signals and rebroadcasts them back to the ground station with the meteorological measurements. This system requires pressure measurements to determine the height of the balloon. An emerging navigation-based technique uses the satellite-based GPS. This technique is very accurate for measuring the position and velocity of the radiosonde, and the GPS has worldwide coverage. However, the cost of GPS technology is high for an expendable application (Dabberdt *et al.*, 2003).

3.7.2   Radiosonde data

Archived radiosonde data in national and international databases begin after 1945 and provide upper-air observations to support a variety of operational and research applications important to hydroclimatology. Spatial patterns of temperature and humidity, and pressure and temperature trends at the surface and in the troposphere and lower stratosphere have been developed from radiosonde data (Dai *et al.*, 2002; Angell, 2003). Constructing temperature and water vapor climatologies and parameterizing water vapor and cloud processes are other areas employing radiosonde data (Ross and Elliott, 2001; Miloshevich *et al.*, 2004). However, utilizing archived radiosonde data for hydroclimatic research requires caution due to the variety of sensors employed globally and changes in instrumentation and observation methods that introduce biases in the data (Elliott *et al.*, 2002; Angell, 2003). Also, radiosonde data are subject to errors due to radiational heating and cooling of the sensor arm, calibration errors, and chemical contamination of the sensors (Hudson *et al.*, 2004). Removal of errors and biases must precede use of the data for hydroclimatic analysis. Metadata, satellite data, and heat transfer models of the instruments are employed to identify and remove errors and biases (Parker *et al.*, 2000).

Each balloon ascent, or sounding, produces about 20 data levels with readings on temperature, pressure, humidity, and related winds. With two soundings each day reported for as many as 2300 upper-air land and marine stations, the magnitude of the raw data is an immense validation and data management problem. The WMO's GCOS establishes the specific record length and homogeneity requirements for upper-air stations. The GCOS Upper Air Network (GUAN) includes the globally distributed stations that meet the GCOS requirements. GUAN strives to ensure the maintenance of key stations and availability of their data (Parker *et al.*, 2000). Quality control and archiving of upper-air data from the GUAN stations is jointly managed by the Hadley Centre of the UK Meteorological Office and NOAA's National Climatic Data Center (NCDC).

Two sets of radiosonde data for GUAN stations are available. Current data from the Global Telecommunications System (GTS) consist of soundings received at NCDC that receive no quality control screening other than review for duplicate records and syntax errors. Historical upper-air data are archived in the Comprehensive Aerological Reference Data Set (CARDS) which is a joint project of NOAA's NCDC and the All-Union Research Institute of HydroMeteorological Information, Russia. CARDS data consist of soundings from multiple data sources that have been processed by the CARDS Complex Quality-Control System (Eskridge *et al.*, 1995). Additional adjustments to the CARDS data to remove inhomogeneities are described by Free *et al.* (2002). CARDS contains over 27 million quality controlled radiosonde observations for the period 1940 to 2000, and this number increases with each revision of the CARDS dataset. Monthly mean data (MONADS) for each CARDS station are available (Elliott *et al.*, 2002). In addition, data sets of monthly and seasonal temperature anomalies and monthly CLIMAT TEMP are provided by the Hadley Centre Radiosonde Temperature (HadRT) products based on monthly average temperature reports from radiosonde operators.

Review questions

3.1 How are instruments that measure solar radiation different from instruments that measure thermal radiation?

3.2 What is the basis for the suggestion that surface air temperature is an indirect indicator of net radiation?

3.3 Where would you expect to find a site with the greatest annual variation in the amplitude of surface temperature?

3.4 What fraction of the mass of the atmosphere lies below the top of Mount Everest (8848 m)?

3.5 What influence does water vapor have on the pressure exerted by a column of air?

3.6 How does a change in air temperature influence the saturation vapor pressure of air?

3.7 In what way was the radiosonde a major advancement in monitoring the atmosphere?

4

Measuring hydroclimate terrestrial components

4.1   A terrestrial focus

Climate of the second kind and the terrestrial branch of the hydrologic cycle embrace the suite of natural processes at or near the land surface that account for the conversion of precipitation into streamflow. The primary variables involved with these processes are fluxes represented by precipitation, evaporation and evapotranspiration, and runoff. These fluxes of energy and mass related to land surface processes are dominantly oriented upward or downward relative to the land surface, and they are aided by energy and mass storages and sinks occurring at or near the land surface. Wind and soil moisture are additional variables that augment or dampen the rate of energy or mass exchange between the land surface and the atmosphere. Wind is a horizontal flow that is a fundamental factor influencing evaporation and evapotranspiration. Soils have non-linear flow properties that transform sudden changes in land surface conditions into gradual changes in subsurface water movement. In addition, soil moisture storage provides water for plant transpiration during periods between precipitation events.

Efforts to generalize the characteristics of the energy and mass fluxes defining climate of the second kind and the terrestrial branch of the hydrologic cycle are complicated by the complexity of the Earth's land surface. The land surface supports an array of heterogeneities and discontinuities related to forcing inputs, state conditions, and land surface properties that influence moisture processing. Heterogeneity of the land surface is obvious in soils, vegetation, and topographic differences at various spatial scales (Becker, 1995). Less evident are the variations in energy and moisture inputs related to surface property differences. These differences may be most apparent when relatively flat terrain is compared to adjacent sloping terrain. The spatial extent of the heterogeneities is revealed by surface observations of variables participating in climate of the second kind and the terrestrial branch of the hydrologic cycle.

Selected hydroclimatic variables at the Earth's surface and instruments used to collect data for these variables are presented in this chapter. Detailed discussions of these instruments are found in the publications cited in Section 3.2.

4.2   Terrestrial hydroclimatic data

Data for hydroclimate analysis involving Earth's surface processes are drawn from both climatology and hydrology. These disciplines share an interest in most of the variables, but the specific characteristics of a variable that are of interest to each discipline may be different. Precipitation illustrates how the perspectives may be similar and different. Both hydrology and climatology are concerned with the quantity and distribution of precipitation and the form in which precipitation occurs. However, the climatological interest in precipitation emphasizes the atmospheric mechanisms responsible for the delivery and distribution of precipitation while the hydrologist focuses on the volume of rainfall received at the surface and its disposition.

One variable inherent to hydroclimatology and clearly the domain of hydrology is runoff or streamflow. Streamflow is observed at a specific location, but the observed data integrate multiple runoff processes operating over an area. Streamflow expresses the depth of runoff over the area contributing to the site where streamflow is measured, but runoff is not measured at discrete locations within this area. This is a distinct contrast to climatic data that in the classic sense are point data based on observations at a specific location representing conditions at that particular site. Streamflow measurement presents different instrument requirements than those needed for observing climate variables, but the significant distinction is the conceptual contrast between point data and areal data.

The areal nature of streamflow measurements emphasizes the need to have areal estimates of other hydroclimatic variables. This can be achieved by employing a network of observation stations or by using information from remote sensing by radar (radio detection and ranging) or satellites. During the last two decades of the twentieth century, a variety of radar, satellite, and data storage and retrieval technologies became available for application to hydroclimatological problems. These technologies provide valuable approaches for bridging differences between point data and spatially distributed data to achieve improved understanding of processes relevant to the terrestrial component of hydroclimatology. Instruments for observing point data relevant to terrestrial hydroclimatology and streamflow are discussed first in the following sections. Techniques for acquiring areal estimates of point data are addressed in

Section 4.8. Radar and satellite data for acquiring areal estimates of hydroclimatic terrestrial components are presented in Chapter 5.

4.3   Precipitation formation

Precipitation is the moisture flux from the atmosphere to the Earth's surface coupling the atmospheric and terrestrial branches of the hydrologic cycle and serving as the downward directed mass flux for climate of the second kind. Rain and snow provide the primary moisture inputs to the land phase of the hydrologic cycle. Dew and fog-drip provide small moisture inputs for specific locations that are very important locally, but the moisture quantity delivered by these processes is small compared to rain and snow. Accurate measurement of rain and snow is essential for hydroclimatic analyses, but precipitation is highly variable in both time and space compared to temperature and pressure. The general problem of data representativeness is particularly challenging regarding precipitation due to its time and space variability, which is a function of the processes responsible for producing precipitation.

Precipitation is produced in a series of stages beginning with supersaturation of ascending and cooling air. Condensation, droplet formation, and successful descent of the droplets or particles to the Earth's surface complete the atmospheric cycling of moisture. The amount of water vapor present in the air is a fundamental factor to precipitation formation. Quantifying the amount of water vapor in the atmosphere is achieved by determining the contribution water vapor makes to total atmospheric pressure. This is the partial pressure due to water vapor or the vapor pressure, and it is typically about 2.5 hPa out of the total atmospheric pressure (Trenberth, 1998). The common convention is to view the process as occurring in a defined volume of air identified as an air parcel. Saturation is the state achieved when evaporation and condensation of water molecules in the parcel are in equilibrium. Saturation is quantified by relative humidity (see Equation 3.4) of 100%.

The saturation vapor pressure is the pressure at which the air parcel is saturated with water vapor, and it varies only with temperature (see Fig. 3.4). This relationship has important consequences relative to the atmosphere and precipitation formation. The first feature is that the saturation vapor pressure of warm air is greater than the saturation vapor pressure of cold air when the quantity of water vapor in the volume is unchanged. The second characteristic is that cooling an air parcel is an efficient means for achieving saturation of air as an initial step in precipitation formation.

A key mechanism for cooling air is to induce vertical displacement of the parcel by lifting the parcel to higher elevations. The vertical ascent of the parcel

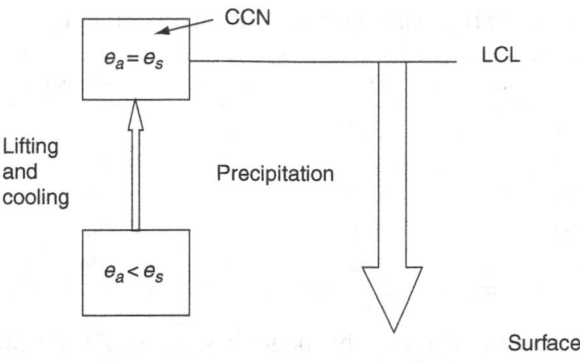

Fig. 4.1. Simplified diagram of the precipitation formation process.

results in cooling as the parcel uses internal energy to expand as it moves into a decreasing free atmospheric pressure environment with increasing altitude. The processes resulting in precipitation are depicted in Figure 4.1. The parcel cools initially at about 9.8 °C km^{-1} of ascent. This cooling is known as the dry adiabatic lapse rate. The free atmosphere cools at a mean vertical rate for the troposphere of 6.5 °C km^{-1}. However, the tropospheric lapse rate varies with time and space.

Three types of lifting mechanisms are recognized as being responsible for triggering vertical atmospheric motion. Convective lifting is related to turbulent air currents initiated by frictional drag due to rough surface features, eddy circulations in the surface air flow, or thermal differences in surface materials that result in differential heating of the air. Frontal lifting occurs due to the contrasting air mass characteristics along the boundary separating expanses of air with different densities and temperatures. The nature of the advancing air mass determines the frontal slope and the horizontal and vertical character of the lifting. Orographic lifting is associated with the forced ascent of air resulting from horizontal flows encountering a terrain barrier such as mountains. The mountains' height influences the character of the lifting, but wind direction relative to the mountains' orientation influences the overall character of the atmospheric lifting.

The lifting condensation level (LCL) defines the height to which a moist air parcel must ascend for adiabatic cooling to produce saturation of the parcel with respect to a plane surface of water. Once saturation of rising air is achieved, condensation can occur if the air is cooled further. Water vapor requires a surface on which to condense because the process would require rarely observed high degrees of supersaturation of air to allow water molecules to collide and stick to produce a droplet. In the free atmosphere, the condensation surface is provided by microscopic impurities suspended in the air and known as cloud condensation nuclei (CCN). These particles have diameters of 0.1 to

Table 4.1. *Selected precipitation types and representative physical characteristics*

Precipitation type	Intensity $(cm\,h^{-1})$	Median diameter (mm)	Fall velocity $(m\,s^{-1})$
Fog	0.013	0.01	0.003
Snow	0.64	3	1
Light rain	1.02	1.24	4.8
Heavy rain	1.52	2.05	6.6
Cloudburst	10.2	2.85	7.9

10 µm and originate from natural and anthropogenic sources. The droplets that form around CCN are relatively small and remain buoyant due to the continuous rising air stream from the surface. These droplets may become dense enough to be visible as clouds. Growth of the cloud droplets is necessary for them to achieve a mass with a terminal velocity great enough to exceed the buoyancy provided by the vertical air currents. Droplet growth depends on the cloud temperature where the growth occurs.

In clouds with temperatures above freezing, droplets can grow by collision and coalescence with other droplets. This growth process is enhanced when droplets of different sizes are involved due to differences in terminal velocities. The greater terminal velocities of larger droplets permit them to overtake smaller droplets. Clouds extending above the freezing level support droplet growth by a three-phase process or the Bergeron process. These clouds contain ice and supercooled water above the freezing level. Since the saturation vapor pressure over ice is less than the saturation vapor pressure over water at the same temperature (see Fig. 3.4), water molecules move from supercooled drops to ice crystals. The ice crystals grow while the water droplets become smaller. The terminal velocity of the ice crystals increases, and they fall toward the Earth's surface. The ice crystals melt and arrive as a water droplet if they encounter air above freezing before they reach the surface. Selected characteristics of different precipitation types are listed in Table 4.1.

4.4   Rainfall

Rain is the most common form of precipitation and is the dominant moisture input to the hydrologic cycle except for latitudes north of 50° N in North America and Eurasia (Peixoto and Oort, 1992). Conceptually, rainfall measurement is straightforward, but there are more than 40 rain gauge designs used throughout the world (Linacre, 1992). These gauges display little agreement on the size of the canister opening or on the height of the gauge above the ground. In addition, a high-density rain gauge network is required to observe total areal

Fig. 4.2. Standard rain gauge used in the United States. (Photo by author.)

rainfall from rain gauge observations, and such high-density rain gauge networks are generally not available (Liu, 2003). Measured rainfall is expressed as a depth of water in cm or mm for a specified time.

4.4.1   Rainfall measurement

Rainfall measurement is typically accomplished by catching rain in a flat bottomed, vertically sided canister. It is assumed that the amount or depth collected per unit area of the gauge aperture is the same as the amount that falls per unit area on the surrounding surface (DeFelice, 1998). The precipitation falling into the cylinder is amplified for ease of measurement by a funnel that directs the precipitation catch to an inside can. The cross-sectional area of the inner can is one-tenth the area of the aperture to facilitate measurements of 1 mm. Depth is measured periodically using a graduated dipstick or continuously using a weighing mechanism or other mechanical devices. Since the rain gauge is basically a cylinder sitting in the airstream, turbulence over the mouth of the canister represents the major accuracy problem (Groisman and Easterling, 1995; Peck, 1997). Other systematic errors in rain gauge measurements are due to wetting of the internal walls of the gauge, evaporation from the gauge, splashing into or out of the gauge, and blowing snow (Legates and Willmott, 1990).

The WMO adopted a standard gauge based on the British design. This gauge has a cylinder diameter of 127 mm. The rim is 1 m above the ground to prevent water splashing into the cylinder (Linacre, 1992). Rain gauges used in the United States have slightly different dimensions than those of the WMO gauge. The standard U.S. National Weather Service rain gauge has a 203 mm diameter opening that is 800 mm above the ground (Fig. 4.2). The majority of rain gauges in the United States are not routinely equipped with wind shields to combat the turbulence known to influence instrument accuracy. Consequently, rain gauge measurements tend to underestimate true precipitation by 5% to 40% with an average bias of 9% (Groisman and Legates, 1994; Duchon and Essenberg, 2001). Wind influences are a particular problem when snow occurs. Snow may accumulate in the receiving canister and then be blown away before it melts or the wind may reduce the amount of snow that collects in the canister producing a bias of up to 50% (Groisman and Easterling, 1995). In general, the standard rain gauge underestimates the snow contribution, and snowfall measurement with the standard gauge is not recommended.

When rain gauges cannot be observed daily, storage or recording gauges are employed. A storage gauge has an enlarged receiving vessel to accommodate greater accumulations of rainfall, but attention must be given to reducing evaporation losses between observations. Recording rain gauges are used in automated or remote weather stations and are classified on the basis of the mechanism used in the recording process (Linacre, 1992). The three major types are weighing, float and siphon, and tipping-bucket. The tipping-bucket type is widely used for real-time data transmissions. Detailed characteristics of storage and recording rain gauges are discussed by WMO (1996), DeFelice (1998), and Guyot (1998).

4.4.2   *Estimating point rainfall*

In the absence of a rain gauge at a selected point, multiple regression and other statistical techniques are employed to estimate point rainfall for the site. However, the predictor variables may be unique for a given site and rainfall accumulation period. Elevation, slope, aspect, latitude, distance from a water body, and distance from an existing rainfall gauge are common predictors. Estimating errors are smallest when the distance to the nearest rain gauge is smallest and when the averaging period for rainfall is longest (Linacre, 1992). Still, it is important to remember that intense and highly localized rainfall can produce enormous differences over small distances.

Correlations between large-scale atmospheric variables and observed surface precipitation have demonstrated encouraging results for estimating point precipitation. Hayes *et al.* (2002) show that 850 hPa wind and humidity radiosonde data satisfactorily estimate station data in mountainous regions of western Washington state, and they suggest interpolation for points between stations should be successful.

4.5   Snowfall

Over much of the Earth's extratropical continental area, a significant portion of precipitation falls as snow. The principal climatic roles of snow relate to its high reflectivity or albedo, low thermal conductivity, water-holding

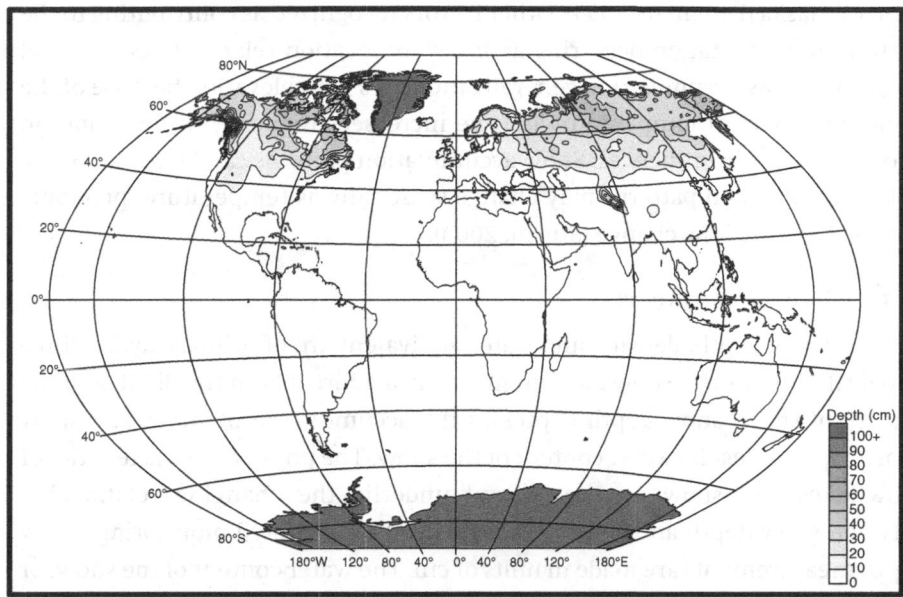

Fig. 4.3. Global annual mean snowcover. (NCEP Reanalysis data courtesy of NOAA/OAR/ ESRL PSD, Boulder, Colorado, USA, from their website at http://www.cdc.noaa.gov/.)

capacity, and thermal inertia (Sturm, 2003). Snow is stored on the surface for periods ranging from hours to months before melting and continuing through the land phase of the hydrologic cycle. It should be immediately apparent that snowmelt is effectively delayed precipitation. Moreover, the high albedo of fresh snow contributes to snow serving as a cold reserve that acts to depress temperatures in the spring when energy is required to warm the snow to 0 °C before melting occurs. In many areas, snowmelt is the main surface-water supply and the main cause of flooding. Snow and snowmelt play an important role in the hydrology of the middle latitudes and of rivers originating in high mountains. For mountainous regions, snowmelt contributes at least 50% of the annual runoff and exceeds 95% in some locations (Rango and Martinec, 1995).

Virtually all land areas above 40° N have a seasonal snowcover of significant duration (Fig. 4.3). This amounts to about 42% of the Northern Hemisphere. However, the perennial cryosphere covers only 8% of the Earth's surface. The annual cycle of snowcover shows the Northern Hemisphere cycle has a larger amplitude than the Southern Hemisphere (Peixoto and Oort, 1992). This is mainly due to the vast extent of snowcover during the northern winter. For the globe, the variation in snow plus ice cover is dominated by changes over the Northern Hemisphere continents. The fraction of average annual precipitation falling as snow shows a steep latitudinal increase and reaches 0.65 on the north coast of Alaska (Dingman, 1994). Other factors recognized as contributing to the portion of precipitation occurring as snow are location relative to oceans and elevation. Snow accumulation generally increases with elevation because of the combined effect of temperature and the increased frequency of precipitation due to orographic influences. Snow accumulation patterns can be complex due to topography, and patterns may change seasonally as temperature, precipitation, and the air flow change (Sturm, 2003).

4.5.1   *Snowfall measurement*

Snow depth, density, and water equivalent are of primary hydroclimatological importance. Snow measurement must address both the depth and the density of snow. Snow depth expresses the accumulation of snow, and snow density quantifies the water content of the snow. The physical characteristics of snowflakes and snow on the ground underlie the changing relationship between snow depth and density and the need for repeated monitoring. Snow depth measurements are made in units of cm. The water content of the snow, or snow water equivalent (SWE), is the vertical depth of water obtained by melting the snowcover. The usual unit of measurement for SWE is mm (WMO, 1996).

Snowflakes are loose aggregates of ice crystals, most of which are branched. Much of the precipitation reaching the ground begins as snowflakes in clouds. Snowflakes commonly melt before reaching the surface and the precipitation assumes a different form as either rain or sleet. In winter the freezing level in the atmosphere may be close to the surface and falling snowflakes have a better chance of surviving to reach the surface. In relatively dry air, snowflakes may reach the ground even when the air temperature is considerably above freezing because the partial melting of the snowflake chills the remainder of the snowflake and retards the rate of melting. Air temperature is a reasonable index of precipitation type, and chances are that snow will occur at temperatures below 0.6 °C (Linacre, 1992).

When snowflakes reach the surface they begin a process of metamorphism that continues until melting is complete. Accumulating snowflakes at the surface constitute the snowpack which builds layer by layer. The initial characteristics of each layer are determined by how much solid precipitation falls, whether the precipitation is accompanied by wind, and the temperature at the time of deposition. After deposition, each layer is subjected to thermal and mechanical metamorphic processes that alter the layer characteristics. New snow layers densify rapidly and settle as the snow crystals fragment and become more rounded (Sturm, 2003). The resulting snowpack is a granular porous medium consisting of ice and pore spaces. When the snow is cold, or its temperature is below the melting point of ice (0 °C), the pore spaces contain only air or water vapor. At the melting point, the pore spaces can contain liquid water as well as air, and the snowpack becomes a three-phase system.

Accurate snow measurement is difficult because falling snow is sensitive to turbulence and to wind depletion that produces large variations in snow depth and density over short distances. For locations that receive occasional snow during the winter, the common method for measuring snow depth is to use a ruler attached vertically to a board placed on the ground or on the previous snow surface. In areas where snow accumulation is substantial, a snow stake or pole with graduated markings is used. In remote areas, aerial snow markers that can be read from airplanes are used to determine snow depth and snow pillows equipped with pressure transducers measure the weight of accumulated snow (WMO, 1996).

The liquid water content is an important physical property of snow. Liquid water content (Θ) is defined as the ratio of the volume of liquid water in the snowpack to the total snow volume

$$\Theta = \frac{V_w}{V_s} \tag{4.1}$$

where V_w is the volume of liquid water and V_s is the volume of snow. Snow density (ρ_s) is the mass per unit volume of snow

$$\rho_s = \frac{\rho_i V_i + \rho_w V_w}{V_s} \tag{4.2}$$

where ρ_i is the density of ice, V_i is the volume of ice, and ρ_w is the density of water. The SWE refers to the amount of water the snow contains. SWE is expressed as the depth of water that would result from the complete melting of the snow in place. SWE is commonly estimated as

$$SWE = \left(\frac{\rho_s}{\rho_w}\right) h_s \tag{4.3}$$

where ρ_s is snow density, ρ_w is the density of water, and h_s is snow depth. In this form, snow density is understood to mean the relative density compared to water. A representative SWE for more than one snow sampling site is estimated by using the arithmetic mean of the snow depth measurements in Equation 4.3 (Seidel and Martinec, 2004).

The density of new-fallen snow is determined by the configuration of the snowflakes. Snowflake configuration is largely a function of air temperature, the degree of supersaturation in the precipitating cloud, and the wind speed at the surface of deposition. Observed densities of fresh snow range from 0.004 to 0.34. The lower values for powder snow occur under calm, very cold conditions, and the higher values for snow occur with higher winds and temperatures (Bras, 1990). It is difficult to measure the density of new snow and an average relative

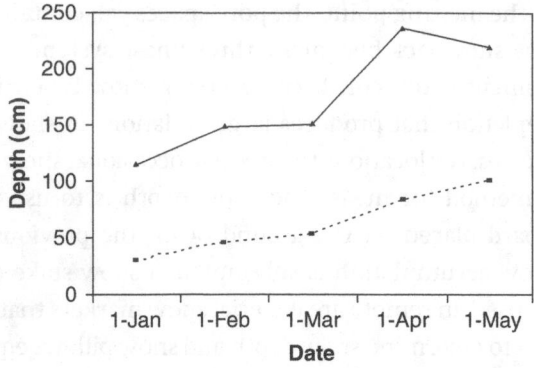

Fig. 4.4. Snow course measurements of snow depth (solid line) and snow water equivalent (broken line) for the first five months of 2006 at Echo Summit, California (39° N, elevation 2258 m). (Data courtesy of the California Department of Water Resources from their website at http://cdec.water.ca.gov/.)

density of 0.1 is commonly assumed. This is the basis for the widely used "rule of thumb" in mountainous areas of North America that 1 cm of snow equals 1 mm of water.

The standard method for measuring SWE is by gravimetric measurement using a tube to obtain a sample snow core. This method is the basis for snow survey procedures used in many countries. The snow sample from the tube is either melted to determine the liquid content or the frozen sample is weighted (WMO, 1996). Specific areas identified as snow courses are designated and visited around the first of each month from January through June. Sites selected as snow courses require the same protection from wind movement as needed by rainfall gauging sites to produce a representative measurement (Peck, 1997). Snow depth is determined and the snow density is measured by weighing an aluminum tube used to extract a core sample of snow. The Mount Rose snow sampler widely used in North America has an inside diameter of 38 mm so that a snow core weighing 28 g is equivalent to 25 mm of water (Bruce and Clark, 1966). Measurements for a snow course in the central Sierra Nevada in California during 2006 are shown in Figure 4.4. Repeat visits are necessary because as soon as snow accumulates on the surface it begins a process of metamorphism that continues until melting is complete. At the beginning of the melt season, the snowpack is typically vertically heterogeneous with several layers of markedly contrasting densities. During melt, density continues to increase and the vertical inhomogeneities tend to disappear due to the formation and drainage of meltwater. A snowpack at 0 °C and well drained tends to have a relative density near 0.35.

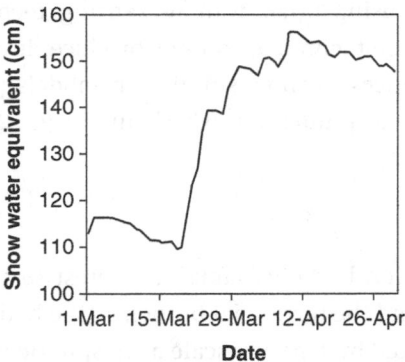

Fig. 4.5. Snow sensor measurements of daily snow water equivalent for the 2005 maximum accumulation period at Meadow Lake, California (40° N, elevation 2182 m). (Data courtesy of the California Department of Water Resources from their website at http://cdec.water.ca.gov/.)

In remote mountain areas where access is difficult or restricted, SWE can be measured automatically by weighing the snow on a pressure-sensing snow pillow made of a metal plate or a flexible bag several meters in diameter filled with an antifreeze liquid. The overlying weight of snow is recorded as changes in pressure using a manometer or a pressure transducer (Ward and Robinson, 2000). Snow pillows are used in selected locations in Switzerland, the United Kingdom, and elsewhere in Europe, and they are widely used in the Western United States. The automated system in the United States called SNOTEL (SNOwpack TELemetry) consists of over 600 sites equipped with at least a pressure-sensing snow pillow, a storage precipitation gauge, and an air temperature sensor to produce daily SWE values. The daily SWE data in Figure 4.5 are from an automated snow sensor. SNOTEL uses VHF radio signals and meteor burst communication technology to collect and communicate data in near-real-time. The data are archived by the Natural Resource Conservation Service, National Water and Climate Center.

4.5.2   Estimating point snowfall

Snowfall displays high variability over short distances in areas of uneven terrain and for areas with diverse vegetation cover. High spatial variability decreases the correlation between adjacent points and reduces confidence in estimating point snowfall from existing point observations. Consequently, it is common to use a snow stake described in the previous section as a representative measurement rather than attempting to develop an estimating equation for other points (Seidel and Martinec, 2004). Snowfall rates are estimated using radar, and snowcover is estimated using other remote sensing techniques described in Chapter 5. Although these techniques do not produce direct estimates of point snowfall depth, advances in snow simulation models provide promising results in estimating snow accumulations (Clark and Vrugt, 2006).

4.6   Wind

Local-scale atmospheric motion is the variable of interest to be measured in climate of the second kind and the terrestrial branch of the hydrologic cycle. Local surface winds are generated by synoptic-scale atmospheric motion, horizontal temperature differences in the planetary boundary layer, topography, and atmospheric instability. These winds advect heat and moisture that can account for spatial and temporal heterogeneities in energy and mass fluxes. Sea breeze, land breeze, and katabatic winds are examples of local surface winds that are embedded in the larger-scale synoptic and global winds of climate of the first kind and the atmospheric branch of the hydrologic cycle.

4.6.1   Wind measurement

In general meteorological and climatological applications, surface wind is considered to be a horizontal, two-dimensional flow represented by its velocity, or rate of movement, and direction. Windfield refers to the combination of air speed and direction in a space-time continuum (DeFelice, 1998). Separate instruments are used to measure these wind characteristics. Anemometers measure wind velocity, and vanes indicate wind direction (Fig. 4.6). Since wind speed and direction are almost always reported as averages, a processing and receiving system is required as part of the equipment installation.

Fig. 4.6. A cup anemometer and wind vane. (Photo by author.)

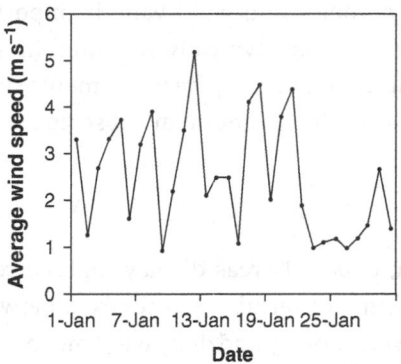

Fig. 4.7. Daily average wind speed for January 2007 at Orland, California (41° N). (Data courtesy of the California Department of Water Resources from their website at http://wwwcimis.water.ca.gov/.)

Anemometers routinely used at climatological stations respond to the kinetic energy of the air. Other types of anemometers (e.g. pressure and sonic) are used for specialized applications. The widely used anemometers have a rotating sensor, and a transmitter that transforms the sensor's rotation into a measurable quantity. The rotating component is three or four small, hollow hemispherical cups or conical cups each attached to arms extending horizontally from a vertical axis. The number of cup revolutions per minute is registered electronically and converted to wind velocity by the transmitter. Wind velocity is usually expressed in $m s^{-1}$ and is often reported as the average for hourly or daily periods (Fig. 4.7).

An alternative design common in the United States is a propeller anemometer resembling a model airplane without wings. The tail swings the propeller into the wind, and the propeller's rotation generates a measurable voltage (Linacre, 1992). Propeller anemometers respond more quickly than cup anemometers to accelerating and decelerating wind. Also, propeller anemometers are more sensitive to low wind velocities (Guyot, 1998).

Wind direction is measured with a vane balanced on a vertical axis to prevent favoring a particular direction. The preferred design is one or two rectangular metal plates attached vertically to the end of a horizontal rod that pivots around a vertical axis. The construction must be heavy enough to smooth out flutter resulting from the smallest wind changes. Vanes will not usually turn when the wind speed is less than $1.5 m s^{-1}$ (Linacre, 1992). The vane direction is observed over a few minutes and recorded as a compass direction. For automated stations, several types of transmitters are available to send vane information electrically in either analog or numerical form (Guyot, 1998).

Appropriate exposure for an anemometer and wind vane is important for reliable measurements. These instruments are typically mounted in an open area at a height of 10 m. Ideally, the land surrounding the instruments should be uniform for a distance equal to 100 times the height of any obstruction (WMO, 1996).

4.6.2   *Estimating wind*

Local surface winds in open, exposed areas display high correlations between two adjacent points when no major terrain features exist between the locations. However, correlations between hourly and daily wind movements are very low for adjacent sites when one site is protected (Peck, 1997). Hills, mountains, valleys, and proximity to the ocean all induce wind variations within short distances.

The standard method for estimating the wind at a point between widely spaced anemometers involves determining a distance-weighted average of the anemometer measurements. The weighting is chosen to make the nearest measurement the most influential in the estimate calculation (Linacre, 1992).

4.7   Soil moisture

Soil moisture is the key state variable in climate of the second kind and the terrestrial branch of the hydrologic cycle. It controls the partitioning of available energy at the surface into sensible heat and latent heat exchanges with the atmosphere, and it links the water and energy balances of the surface through the moisture and temperature states of the soil. Soil water is the immediate source of moisture that evaporates and transpires from the soil and vegetation into the atmosphere (Robock, 2003), and it is the switch that controls the proportion of precipitation that percolates vertically into the soil, evaporates from the land, or eventually becomes runoff. Soil moisture integrates precipitation and evaporation over extended periods, and it, along with snowcover, provides a significant memory component for the atmosphere–land system.

Soils form in response to physical, chemical, and biological processes modifying parent material on the land surface. The soil resulting from these natural processes is a mixture of solid, liquid, and gaseous materials. The solid materials include particles of different sizes, shapes, and mineral composition along with organic matter. Solid particles commonly make up the majority of the mass volume of the soil and constitute the soil matrix. Soil water and soil air occupy the void spaces or pores between the solid particles. The volume of air and water in pores is complementary in that as one increases the other decreases. Porosity is an index of the relative pore space in a soil, and the size of pores varies depending on soil texture and soil structure. Soil texture is determined by the relative amounts of the mineral particles of sand, silt, and clay making up the soil matrix. Soil structure is the arrangement of the soil particles into granules or soil aggregates of different shapes, sizes, and volumes of pore spaces. General values of porosity range from 30% to 60%. The wettest possible soil condition in which all the pores are filled with water is saturation (Hillel, 2004).

Soil moisture is commonly regarded as the amount of water in the upper layer of the soil typically down to a depth of one meter that interacts with the atmosphere. The amount of moisture in this soil layer is a function of precipitation, soil texture, soil structure, soil porosity, organic matter content, and the rate of moisture withdrawal from the soil, and soil moisture varies considerably on scales of only a few meters (Miller *et al.*, 2005). Fine sands hold much less water per unit depth than silt or clay soil, but moisture storage for most soils is around 15% of the bulk soil volume of the root zone, which is the layer from which plant roots can extract water during transpiration (Dingman, 1994). Precise *in situ* soil moisture measurements are sparse, each value represents a small area, and the available data are relatively recent. The Global Soil Moisture Data Bank contains soil moisture observations for over 600 stations, but all of the stations are in the Northern Hemisphere (Robock *et al.*, 2000). Details of soil-water relationships and direct and indirect field methods for determining soil moisture appear in the soil science and hydrology literature (e.g. Maidment, 1993; Ward and Elliot, 1995; Hillel, 2004; Rose, 2004). An overview of soil properties relevant to soil moisture and selected soil moisture measurement techniques are presented here.

4.7.1   Soil water properties

Soil water in the soil profile is important for moisture storage and for influencing energy and moisture fluxes in land surface hydroclimatic processes. The soil profile is a vertical cross-section through the soil commonly comprising a number of soil layers having different physical characteristics. Soil texture and soil structure are the physical characteristics most relevant to the soil's ability to store and release water. In the following discussion, unsaturated soil conditions are assumed.

Two basic approaches aid in characterizing and measuring soil water. A physically based approach emphasizes the moisture status of the soil resulting from the interaction of forces related to the solid, liquid, and gaseous soil components. A second approach employs the energy state of soil water. Although the energy state is most often associated with soil water movement, it is relevant to soil wetness (the contemporary term for soil water content) because it is used to

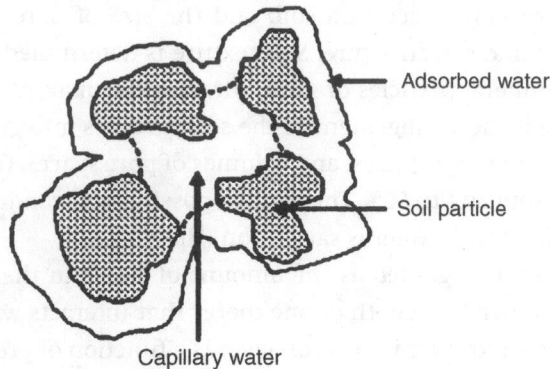

Fig. 4.8. Simplified and magnified view of water in soil pores.

estimate soil water storage and to define the work required to remove water from the soil by evaporation and plant transpiration (Hillel, 2004).

Soil wetness

Under natural conditions, soil moisture is dependent on time since the last precipitation event. Water entry into the soil through the soil surface occurs by infiltration as discussed in Section 4.7.4.

Water molecules adhere to soil particles as a film of water in response to forces related to capillarity, adsorption, and osmosis. Capillary forces are the result of surface tension between soil air and soil water and the attraction of water molecules to compatible surfaces provided by soil minerals with oxygen atoms to share with water's hydrogen atoms. Liquid molecules are attracted more to each other than to water vapor molecules in the air, and this causes contraction of the liquid surface. A curved liquid surface results from the pressure differences experienced by the liquid and vapor molecules. The pressure on the liquid molecules is likely to be lower than the atmospheric pressure acting on the vapor molecules (Hillel, 2004). Withdrawing water from the soil increases the pressure difference as the liquid surface becomes more curved and is maintained only in the smaller pores (Fig. 4.8). Soils with larger pores retain less water than soils with smaller pores because the smaller pores act as narrower capillaries and exert greater force on soil water than the force exerted by larger pores.

Capillary water is augmented by water molecules adsorbed upon the surface of soil particles largely due to electrostatic forces. The forces involved are only effective very close to the soil particle surface and only a thin film of water is held in this way. Adsorptive forces are greatest for the first layer of water molecules, and the second layer is attached to the first by hydrogen bonding so that the attractive force diminishes rapidly with distance as each additional layer is added (Hillel, 2004). Differences in soil particle size are important in determining differences in adsorbed water. Large soil particles are able to adsorb more water than small particles due to their greater surface area. Adsorbed water has a degree of mobility that becomes important at low soil wetness (Catriona *et al.*, 1991).

Capillary and adsorptive forces are regarded as being in equilibrium and are not easily measured separately (Ward and Robinson, 2000). The water pressure in pores is less than atmospheric pressure in unsaturated soils, and both capillary and adsorption forces are regarded as exerting tension or suction on the soil water. Therefore, it is usual to consider their combined effect on water in the soil matrix as matric suction or matric potential. The terms tension, suction, and pressure are used interchangeably in soil water studies. Tension or suction is negative pressure in that the pressure is less than atmospheric pressure (Hillel, 2004).

Osmotic pressure is a third force acting to retain soil water. This force is due to the presence of solutes in soil water and the dipolar nature of water molecules. Ions in solution are attracted by the electric field around individual water molecules. The influence of osmotic pressure is often ignored unless there is a difference in solute concentration across a permeable membrane. In addition, osmotic pressure applies to water alone and not the soil solution (Hillel, 2004).

The total force holding water in the soil is the sum of the matric and osmotic forces. These forces vary with soil wetness, and soil wetness varies with differences in the shapes and sizes of soil particles. Soils with large pores empty at low suction, but soils with small pores empty at higher suctions. The relationship between soil wetness and soil moisture suction is fundamental to understanding soil moisture behavior and measurement (Ward and Robinson, 2000).

In theory, the capillary/adsorptive attraction is greater than the force of gravity and this water does not drain away. Additional water entering the plant root zone percolates vertically through the soil once the film of capillary water has reached a maximum diameter. The soil water remaining after internal drainage ceases is traditionally known as field capacity, but this is a subjective concept that lacks a universal physical basis since drainage typically continues for long periods (Hillel, 2004). The water moving vertically in response to gravity passes through the zone of air and capillary water storage and continues downward as long as void spaces in the strata are available to provide transmission routes. Eventually, this gravity water may fill all of the void spaces in the strata. The subsurface water driven by gravity is groundwater that commonly discharges into rivers and streams and sustains streamflow between rainfall events.

Capillary water adhering to soil particles is vital to plants that are able to extract the water by the suction action created by their vascular structure and transmitted to their roots. The root depth of plants varies from a few millimeters to tens of meters, and the water accessible to plants varies with the root depth and the physical characteristics of the soil. A capillary water flow toward the surface can occur under extreme drying conditions at the surface, but the quantity of water responding in this manner is very small. Consequently, the common convention is to assume that capillary water is available to plants only, and the quantity of this water is estimated by the plant's rooting depth. Within the root zone, plants extract 40% of their needed water from the top quarter of the root zone, and only 10% comes from the bottom quarter of the plant root zone.

Soil wetness can be expressed either on a mass or volume basis. In either case, the derived value is dimensionless and can be regarded as a fraction or expressed as a percentage. Mass wetness (θ_g) is expressed as

$$\theta_g = \frac{M_w}{M_s} \tag{4.4}$$

where M_w is the water mass in g and M_s is the solids mass in g for a soil sample. The mass of air is negligible and the total mass is taken as the sum of the soil and water masses in determining the bulk density, which is the ratio of solids to the total soil volume. Bulk density is a factor in determining θ_g, which ranges between 25% and 60% in different soils (Hillel, 2004).

Volume wetness (θ_v) is expressed as

$$\theta_v = \frac{V_w}{V_t} \tag{4.5}$$

where V_w is the volume of water in cm^3 and V_t is the total volume in cm^3 of a soil sample. At saturation, θ_v is equal to the porosity and ranges between 40% and 60%. For many purposes, θ_v is multiplied by the soil depth to express soil water as a depth of water that is compatible with the units used to measure rainfall, evaporation, transpiration, and runoff (Hillel, 2004).

Soil water energy

Soil water potential energy expresses the ability of soil water to perform work and is of primary importance in determining the state and movement of soil water. Kinetic energy is considered negligible due to the slow movement of soil water. The concept of soil water potential energy expresses the specific potential energy of soil water relative to that of water in a standard reference state. Differences in soil water potential energy occur over a wide range of time and space scales, and knowledge of the relative soil water potential energy allows determination of the condition of soil water compared to the soil system equilibrium state. Consequently, the soil water potential energy state serves as the basis for estimating how much work must be expended by plants to extract a unit amount of water (Hillel, 2004).

Soil water is held in soil pores under surface tension forces and a concave surface extends from particle to particle across each pore channel. The radius of curvature reflects the surface tension on the individual, microscopic air–water interface of a specific pore. The potential energy of water becomes greater as the water content of the pore increases and the radius of curvature of the water film in the pore becomes greater. A water molecule adsorbed in soil is less free to move than a water molecule without adhesion to a surface. A potential energy gradient results, and the related force is directed from higher to lower potential resulting in the energy potential of soil moisture being taken as negative (Hillel, 2004). This concept is applied to both soil wetness and soil water flow in that it defines a gradient indicating a direction of water movement toward a point having the lowest potential. Negative soil water potential is the work required to remove water from the soil by evaporation or transpiration.

The relative concentration of potential energy in soil water at different locations is the characteristic of interest. The major constraint on defining soil water potential is that the absolute potential energy of water cannot be measured. Recognized factors that can change the potential energy of soil water are the adsorption of water onto soil particles, solutes dissolved in soil water, elevation of soil water in the Earth's gravitational field, and applied pressure. The total soil water potential (ψ_t) is the net result of all of these factors and is expressed as

$$\psi_t = \psi_z + \psi_m + \psi_o \tag{4.6}$$

where ψ_z is the gravitational potential expressing the force gravity has on water, ψ_m is the matric potential, and ψ_o is the osmotic potential (Hillel, 2004). The unit for expressing the soil water potential is the kilopascal (kPa).

The separate potential forces in Equation 4.6 act in different ways for unsaturated soil. An arbitrary reference level is used for determining ψ_z, and selection of the soil surface as the reference pool results in a negative value. Work is needed to withdraw water against the soil matric forces and this produces a negative value for ψ_m. Interacting capillary and adsorptive forces between water and the soil matrix bind water in the soil and lower its potential energy below that of bulk water. Thus, soil water has a pressure lower than atmospheric, a condition known as tension or suction, and a negative pressure potential (Hillel, 2004). The energy differences between pure water and water containing dissolved salts determine ψ_o. It is a negative potential that increases as the solute concentration increases as ψ_o acts to retain or draw water into the soil and to lower the total soil water potential. For unsaturated soils not involving areas with high evaporation and high solute concentrations in the soil water, ψ_o is taken as zero and ψ_t is reduced to two terms.

4.7.2   Measuring soil moisture

Numerous instruments are available for measuring either soil wetness or soil water potential. The spatial and temporal resolutions required for soil moisture measurements vary with the data application, and there is no universally recognized standard method of measurement (Hillel, 2004). Soil wetness is determined by direct and indirect methods. Soil water potential is measured by indirect techniques. For hydroclimatic purposes, soil wetness, soil water potential, or both measurements may be important. A brief description of selected measurement methods is presented to illustrate how the data are acquired.

Direct measurement of soil wetness

The simplest and most widely used method for measuring soil wetness (θ_g) is the gravimetric or thermostat-weight technique (Robock *et al.*, 2000; Hillel, 2004). This direct measurement technique is the standard method to which all other methods are compared. A soil sample is removed from the field by coring or augering and transported to a laboratory in a leak-proof container. In the laboratory, the sample is weighed and then dried in an electrically heated oven at 105 °C until the mass stabilizes at a constant value. The difference between the sample weight before and after drying is attributed to the mass of water in the original sample.

The methodology requires simple equipment, but it is destructive to the sample site which becomes unsuitable for a large number of repeated measurements over time. The laboratory procedure is laborious, and a period of at least 24 hours is usually considered necessary for complete oven drying (Hillel, 2004).

Lysimeters are a non-destructive direct method to measure soil moisture, but they are limited to research facilities due to their expense and maintenance requirements. These large soil-filled containers are weighed to determine changes in mass for a specific time and the change is attributed to changes in soil moisture. This procedure requires establishing the wetness of the soil in the container as an initial benchmark for determining later soil wetness. Lysimeters are discussed in more detail in Section 4.8.3 regarding measuring evapotranspiration.

Indirect measurement of soil wetness

The neutron probe is widely used as an indirect method for measuring soil wetness due to the high efficiency and reliability of the technique. A radioactive source of high-energy neutrons is lowered into an access tube in the soil that provides standardized measuring conditions (Hillel, 2004). The number of neutrons slowed or thermalized by collisions with hydrogen nuclei in the soil water is measured by a detector. There is a fairly linear relationship between the detector count rate and the soil wetness, but the relationship varies from soil to soil. Neutron probe readings are usually calibrated for a given soil against the gravimetric method, but inherent uncertainties with the neutron probe usually limit its utility to measuring moisture differences rather than absolute moisture content (Ward and Robinson, 2000). However, the method is non-destructive and it has the advantage that data are acquired from the same location and depths at each observation.

A number of other instruments are used for indirect measurement of soil wetness. The capacitance probe uses the dielectric constant of the soil as a measure of soil moisture content and provides a non-radioactive alternative to the neutron probe. Time-domain reflectometry (TDR) is another technique for determining soil wetness by measuring the dielectric constant, but this technique employs the soil water response at microwave frequencies. The presence of soil water slows the speed of the electromagnetic wave emitted by the probe (Hillel, 2004). Frequency-domain reflectometry is similar to TDR except that it employs changes in the frequency of microwave signals due to differences in the dielectric properties of the soil with different water content. Gamma densitometry derives soil moisture content based on the relatively greater gamma radiation attenuation of water compared with other soil components.

Measuring soil matric potential

Tensiometers are the oldest and most widely used technique for measuring soil matric potential. This technique allows the soil solution to come into equilibrium with a reference pressure indicator by using a liquid-filled permeable ceramic cup connected to a pressure measuring device. The cup is inserted into the soil and water flows between the cup and the soil until the pressure potential inside the cup equalizes that of the soil water. Tensiometers are not affected by the osmotic potential of the soil solution, and they measure the matric potential with good accuracy in the wet range. However, they provide only a point measurement of matric potential and operate between saturation and -80 kPa (Hillel, 2004).

Other approaches for measuring soil matric potential include burying a resistance block of gypsum, nylon fabric, or fiberglass in the soil. Two electrodes are embedded in the porous block. The matric suction of the block comes into equilibrium with the soil water and the resistance across the electrodes varies with the resulting water content of the block (WMO, 1996). Thermocouple psychrometers are used to measure the vapor phase of the soil environment which is assumed to be in equilibrium with the soil matric potential. Neither the resistance block nor the psychrometer is sensitive to wet conditions, but both are well suited to a dry soil environment.

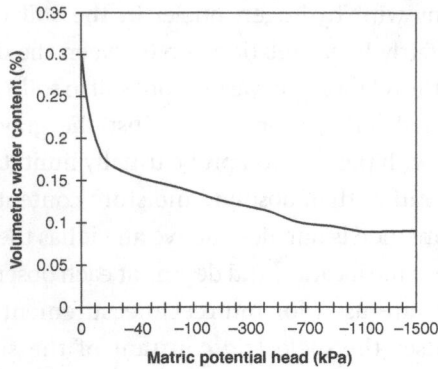

Fig. 4.9. Soil moisture characteristic curve.

Schneider *et al.* (2003) demonstrate that a network of heat dissipation sensors can provide a continuous, inexpensive system of soil water profile observations across a range of soil conditions and sites in the Southern Great Plains. The heat dissipation sensors measure soil matric potential from which volumetric water content is estimated.

4.7.3   Estimating soil moisture

Challenges encountered in measuring soil wetness and matric potential have encouraged development of techniques for estimating unsaturated soil water conditions. The influence of soil structure and pore size distribution in determining soil moisture characteristics is the physical basis for expecting moisture differences among soil types. The relationship between soil wetness and matric potential is determined experimentally in the laboratory by placing a saturated soil sample on a porous plate and subjecting the sample to a series of suctions or negative pressure heads. Water content is obtained by weighing the soil sample before and after oven drying to determine how much water is retained by the soil at the specific suction (Ward and Robinson, 2000; Hillel, 2004). Soil wetness and matric potential data are used to construct soil moisture characteristic curves that are a graphical representation of the relationship between these variables for a specific soil (Fig. 4.9).

Soil moisture characteristic curves are not unique curves for a given soil type because the equilibrium soil wetness at a given suction is greater during drying than during wetting (Hillel, 2004). The current suction and prior soil conditions are primary factors influencing the equilibrium soil wetness for a given suction, and this dependence on the previous state of the soil water prior to the current equilibrium condition is known as hysteresis. The main contributors to hysteresis are pore irregularity, contact angle between water and the solid walls, entrapped air, and shrinking and swelling (Hillel, 2004). The result is that the wetting curve is drier than the drying curve over a wide range of matric potentials. Hysteresis is a serious practical problem for soils subject to change by both wetting and drying influences. In practice, the problems associated with measuring the moisture characteristic accurately commonly lead to the hysteresis phenomenon being ignored. The drying curve is easier to establish experimentally and is the one most often reported in the literature as the soil moisture characteristic curve or the moisture retention curve (Hillel, 2004).

At the field scale, soil moisture (SM) in the active soil layer is estimated using an alternative form of the water balance (Equation 2.13) as suggested by Robock (2003). Temporal soil moisture variations are expressed as

$$\frac{\partial \text{SM}}{\partial t} = P_\text{r} + P_\text{sn} + \text{WT} - \text{ET} - R \qquad (4.7)$$

where P_r is rainfall, P_sn is snowmelt, WT is water table contributions, ET is evapotranspiration, and R is runoff. All units are in mm. Soil moisture retention characteristics are incorporated in the algorithm that quantifies evapotranspiration. Water balance soil moisture estimates can be applied to single or multiple layer soils, and the soil moisture estimates are easily incorporated in land surface models. Formulations such as Equation 4.7 quantify the soil moisture component driven by atmospheric forcing, but they are less successful at quantifying small-scale soil moisture field variability that appears as a stochastic response to topography, soils, and vegetation (Robock et al., 2000). Nevertheless, Yamaguchi and Shinoda (2002) report success with simple models using limited daily meteorological data that estimate absolute soil moisture amounts in the upper 50 cm soil layer in Sahelian Niger. Regional soil moisture is controlled principally by rainfall, and the models indicate soil moisture variations from intraseasonal to interannual time scales. Braud et al. (2003) use a simplified in situ method to estimate regional soil hydraulic property variability in central Spain. Their method employs a two-layer soil model and representative particle size distributions and infiltration rates based on sample data medians to characterize spatial soil variability. A statistical analysis approach employed by Cosh et al. (2004) reveals the underlying soil type distribution is the most important parameter in accounting for temporal soil moisture variability in southwestern Oklahoma.

4.7.4   Water infiltration into soils

Excluding extreme conditions, a fraction of precipitation enters the soil surface by infiltration and is redistributed to successively deeper layers of the soil profile. The infiltrating precipitation replenishes the water stored in the soil that is available to plants, and it determines the water eventually recharging

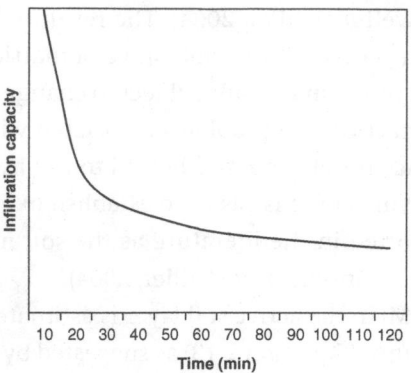

Fig. 4.10. Infiltration rate shown as a function of time.

groundwater. Soil water moves in response to a number of forces, and infiltration may involve water movement in one, two, or three dimensions. However, soil water is commonly conceptualized as one-dimensional vertical flow, and this perspective of infiltrated water in the unsaturated zone is emphasized here. Comprehensive treatments of infiltration are found in Ward and Robinson (2000), Hillel (2004), and Rose (2004).

The infiltration rate is the volume flux of water flowing into the soil profile per unit of soil surface area (Hillel, 2004). Horton (1940) defined the soil's infiltration capacity as the condition when the rainfall rate exceeds the infiltration rate. When the rainfall delivery rate is smaller than the infiltration capacity, water infiltrates as fast as it arrives and infiltration is supply controlled. When the delivery rate exceeds the soil's infiltration capacity the infiltration rate determines the flux and the process is soil controlled. The soil can limit the infiltration rate either at the surface or within the soil profile (Hillel, 2004). The infiltration rate generally changes systematically with time during a given rainfall event in response to changes in rainfall intensity and duration and to changing conditions of the soil's surface properties (Fig. 4.10). Surface properties influencing infiltration include soil texture, soil structure, initial soil moisture content, soil chemistry and temperature, the nature of the vegetation cover, and the slope angle of the land. These factors influence the ability of air and water to move through the soil, and they collectively determine the soil's permeability. In addition, conditions affecting infiltration change between rainfall events and this may magnify the complexity of the surface response to infiltration.

Infiltration rates vary from soil to soil and with the presence or absence of vegetation. Coarse-textured soils generally have larger pore spaces than fine-textured soils. Infiltration rates for coarse sand with good structure range as high as $25\,\mathrm{mm\,hr^{-1}}$ while the infiltration rates for clay loams are $8\,\mathrm{mm\,hr^{-1}}$.

Table 4.2. *Soil permeability and infiltration rates for selected surface conditions*

Permeability class	Infiltration rate (mm hr^{-1})
Slow	1.5–5.1
Moderate	15–51
Rapid	152–508

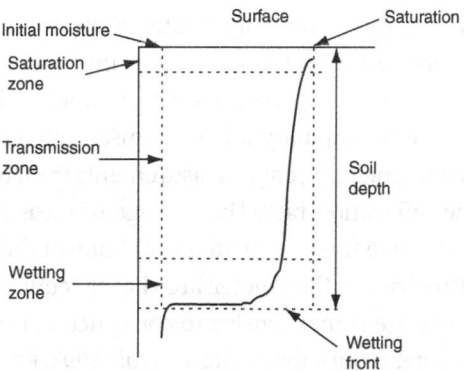

Fig. 4.11. Simplified diagram of the decrease in soil matric suction with the curve of water content versus soil depth. (Adapted from Hillel, 2004, Fig. 14.3. Used with permission of Elsevier Science.)

Infiltration rates for general soil permeability classes are shown in Table 4.2. Poor structure can reduce infiltration rates by 50%. In addition, infiltration rates tend to be high at the beginning of the process and gradually decrease to a nearly constant value as soils become wetter. The decreasing infiltration rate is due to the closure of pore spaces as some soil granules expand in response to wetting and the delay in vertical water movement related to frictional resistance that increases with depth. Final infiltration into sandy soils is about 20 mm hr^{-1} while the expected rate into heavy clay soils is less than 1 mm hr^{-1} (Hillel, 2004).

Water infiltrates the surface at a given rate as long as the underlying soil profile can conduct the infiltrated water away at a corresponding rate. Soil texture, structure, stratification, and the initial soil water potential gradient are all important contributors to maintaining the necessary water redistribution rate. However, a decrease in the matric suction gradient occurs as infiltration proceeds and the wetted zone deepens (Fig. 4.11). At any given time, wetness decreases with depth at a steeper gradient down to the wetting front, which is a sharp boundary between moistened soil above and initially dry soil at greater depth (Hillel, 2004).

Measuring infiltration

Point infiltration measurements for determining infiltration rate variations with time are made with ring, sprinkler, tension, and furrow infiltrometers. These instruments are described by Rawls *et al.* (1993) and Rose (2004). The purpose of each device is to define the infiltration capacity for a small area by applying a known quantity of water at a specific rate and monitoring changes in soil-water relationships.

The ring infiltrometer illustrates the basic concepts for measuring infiltration. A metal ring with a diameter of 30 to 100 cm and a height of 20 cm is driven into the ground about 5 cm to form an impermeable boundary. Water is applied inside the ring, and intake measurements are recorded until a steady infiltration rate is observed. A double-ring arrangement is used to create a buffer zone to eliminate the effect of lateral spreading of water by capillary forces in unsaturated soils or a correction factor can be applied to the singe-ring data. Measurements from the inner ring are used for determining the infiltration rate. The average of measurements from several sites is recommended due to the high spatial variability of the infiltration capacity. The advantages of the ring infiltrometer are that it requires only a small area for measurements, the device is inexpensive to construct and simple to operate, and it does not have high water requirements (Rawls *et al.*, 1993; Rose, 2004).

Estimating infiltration

Physical, approximate, and empirical models are used to estimate infiltration. Physical models are the most data intensive and have benefited from advances in computer software. Richards' (1931) non-linear equation describing water flow in soils is a physically based infiltration equation that combines Darcy's equation for vertical unsaturated flow in a homogeneous porous medium with the continuity equation. Richards' equation expresses the time-dependent infiltration rate in terms of antecedent soil profile soil moisture conditions, the rate of water delivery at the surface, and the conditions at the bottom of the soil profile (Rawls *et al.*, 1993). The partial differential equation of one-dimensional vertical flow takes the form

$$\frac{\partial \theta}{\partial t} = \frac{\partial}{\partial z}\left[\frac{K(\theta)}{\left(\frac{\partial \psi}{\partial z} - 1\right)}\right] \tag{4.8}$$

where θ is the downward water content gradient, t is time, z is downward vertical flow, K is the hydraulic conductivity, and ψ is the matric potential gradient. All rainfall infiltrates into the soil when the rainfall intensity is less than or equal to the saturated hydraulic conductivity of the soil profile. High rainfall intensities that exceed the infiltration rate produce infiltration during the early stages of the event until the soil surface becomes saturated.

Approximate models strive to represent physically based infiltration conditions based on major parameters and a holistic view of the infiltration process. However, parameter characterization is commonly difficult. A widely used approximate model based on the total amount of water infiltrated was introduced by Green and Ampt (1911) and in its simple form is

$$f = i_c + \frac{b}{S_s} \qquad (4.9)$$

where f is the infiltration rate, i_c is the eventual steady-state flow rate driven by the soil water potential gradient between the wet surface and the drier soil below, b is diffusivity flow of water that moves into the drier soil ahead of the advancing wetting front, and S_s is the volume of water stored in the depth of soil saturated by infiltration. Terms i_c and b are constants for given soil texture and moisture conditions, and Equation 4.9 is often used in computer models of infiltration (Ward and Robinson, 2000). The Green and Ampt model provides a reasonable approximation of reality for one-dimensional vertical flow with water supplied under positive hydrostatic pressure to coarse-textured soils that are initially dry and without entrapped air (Hillel, 2004). The time-based equation developed by Philip (1957) is an extension of the Green and Ampt model (Ward and Robinson, 2000).

Empirical infiltration models use parameters estimated from measured infiltration rate–time relationships for a given soil condition. This is a common limitation that hinders the wide-scale application of these models. The Horton (1940) model is widely used in hydrologic modeling and illustrates the nature of empirical infiltration models. It is founded on the concept that infiltration capacity passes through a cycle for each storm starting with a maximum value and decreasing rapidly at first as the surface becomes sealed. Horton's three-parameter model is expressed as

$$f_p = f_c + (f_o - f_c)e^{-kt} \qquad (4.10)$$

where f_p is the infiltration capacity, f_c is the minimum constant infiltration capacity, f_o is the maximum infiltration rate at the beginning of the storm event, k is the rate of decrease in the infiltration capacity, and t is time (Rawls et al., 1993). Numerous field experiments have found this relationship is most likely to occur where bare soils are exposed to rainfall as happens in arid and semi-arid areas (Ward and Robinson, 2000).

Rainfall excess models lump infiltration with other surface processes. These models are commonly used for estimating runoff and are discussed in Section 6.10.3.

4.8   Evaporation and evapotranspiration

Evaporation from free water surfaces and bare soil and evapotranspiration from vegetated surfaces support upward directed energy and mass fluxes that complement downward directed precipitation in climate of the second kind and the terrestrial branch of the hydrologic cycle. Furthermore, evaporation and evapotranspiration have an important role in determining surface temperatures, surface pressure, rainfall, and atmospheric motion. The upward directed energy and water vapor fluxes from the Earth's surface involve the passage of water from the liquid to the gaseous state. It may be thought of as the mass flux that is the opposite of precipitation but because it is water vapor it is not visible in the same way as precipitation. The water vapor flux occurs as evaporation from free water surfaces (i.e. lakes, rivers, and the oceans) and moist soil surfaces and as transpiration from living plants. Evaporation from free water surfaces and moist soil surfaces occurs when the ambient air vapor pressure gradient is less than the vapor pressure at the evaporating surface and an external source of energy is present.

Transpiration is a more complex process and varies considerably from one plant species to another largely in response to the plant rooting depth and the vegetative area. Water is transported by the plant vascular system from the roots to the stomata on the underside of the leaves where it is extruded. Water exiting the stomata is evaporated by available energy, and this loss of water is plant transpiration. The combined evaporation from the soil, plant, and other surfaces and the transpiration from plants covering the ground constitute evapotranspiration. When the ground is well covered with plants, transpiration is the dominant moisture exchange process and evaporation is a small fraction of the total moisture flux (Szilagyi and Parlange, 1999). The physical process is the same in either case, and it is common to use either evaporation or evapotranspiration to embrace both processes. Evaporation is favored in Europe and evaporation and evapotranspiration are both widely used in the United States (Ward and Robinson, 2000). In the discussion that follows, evaporation and evapotranspiration are treated separately in an effort to maximize clarity.

The evaporation process can be described by a relatively simple generic equation analogous to Ohm's characterization of electric current expressed in terms of electric potential and resistance. The potential for evaporation is directly proportional to the concentration of water vapor at the surface and in the air above the surface or the vapor pressure gradient. Resistance to the vapor transfer occurs in the form of the relative difficulty the surface and the atmosphere present to the net movement of water vapor away from the surface. Within this framework, evaporation (E) from a surface into the atmosphere is expressed as

$$E = \frac{\rho_a \varepsilon^*}{P} \left\{ \frac{e_s^* - e_a}{r_s + r_a} \right\} \tag{4.11}$$

where ρ_a is air density, ε^* is the ratio of the molecular weights of water vapor and air, P is atmospheric pressure, e_s^* is the saturation vapor pressure of the surface, e_a is the vapor pressure of air, and r_s and r_a are the resistance to evaporation by the surface and the air, respectively. This expression clearly illustrates that the highest potential evaporation rate for any given atmospheric state will occur when the surface is completely wet, and the constraint on achieving that potential rate is due to the resistance to water vapor transfer presented by the environment (Willmott, 1996).

Free water surfaces present the least resistance to water molecule exchange with the air since the water-surface vapor pressure is a maximum for any given temperature. Snow and ice surfaces present greater resistance relative to a free water surface because water molecules in snow and ice are bound in a more rigid molecular structure. Land surface resistance to evaporation varies with the surface roughness and the structure of the plant canopy related to the nature of the vegetation, with the soil type and texture, and with the degree to which the vegetation and soil are wet. Wind is the primary atmospheric resistance factor, but surface geometry and atmospheric stability also contribute. In essence, wind promotes efficient transport of water molecules away from the surface and maintains a vapor pressure gradient that supports the maximum evaporation determined by available energy.

The fundamental role evapotranspiration plays in the hydrologic cycle is evident in that about two-thirds of the land surface precipitation is allocated to evapotranspiration (Szilagyi and Parlange, 1999). However, the importance of the energy flux that accompanies evapotranspiration must be acknowledged. Since this vapor flux involves a phase change of water, evapotranspiration accounts for a significant transfer of energy from the Earth's surface to the atmosphere. Both the energy and vapor transfers influence atmospheric behavior and Earth's climate.

Although the entire troposphere is influenced by solar radiation absorbed at the Earth's surface, atmospheric responses to energy and mass fluxes at the Earth's surface are particularly evident in the atmospheric boundary layer. The atmospheric boundary layer is the lowest 1 km of the atmosphere above the Earth's surface (Andrews, 2000). Atmospheric motion in this region is directly influenced by frictional effects related to the nature of the underlying Earth's surface. Equally important is the boundary layer's role in providing an environment that encourages or inhibits energy and mass fluxes at the surface. Interactions between the Earth's surface and the boundary layer provide the

Fig. 4.12. A galvanized sheet-metal evaporation pan known in the United States as a National Weather Service Class A pan. (Photo by author.)

atmosphere with water, provide the major cause of diurnal temperature change, and contribute to atmospheric mixing and turbulence.

4.8.1   Measuring evaporation

Evaporation is measured using evaporation pans or tanks of various sizes, atmometers, and inflow and outflow measurements of lakes. These techniques are constrained by a number of physical factors that influence evaporation and make extrapolation of data problematic. Only selected instruments are presented here to illustrate common characteristics.

The U.S. National Weather Service Class A pan (Fig. 4.12) is 120.7 cm in diameter, 25.4 cm deep, is supported 4 cm above the ground surface, and is initially filled with water up to 5 cm of the rim. The unpainted, galvanized steel pan is set on a wooden platform to allow free air circulation under the pan. Water is added to the pan each day until the surface is level with the spike tip in the stilling cylinder. Evaporation is the daily difference between observed water levels corrected for any precipitation measured by a nearby rain gauge. The evaporation data in Figure 4.13 show variations in amounts related to daily changes in available energy and to a generally decreasing energy availability over the course of the month.

Turbulence created by the pan, heat transfer through the sides of the container and small heat storage in the limited water volume all tend to cause the water loss from the pan to be different from that over an open water surface. Use of evaporation pan data to estimate lake evaporation or evaporation from a land surface requires application of an empirical pan coefficient to reduce values to realistically represent lake or vegetation evaporation (Ward and Robinson, 2000). Nevertheless, evaporation pans are the most commonly used device for measuring evaporation.

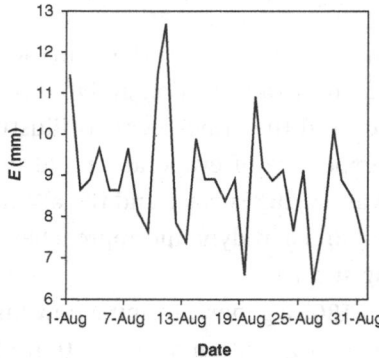

Fig. 4.13. Daily evaporation for August 2006 at Davis, California (39° N). (Data courtesy of the California Department of Water Resources from their website at http://wwwcimis.water.ca.gov/.)

The Russian $20 \, \text{m}^2$ evaporation tank is recommended by the World Meteorological Organization as the international standard reference evaporimeter. This flat-bottomed cylinder has a diameter of 5 m, a surface area of $20 \, \text{m}^2$, and is 2 m deep. It is buried in the soil with its rim 7.5 cm above the ground. The tank has a replenishing vessel and a stilling well with an index pipe on which a volumetric burette is placed to measure the water level in the tank. A needle point inside the stilling well indicates the height to which the water level is to be adjusted to determine the moisture loss (WMO, 1996).

Atmometers measure water loss from a standard moistened surface (Guyot, 1998). These devices use a small water-filled porous porcelain bulb or wetted pieces of paper to serve as the evaporating surface. Either a tube filled with water or an attached container of water maintains the wetness of the evaporating surface mounted on top of the cylindrical water reservoir. Utilization of atmometer data is constrained in that data are not comparable unless they come from identical instruments exposed in similar conditions (Guyot, 1998). In addition, atmometers indicate the drying power of the air, but the water loss cannot be converted to a depth of water loss over the ground (Ward and Robinson, 2000).

The determination of evaporation using lake or reservoir levels and inflow and outflow data is theoretically possible, but in these cases evaporation is not actually measured. Evaporation from water bodies is a residual in a mass balance equation and is subject to considerable uncertainty and error if the evaporation quantity is small compared to the other variables. Seepage is an especially troublesome variable that introduces errors in mass balance equations for lakes and reservoirs. Under the best circumstances, the mass balance computation contains the net sum of all measurement errors as well as evaporation.

4.8.2   Estimating evaporation

The physics of the evaporation process relate to the climatic forcing that supplies the energy for the latent heat of vaporization and to diffusion of water vapor between the Earth's surface and the atmosphere as illustrated in Equation 4.11. The thermodynamic perspective of evaporation addresses the energy balance characteristics of the evaporating surface and the allocation of energy to latent heat. The mass transport and aerodynamic approaches emphasize the vapor flux from the evaporating surface.

Dalton is credited with introducing in 1802 a general formula now known as the mass transport method that predicts evaporation as a function of vapor pressure. Dalton recognized that the rate at which water molecules leave a water surface is dependent on the temperature of the water surface and the atmospheric partial pressure due to water vapor. For the evaporation process to continue efficiently, atmospheric motion or wind is needed to carry away the water molecules from the evaporating surface. Evaporation from open water (E_o) is expressed as

$$E_o = u(e_o - e_a) \tag{4.12}$$

where u is an empirically determined constant involving some function of windiness, e_o is the vapor pressure at the surface and e_a is the actual vapor pressure in the air at some point above the surface. This method offers the advantage of simplicity in calculation once the empirical constants are developed. Numerous applications have shown that the mass transport method works well when u is locally determined in advective and non-advective cases.

The evaporation of water from a given surface is greatest in warm, dry conditions because the air is warm, the saturation vapor pressure of water is high, and the actual vapor pressure of the air is low. Aerodynamic or profile techniques for estimating evaporation are based on the physical processes governing the turbulent diffusion of momentum, sensible heat, and water vapor in the atmosphere. Each of these fluxes is estimated by analogous equations using turbulent exchange coefficients assumed to be identical. The water vapor flux (E) is estimated by

$$E = \frac{(M_w/M_a)}{P} \rho_a K_w \frac{\partial e_a}{\partial z} \tag{4.13}$$

where M_w and M_a are the mole weights of water vapor and air, respectively, P is atmospheric pressure, ρ_a is air density, K_w is the turbulent exchange coefficient for water vapor, and $\partial e_a/\partial z$ is the vertical gradient of water vapor (Rosenberg et al., 1983). However, research has shown that the assumption of identity for the exchange coefficients is only valid under neutral atmospheric stability, and the effects of non-neutral conditions must be determined experimentally. The like-lihood of large errors occurring in determining each of the fluxes limits the use of this technique (Guyot, 1998).

Penman (1948) introduced a method for estimating evaporation from open water surfaces based on both energy budget and aerodynamic approaches and eliminating terms not commonly measured. In essence, this formulation is a weighted sum of an evaporation rate due to net radiation as defined in Equation 2.11 and an evaporation rate due to turbulent transfer. The formulation assumes no advected energy and no heat storage effects. Only meteorological measure-ments made at one level above the surface are required, and E_o is expressed as

$$E_o = \frac{\Delta R_{no} + \gamma E_a}{\Delta + \gamma} \tag{4.14}$$

where Δ is the slope of the saturation vapor pressure curve at the mean wet bulb temperature of the air, R_{no} is the net radiation over open water, γ is the psychrometric constant expressed as $\gamma = PC_p/L_{\varepsilon*}$ where P is the atmospheric pressure, C_p is the specific heat of air at constant pressure, and $\varepsilon^* = M_w/M_a$ where M_w and M_a are defined above. In this form, evaporation due to the energy balance is represented by R_n and evaporation due to the aerodynamic compo-nent is represented by $E_a = f(U)(e_a^* - e_a)$ in which $f(U) = 0.27(1 - U/100)$, U is the wind run in km per day at 2 m, and e_a and e_a^* are the vapor pressure and saturation vapor pressure of the air in hPa at some height above the surface, respectively. The required weather variables are measured routinely at most weather stations except R_n and it can be estimated. The Δ and γ terms are weighting factors to determine the relative importance of the energy balance and aerodynamic terms in accounting for total evaporation.

4.8.3   Measuring evapotranspiration

Measuring evapotranspiration is difficult because of its upward direc-ted orientation and the influence of various surface materials on the nature of the process. It is measured in equivalent depth of water returned to the atmo-sphere over a specified period to make it comparable with precipitation. Although commercial gauges with a surface that simulates a well-watered leaf are available for estimating local evapotranspiration, the most accurate evapo-transpiration measurement is achieved using lysimeters or soil-filled tanks covered with vegetation. In general, lysimeters are categorized as weighing or non-weighing types. There is no universal international standard lysimeter for measuring evapotranspiration (WMO, 1996).

A lysimeter is an instrument that integrates the influences of temperature, wind speed, solar radiation, and available moisture on the evapotranspiration

Fig. 4.14. Weighing lysimeter with short grass surrounded by a field being prepared for planting. (Photo by author.)

process. Weighing lysimeters consist of a soil-filled container several meters in diameter and 1 or 2 m deep. The container is buried at ground level and the surface is planted with the same vegetation as the surrounding area (Fig. 4.14). The vegetation can include orchard and vine crops as reported by Johnson *et al.* (2005). The bottom of the container is positioned on highly sensitive hydraulic weighing systems or on load cells with strain gauges of variable electrical resistance (WMO, 1996; Barani and Khanjani, 2002). The container is constantly weighed and weight changes are equated to the change in moisture content of the soil-filled tank. An alternative design is to float the lysimeter tank in water and use the change in liquid displacement as a measure of the weight gain or loss from the lysimeter tank related to changes in soil water. Any precipitation or irrigation on the lysimeter is measured at ground level and any percolation from the soil in the lysimeter is recorded. By this method, evapotranspiration equals precipitation plus irrigation minus percolation minus change in soil storage. A weighing lysimeter can determine evapotranspiration accurately for periods as short as one hour using mechanical scales or for periods of 24 hours using hydraulic weighing systems (Barani and Khanjani, 2002). Units are expressed in mm of water for the appropriate time interval to correspond with measures of precipitation and irrigation. Unfortunately, the expense of installing and operating lysimeters limits their use to research centers.

Evapotranspiration is controlled by the energy demand for moisture, water availability, and the extent of plant cover and its growth stage. Operated under natural conditions allowing moisture to be replenished by precipitation only, the weighing lysimeter provides a measure of the physical water loss from the surface called actual evapotranspiration (ETa). Natural moisture loss depends on climatic factors related to net radiation, wind velocity, and humidity and to other physical influences such as soil type, soil moisture content, vegetation rooting depth, and land management practices. Application of a constant water supply to the lysimeter by surface or subsurface irrigation so that the water loss is never restricted by a lack of available water permits the lysimeter to measure the moisture loss limited only by available energy. Under these conditions, the lysimeter provides a measure of the maximum evapotranspiration flux or potential evapotranspiration (ETp). This is the moisture flux occurring from an extended surface of short green vegetation that fully shades the ground, exerts little or negligible resistance to the flow of water, and is always well supplied with water (Rosenberg et al., 1983). The possibility that evapotranspiration may not be satisfied by the available moisture supply creates the need for the two designations for evapotranspiration. ETp is a climatically defined quantity independent of surface characteristics. ETa is the moisture flux occurring from a vegetated surface to the atmosphere in response to the coincidence of energy and moisture. Due to the defining role moisture plays in these two expressions, ETa can be equal to or less than ETp but it can never exceed ETp.

Reference evapotranspiration (ETo) is a term used to relate the evapotranspiration concept to crop water requirements. ETo is essentially equivalent to ETp with the exception that the leaf surfaces are typically not wet and a reference crop of short grass or alfalfa is specified (Jensen et al., 1990). Even though ETp and ETo have technical distinctions, McKenney and Rosenberg (1993) assert they are basically similar conceptually. ETp is emphasized in the following discussions because natural landscapes are the dominant focus of hydroclimatological studies. ETo is used in those instances where agricultural applications are the major focus.

Non-weighing or percolation-type lysimeters measure ETa by indicating volumetric change in the container water balance. A simple design is a petroleum barrel filled with soil and placed in a hole so that its top is level with the surrounding surface. Excess water draining to a pebble layer at the bottom of the barrel is removed through a small tube either by a gravity drain or by a suction hand pump (WMO 1996). A percolation lysimeter uses direct sampling of soil moisture changes or indirect sampling with a device such as a neutron probe along with water drainage out of the bottom of the root zone to determine ETa. In locations with a high water table, the water table in the lysimeter is maintained at a constant level and the water added to maintain the water level is a measure of ETa (Nokes, 1995). Non-weighing lysimeters can be used only for long-term measurements unless the soil moisture content is measured by some independent technique (WMO, 1996).

4.8.4   Estimating evapotranspiration

The difficulty in measuring evapotranspiration has promoted the use of various approaches for estimating evapotranspiration. The specific application of evapotranspiration information is an important determinant in choosing an estimating technique because not all techniques perform equally well in all settings. The space and time purposes of the evapotranspiration data are important in defining the nature of the required meteorological data and the likelihood of it being available. Estimating techniques that provide excellent results for 30-minute time steps are not readily extrapolated for decadal time-series studies. Techniques providing excellent evapotranspiration assessments for irrigated agriculture may not perform well for heterogeneous natural vegetation experiencing soil drying. Because application characteristics vary, estimating techniques favored by agriculture, climatology, engineering, forestry, hydrology, and meteorology differ. However, the physical basis for estimating the evapotranspiration process is the same.

Evapotranspiration is estimated with considerable accuracy using energy balance concepts, highly sensitive and efficient instrument arrays, and a variety of indirect techniques for computing evapotranspiration based on the physics of the phase change process. These techniques are described in detail in the micrometeorological and hydrological literature (e.g. Rosenberg et al., 1983; Shuttleworth, 1993; Guyot, 1998), and selected examples are discussed here to illustrate the nature of these techniques.

There are a number of equations in common use for estimating evapotranspiration that share a focus on the energy balance at the Earth's surface. All of these approaches have some basis in theory with one or more experimentally determined or estimated variables. The form of the equation is dictated by its application to a type of land surface, climatic setting, or vegetation cover. The common element linking these equations is their focus on the energy balance at the Earth's surface as shown in Equation 2.12, which expresses the energy balance as the algebraic sum of the energy fluxes. The energy balance approach is commonly used for daily estimates of evapotranspiration, and it enjoys a considerable degree of success when applied locally. However, detailed meteorological measurements are needed for these estimates, and the effect of atmospheric stability is important and must be included in the computation for short-term estimates.

The energy balance approach to estimating evapotranspiration is concerned with estimating the portion of net radiation available for conversion into latent heat. A simple expression of this condition is the ratio of the sensible heat flux (H) to the latent heat flux (LE). This relationship is commonly termed the Bowen ratio (β) and is expressed as

$$\beta = \frac{H}{LE} = \frac{\gamma(T_s - T_a)}{(e_s - e_a)} \tag{4.15}$$

where γ is the psychrometric constant which varies weakly with temperature, T_s is the mean surface temperature, T_a is mean air temperature at a given height, e_s is the saturation vapor pressure of the evaporating surface at temperature T_s, and e_a is the actual vapor pressure of the air at the height where T_a is measured. Accurate measures of net radiation, soil heat flux, and vertical profiles of temperature and humidity are needed for this approach. These measurements are relatively simple and rapid response instruments are not needed for most research purposes, but routine field measurements for these evapotranspiration calculations may be difficult to acquire (Ward and Robinson, 2000). The sign of H often changes in the evening and morning so that Equation 4.15 is not defined at these times. The Bowen ratio estimates ETp when the surface is well watered and LE is limited only by R_n. When moisture is a limiting factor for energy partitioning at the surface, then the Bowen ratio estimates ETa.

The energy balance can be coupled with empirical relationships depicting various surface conditions to estimate evapotranspiration. Priestly and Taylor (1972) proposed a method using an empirically derived constant as an advective term so that evapotranspiration from a moist surface (ETp) depends only on measured temperature and available energy. The equation takes the general form of

$$ETp = c \left(\frac{\Delta}{\Delta + \gamma} \right) (R_n - G) \tag{4.16}$$

where c is an empirically derived constant, Δ is the slope of the saturation vapor pressure curve at the mean wet-bulb temperature of the air, γ is the psychrometric constant, R_n is net radiation, and G is the soil heat flux. The Priestly–Taylor method is most reliable in humid areas using a constant value of 1.26. In arid climates, a constant of 1.74 is recommended (Shuttleworth, 1993). Other empirical radiation-based equations have been proposed for use in either humid climates (Turc, 1961) or arid climates (Doorenbos and Pruitt, 1977).

Parlange et al. (1995) conclude that the fundamental difference among energy balance equations for estimating ETa is in their bases for partitioning available energy to determine evaporation as a residual of the measured terms. They propose a general form of the various equations is

$$LE = B' \left[A \left(\frac{\Delta}{\Lambda + \gamma'} \right) (R_n - G) + B \left(\frac{\gamma}{\Lambda + \gamma'} \right) f(u) (e_a^* - e_a) \right] \tag{4.17}$$

where LE is the latent heat flux in $W\,m^{-2}$; L is the latent heat of vaporization with a value of $2.45 \times 10^6\,J\,kg^{-1}$ at typical temperatures; E is the evaporation rate in $kg\,m^{-2}\,s^{-1}$; Δ is the slope of the saturation vapor pressure curve taken at the temperature of interest; R_n is net all-wave radiation incident on the surface; G is the soil heat flux into the ground; e_a and e_a^* are the vapor pressure and saturation vapor pressure of the air at some height above the surface, respectively; γ is the ratio of the specific heat of air at constant pressure to the latent heat of vaporization generally taken to be a constant of $0.67\,hPa\,K^{-1}$ at standard temperature and pressure; $f(u)$ is some function of the wind velocity historically taken to be of the linear form $f(u) = a + bu$ with a and b being constants; A, B, and γ' are parameters that take on various values depending on the particular formulation; B' is the Budyko–Thornthwaite–Mather parameter, which is a function of the availability of surface water and is generally taken to be 1.0 until some measure of field capacity is reached and then allowed to decrease to zero with limited water availability (Parlange et $al.$, 1995). Since ET is linked to the energy balance through LE, ETa is derived from Equation 4.17 by

$$ETa = LE/L \tag{4.18}$$

where all of the variables are defined above.

Penman's E_o is converted into ETp for a vegetated land surface by employing empirically derived coefficients for a particular site and time of the year using a modified form of Equation 4.13. A widely used formulation by Monteith (1965) incorporates vegetation-canopy conductance, and it characterizes the vegetation surface as one big leaf. ETp from land surfaces by the Penman–Monteith method is estimated by

$$ETp = \frac{\Delta(R_n - G) + \rho_a c_p (e_a^* - e_a) r_a^{-1}}{\Delta + \gamma(1 + r_s r_a^{-1})} \tag{4.19}$$

where Δ is the slope of the saturation vapor pressure curve at the mean wet-bulb temperature of the air; R_n is net radiation incident on the surface; G is the soil heat flux into the ground; ρ_a is air density $(kg\,m^{-3})$; c_p is specific heat at constant pressure $(J\,kg^{-1}\,K^{-1})$; e_a and e_a^* are the vapor pressure and saturation vapor pressure of the air at some height above the surface, respectively (hPa); r_a is the aerodynamic resistance to vapor transfer $(s\,m^{-1})$; γ is the psychrometric constant or the ratio of the specific heat of air at constant pressure to the latent heat of vaporization, generally taken to be a constant of 0.67 at standard temperature and pressure $(hPa\,K^{-1})$; and r_s is minimum canopy resistance $(s\,m^{-1})$. This equation is remarkably successful in estimating ETp in many environments and for periods as short as 20 minutes. However, the surface resistance must be determined by botanical methods that are difficult to extrapolate to other areas or types of vegetation, and the surface resistance can vary dramatically over a short period in response to environmental variables. Nevertheless,

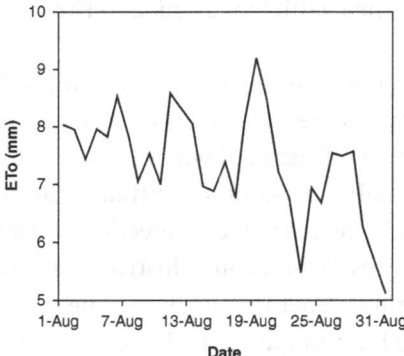

Fig. 4.15. Daily reference evapotranspiration (ETo) for August 2006 at Indio, California (34° N). (Data courtesy of the California Department of Water Resources from their website at http://wwwcimis.water.ca.gov/.)

this approach produced the best results in an extensive comparison by Jensen *et al.* (1990) of 19 methods for estimating potential evapotranspiration compared to lysimeter and well-watered alfalfa data.

The Food and Agriculture Organization of the United Nations (FAO) recommends a modified form of the Penman–Monteith equation for ETo whose foundation is based on water needs associated with well-managed irrigated agriculture. The reference surface is defined as a hypothetical grass reference crop with an assumed crop height of 0.12 m, a fixed surface resistance of $70\,\mathrm{s\,m^{-1}}$, and an albedo of 0.23. This reference surface closely resembles an extensive green, well-watered grass of uniform height, actively growing, and completely shading the ground. The fixed surface resistance of $70\,\mathrm{s\,m^{-1}}$ implies a moderately dry soil resulting from a weekly irrigation frequency (Allen *et al.*, 1998). The FAO Penman–Monteith equation applied on a daily basis and resulting in $\mathrm{mm\,d^{-1}}$ has the form

$$\text{ETo} = \frac{0.408\Delta(R_n - G) + \gamma\left(\dfrac{900}{T+273}\right)u_2(e_a^* - e_a)}{\Delta + \gamma(1 + 0.34u_2)} \tag{4.20}$$

where T is the mean daily air temperature at 2 m height (°C), u_2 is wind speed at 2 m height ($\mathrm{m\,s^{-1}}$), and all other variables are defined for Equation 4.19. An example of daily ETo for a site in California calculated with the Penman–Monteith equation is shown in Figure 4.15.

The FAO Penman–Monteith equation is generally limited to applications for agricultural crops, but it can be applied to natural vegetation. However, greater variations in plant density, leaf area, and water availability in natural vegetation increase uncertainty in the nature of the coefficient applied to ETo to estimate the actual water loss (Allen, 2000).

The eddy covariance technique assumes that the moisture flux away from the surface can be estimated by simultaneous measurements of vertical wind velocity and atmospheric humidity. Upward diffusion of water vapor from the surface occurs when upward directed small parcels of air entrained in turbulent eddies are more moist than downward directed eddies. Detection of the upward transport by turbulent motions requires fast-response instrument arrays that record wind, temperature, and humidity fluctuations with time constants of seconds or shorter. In addition, these instruments must be sensitive enough to detect fluctuations caused by the rapid passage of different eddies. However, advances in instrumentation, data acquisition, and data archiving have accelerated use of this technique for quasi-continuous measurements (Baldocchi, 2003). The vertical water vapor flux (ETa) is given by

$$\text{ETa} = \frac{\overline{w' \rho_v'}}{\rho_w} \tag{4.21}$$

where w' is the vertical wind speed component departure from the mean, ρ_v' is concurrent variation of absolute humidity about the mean value, ρ_w is the mass density of water, and the overbar represents a time average usually of 30 minutes. Eddy covariance is part of the FLUXNET program involving an international network of sites assessing latent and sensible heat flux variability across different climates and ecosystems (Wilson et al., 2002; Baldocchi, 2003).

Surface renewal is an indirect and inexpensive method for estimating evapotranspiration by measuring high-frequency temperature or humidity variability within the plant canopy considering both vertical turbulence and horizontal advection. The basic premise of surface renewal is that major canopy–atmosphere energy or mass exchange eddies create ramp patterns in the scalars being exchanged. The typical temperature or humidity ramp has a gradual rise (or decrease) in temperature or humidity followed by a relatively sharp drop (or rise) (Paw U et al., 2004). The ramp pattern occurs because the turbulent exchange is driven by the regular replacement of an air parcel in contact with the sinks and sources in the surface where the exchange occurs. As one air parcel sweeps down to the surface it replaces another parcel that is ejected from the canopy (Castellvi, 2004). When high-frequency air temperature or humidity is plotted against time, the constant air renewal is inferred from the sawtooth pattern exhibiting ramp-like shapes related to coherent eddy structures (Paw U et al., 1995).

High-frequency measurements of 1–20 times per second permit quantification of the mean ramp amplitude and inverse ramp frequency of temperature or humidity used to estimate the sensible heat flux or the latent heat flux in the surface renewal method (Paw U *et al.*, 1995). Fine-wire thermocouples are used for high-frequency temperature measurements, and fast-response hygrometers are used for high-frequency water vapor measurements. However, the relatively greater expense of fast-response hygrometers makes it more practical to use the more indirect method based on temperature measurements. The surface renewal temperature ramp amplitude and inverse ramp frequency are used to estimate the sensible heat flux using an energy conservation equation (Paw U *et al.*, 1995). The latent heat flux attributed to evapotranspiration is estimated as a residual using measured net radiation and soil heat flux measurements (Spano *et al.*, 2000).

A number of empirical techniques for estimating evapotranspiration have been developed to overcome the demanding data requirements of other approaches. Unfortunately, simplified estimating procedures sacrifice some facets of the true physical aspects of the evapotranspiration process. The major advantage of empirical approaches is that they can be applied in areas where only standard climatic data are available, and they are often applied successfully to areal estimates. Two empirical techniques are discussed to illustrate the characteristics of this method for estimating ET.

One of the best known of the empirical approaches recognizing both the physical and biological roles in evapotranspiration was developed by Thornthwaite (1948). He proposed an expression for evapotranspiration requiring only information on temperature and day length. Thornthwaite's expression for ETp is

$$ETp = 1.6d(10T/I)^a \tag{4.22}$$

where d is a correction factor based on the number of daylight hours per day compared to 12 and the number of days in a month compared to 30, T is mean monthly temperature (°C), I is an annual heat index or the sum of 12 monthly heat-index values (i) determined from $i = (T/5)^{1.514}$, and the exponent a is the non-linear function of the heat index approximated by the expression $a = 0.49239 + 1.79 10^{-2}I - 7.71 10^{-5}I^2 + 6.75 10^{-7}I^3$. Since the daylength factor is a function of latitude, Thornthwaite's ETp can be solved for any location where monthly temperature data are available.

This method is often criticized because it uses temperature as the basis for estimating evapotranspiration rather than radiation values. However, Bailey and Johnson (1972) demonstrate that the heat index term is a parameterization of radiation, humidity and wind, and Mather and Ambroziak (1986) show that Thornthwaite's method does account for insolation. In addition, it has been used successfully in a variety of regional applications to solve hydroclimatic problems where only temperature data are available. Jensen *et al.* (1990) caution that the Thornthwaite method is only applicable in humid regions similar to the locale where it was developed, but Shelton (1978) presents evidence that an adjusted form of monthly Thornthwaite ETp is valid for semiarid regions. Application of the Thornthwaite method to periods of less than a month often leads to errors because short-term mean temperature is not correlated with incoming radiation in the same manner as long-term radiation (Rosenberg *et al.*, 1983). Guyot (1998) concludes the Thornthwaite formula is well suited for drainage-basin-scale analyses using cumulative values.

Blaney and Criddle (1950) developed a temperature-based equation for estimating ETa specifically designed for agricultural applications. The original relationship was intended for seasonal estimates of crop water requirements for irrigated agriculture in the western United States. The Blaney–Criddle formula is based on the assumption that crop evapotranspiration varies with the sum of the products of mean monthly air temperature and the monthly percentage of daylight hours. An empirically determined and crop specific consumptive use coefficient is applied as the other variable. The Blaney–Criddle method is attractive because it is relatively easy to use, and the required data are readily available. The original equation expressing monthly consumptive use or crop specific ETa in mm d^{-1} is defined as

$$u = k_m f \tag{4.23a}$$

where k_m is an empirically defined monthly consumptive use coefficient specific for a given crop, and f is a monthly consumptive use factor determined by $f = 0.01(1.8T_a + 32)p$, where T_a is the monthly mean temperature in °C, and p is the monthly percentage of total annual daylight hours. The total consumptive use for the season (U) is expressed as

$$U = \sum_{i=1}^{n} u = \sum_{i=1}^{n} k_m f \tag{4.23b}$$

where U is accumulated over any period, and n is the number of months in the period (Rosenberg *et al.*, 1983).

A number of alternative forms of the Blaney–Criddle formula are employed internationally. The U.S. Soil Conservation Service (USCS,1970) suggested a modified form of the original formula as

$$U = k_t k_m f \tag{4.23c}$$

where k_t is a climatic coefficient related to mean air temperature and given by $k_t = 0.0173t - 0.314$ when $k_t > 0.3$ (McKenney and Rosenberg, 1993). A version of the formula commonly called the FAO Blaney–Criddle equation (Doorenbos and Pruit, 1977) for monthly ETa in mm d^{-1} has the form

$$\text{ETa} = c[p(0.46T + 8.13)] \qquad\qquad (4.23d)$$

where c is an adjustment factor based on minimum relative humidity, sunshine hours, and daytime wind estimates, p is the ratio of actual daily daylight hours to annual mean daily daylight hours expressed as a percentage, and T is mean air temperature in °C. Graphs and tables are provided by Doorenbos and Pruitt (1977) to determine c, and equations for calculating both p and c are provided by Shuttleworth (1993).

4.9   Streamflow

Streamflow is the flow rate or discharge of water at a specified location on a natural stream channel. In general terms, streamflow is the residual of the precipitation that falls on the area upstream of the selected stream channel reference point. The dominantly horizontal character of streamflow establishes its roles in climate of the first kind and the terrestrial branch of the hydrologic cycle as the transfer function for returning excess continental liquid moisture to the oceans. Streamflow at any particular time integrates all of the hydroclimatic processes and storages occurring upstream relative to the selected reference point.

Modern methods for measuring streamflow developed in the nineteenth century are described in Chapter 1. The nature of streamflow measurements is responsible for streamflow data being very expensive, difficult to obtain, and commonly restricted to selected rivers and streams (Ward and Elliot, 1995). An additional consideration is that many gauged streams are affected to some extent by reservoir regulation and water diversions. Although the number of hydrological stations expanded greatly at the international level in the closing decades of the twentieth century, all continents other than Europe and North America have relatively low density hydrological networks. Europe has a high density hydrological network that supports the longest continuous records. Streamflow was recorded on many large European rivers in the nineteenth century, and in Britain the longest continuous flow measurements started on the River Lea in 1879 (Arnell, 1996). However, the earliest continuous streamflow records for North American rivers begin around 1900. An extensive network of streamflow measurements for the United States is maintained by the U.S. Geological Survey in cooperation with state water agencies. Many national computer archives of hydrologic data can be accessed through the Hydrological Operational Multipurpose System of the WMO (Mosley and McKerchar, 1993).

The hydrologic balance (Equation 2.13) illustrates that the single parameter unique to the hydrologic cycle is runoff. Runoff is that part of precipitation that eventually becomes the water in river and stream channels, but it is generated over an area. Consequently, runoff is never measured directly and it is unlikely that it ever will be measured directly. Rather runoff is determined using a series of intermediate steps. First it must be recognized that runoff refers to a depth of water spread evenly over a specified area. This is the major basis for explaining why runoff is not measured directly. This point may also aid in clarifying why confusion exists over the difference between runoff and stream discharge or streamflow. The water in a river or stream channel is both runoff and discharge. Streamflow is runoff from an area contributing water to the stream channel, and this area is called the watershed, drainage area, drainage basin or catchment. Streamflow is also the stream discharge at a specific point on the river or stream. The runoff phenomenon is treated in detail in Chapter 6. The quantification of streamflow and runoff is addressed here.

Stream discharge is a volumetric measure for a specific time interval and is commonly reported as $m^3 s^{-1}$. For periods longer than a day, the meaning becomes less precise. Discharge for longer periods is expressed as km^3.

Quantification of runoff is aided by the natural collection system represented by the watershed or the area contributing the water that passes through a given river or stream cross-section. This area is commonly topographically defined, and the boundary delimits a watershed divide. Delineation of the watershed area is important for defining the paths and rates of water movement between the time it arrives as precipitation on the watershed and when it arrives at the watershed outlet. A complex array of space and time variations in physical watershed variables is integrated by the runoff process. The result is that runoff provides the space and time averaging that is difficult to achieve when individual hydroclimatic parameters are measured and averaged over an areal unit. Runoff is the volume of water quantified by streamflow, but it is also the depth of water that would occur by spreading the streamflow volume evenly over the area upstream of the stream gauge.

Streamflow is the only phase of the hydrologic cycle for which reasonably accurate measurements are made of the volumes involved. However, streamflow at a given location is seldom measured in a single, relatively simple manner, and the measurement procedures are subject to introduced errors over time. For most rivers and streams, there is a specific relationship between the flow at a particular location along the channel and the water level in the channel. This relationship is due to the physical characteristics of the watershed and the stream channel. Measurement errors are commonly related to changes in these physical characteristics. Nevertheless, the assumed relationship between the water level and the magnitude of discharge is the underlying basis for determining streamflow. Typically, streamflow data are available for a much shorter duration than climate data.

Fig. 4.16. Staff gauge on a small stream that has seasonal flooding. The streamflow depth at this site is used as a benchmark for flood forecasting at this point and further downstream. (Photo by author.)

In practice, three steps followed in developing streamflow data are measuring the water level, measuring stream discharge, and identifying the relationship between water level and discharge. River stage refers to the water-surface elevation or water level at a point along a stream relative to an arbitrary datum or fixed reference point of known elevation. The principal devices for determining the water level are the staff gauge and a variety of mechanical instruments for measuring, recording, and transmitting water-level data (Mosley and McKerchar, 1993; Herschy, 1997). The vertical staff gauge is the simplest type of manual gauge, and it represents the river stage concept clearly (Fig. 4.16). The gauge is fixed to a bridge pier or piling and has a graduated scale that includes all possible water levels that may occur at the location. An observer determines visually the water level at the time of observation. This method is most appropriate for large rivers where the river stage changes slowly. Since observations are done only once a day, a major disadvantage of the manual gauge is that the water level may change significantly between gauge observations.

Recording gauges overcome the problem of water-level changes between observations by providing a continuous water-level record. Both float-actuated and pressure-actuated recording gauges can be installed in a shelter over a well that is connected by open pipes to the river or stream. Water-stage fluctuations in the channel coincide with the water level in the stilling well, and a continuous water-level record is produced. Modern telemetry permits real-time retrieval of data. Figure 4.17 portrays daily streamflow for a large river at the beginning of the snowmelt season.

Stream discharge is the basic measurement for a watershed, but discharge is a derived term expressing the volume rate of water movement past a given point

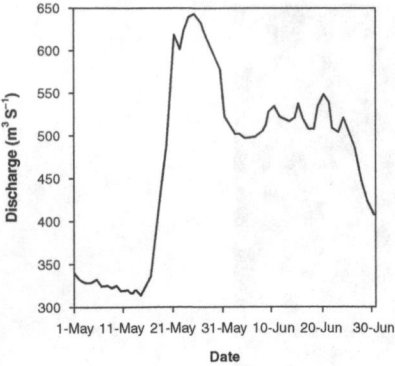

Fig. 4.17. Daily streamflow for the Columbia River at the International Boundary for two months during the 2006 high flow season. (Data courtesy of the U.S. Geological Survey from their website at http://waterdata.usgs.gov/nwis/.)

with respect to time. The actual measurements involved in determining stream discharge are stream velocity and the cross-section area of the stream through which the water is passing. Stream velocity is the linear rate of water movement, but the highly variable conditions of streamflow prevent the adoption of a universal velocity measurement method. However, various mechanical current meters are the most widely used method for measuring stream velocity. The stream cross-section at the point where velocity is measured defines the area for deriving discharge. Therefore, stream discharge (Q) is calculated as

$$Q = vA_c \tag{4.24}$$

where v is stream velocity and A_c is the cross-sectional area of flow.

For all but the smallest streams, multiple measurements of stream velocity and a corresponding cross-section area are needed. The river cross-section is divided into a number of predetermined equal-area subsections ranging from 20 to more than 30 depending on the physical characteristics of the stream channel and the measurement method employed. No subsection should account for more than 5% of the total discharge. Therefore, the spacing of the subsections will be closer together where the flow is fast and deep and farther apart where flow is slow and shallow.

Cross-sections are negotiated by wading, by specially constructed cable ways, by boat, or from bridges. The flow in each subsection is measured with a current meter and a sounding is taken to determine the water depth. Because of expected velocity distributions with depth across a stream channel, velocities are usually measured at 0.2 and 0.8 of the total depth in each river subsection when the depth exceeds 0.76 m (Mosley and McKerchar, 1993). The mean of these two velocities closely approximates the average velocity in a vertical section in most free-running streams. When only one measurement is made, the velocity is measured at 0.6 of the subsection depth. Total discharge is the sum of the individual subsection discharges. Alternative discharge measurement methods are described by Moseley and McKerchar (1993) and Herschy (1997).

The direct measurement of stream discharge is difficult and time consuming. However, a plot of measured discharge against water level at the time of discharge measurement usually defines a smooth curve known as the stage–discharge relation or a rating curve. The rating curve and a continuous water-level record at the gauge serving as the basis for the curve permits computation of a continuous streamflow record. Periodic discharge measurements are employed to verify that the stage–discharge relation has not changed. The rating curve can be altered by changes in the stream bed, stream velocity or aquatic vegetation. It is critical to recognize that discharge is an instantaneous measure that is related to a specific day and hour. A continuous record of the river stage permits quantification of discharge for periods as long as 24 hours.

4.9.1   Calculating runoff depth

Streamflow quantification is a necessary first step in determining the depth of runoff. The volumetric expression of stream discharge represents both streamflow and runoff. Recall that runoff is the volume of water produced by a specific drainage area. Runoff and streamflow are both expressed in km^3, but runoff is also expressed in cm or mm depth over the drainage area. Consequently, the computation of stream discharge produces the volumetric expression of runoff, but runoff expressed as water depth over the drainage basin requires an additional computation.

Runoff in cm (R_{cm}) at the gauge site is calculated by

$$R_{cm} = Q/A_d \times 0.000\,01 \qquad\qquad (4.25)$$

where Q is discharge in km^3, A_d is the area above the streamflow measurement point in square kilometers (km^2), and 0.000 01 is the km^3 of water produced by 1 cm of water covering 1 km^2. Runoff expressed in cm or mm has the advantage that it is comparable to the commonly used measures of precipitation and evapotranspiration.

4.9.2   Estimating streamflow

The streamflow response to precipitation at varying time scales is determined by the manner in which watershed processes partition precipitation into runoff and streamflow. The varied nature of these individual processes is examined in Chapter 6. Estimates of the quantity and volume of runoff are derived using models and equations intended to represent these processes and transform precipitation into a quantitative description of the stream response. The underlying basis for the models and estimating equations is the mass conservation of water expressed in the hydrologic balance (Equation 2.13). Watershed models are discussed in Section 6.15.

4.10   Estimating areal hydroclimatic data

For many hydroclimatic applications, a significant problem arises concerning the use of point sensors for climate data and data extrapolation away from the instrument site. This is often expressed as the representative area and is of considerable importance in hydroclimatology. Different physical factors influence determination of the representative area for a given variable. Gridding and remote sensing techniques provide alternative approaches for developing areal hydroclimatic data estimates, but point data are the benchmarks for verifying these estimates. Precipitation and evapotranspiration illustrate the nature of the challenges encountered in determining areal data from point observations.

4.10.1   Estimating areal precipitation

Precipitation is likely to be variable and difficult to extrapolate with confidence, especially in mountainous regions (Basist et al., 1994). The relationship between rainfall at two locations depends on the distance separating the points, the nature of the terrain and rainfall, and the time step employed (Linacre, 1992). Atmospheric conditions are responsible for much of the uncertainty surrounding the spatial pattern of precipitation because precipitation-producing conditions in the atmosphere are constantly in motion and the size and life span of atmospheric precipitation areas constantly changes. Low-latitude precipitation in summer is convective in nature and related to cumuliform clouds. Individual clouds typically exhibit a horizontal extent of only a few kilometers. The spatial distribution of precipitation even over flat terrain or the ocean can be quite variable. In extreme cases, rainfall at a rate of a few tens of millimeters per hour can occur at one location with no precipitation occurring only a few kilometers away. Typical rain-gauge networks do not have a density sufficient to measure variations at 1 km except for research sites. The spatial pattern of precipitation is typically determined from measurements made at several rain gauges within an area. On scales of a few tens of kilometers, spatial correlation of rainfall is directly related to latitude. Correlations are stronger during the cold season, and this implies rainfall is related to frontal activity. Frontal rainfall is more organized than convective systems common in the summer.

Techniques and algorithms for estimating areal precipitation by interpolating point measurements are described in many sources (e.g. Dingman, 1994; Elliott, 1995; Mays, 2005), and these approaches are supplemented by remote sensing technologies addressed in Section 5.7. Precipitation models provide another approach for estimating areal precipitation. Statistical models incorporate diverse information for estimating precipitation especially in mountainous areas. Basist *et al.* (1994) use six topographic variables in a multivariate analysis to estimate mean annual precipitation for ten mountainous areas representing five global climatic regions. Hayes *et al.* (2002) use twice-daily 850 hPa wind and humidity radiosonde data to estimate point and areal precipitation in the central Cascade Mountains of Washington State. Daly *et al.* (2002) present a hybrid statistical-geographic approach called PRISM (Parameter-elevation Regressions on Independent Slopes Model) that estimates areal precipitation using point precipitation measurements and a digital elevation model (DEM). PRISM was originally developed for precipitation estimates, but it has been applied to temperature and snowfall. High-resolution numerical precipitation models are widely used to estimate areal precipitation at various scales. At the global scale, Chen *et al.* (2002) provide monthly precipitation on a 2.5° latitude/longitude grid over land areas, and Legates and Willmott (1990) present monthly precipitation on a 0.5° latitude/longitude grid over both land and oceans.

4.10.2   Estimating areal evapotranspiration

Areal estimates of evapotranspiration are essential in hydroclimatic analysis. A complex set of problems underlies efforts to estimate evapotranspiration for areal units, such as river basins, because the most rigorous equations for point evapotranspiration estimates are difficult to extend with confidence beyond the site where the detailed measurements are performed. At the watershed scale, the spatial distribution of the soil moisture content and the soil–plant–atmosphere relationship display continuous and discontinuous properties that create a mosaic of landscape elements with variable properties.

Development of areal estimating methods requires incorporating the numerous processes and feedback mechanisms involved in evapotranspiration and the network of instruments for collecting adequate data. Consequently, there is no agreement on a standard methodology for determining areal average evapotranspiration (Parlange *et al.*, 1995). The problem of estimating areal evapotranspiration is exacerbated when complex topography and land use patterns are present, and this further contributes to areal evapotranspiration being the least understood aspect of the hydrologic cycle. One approach for estimating regional evapotranspiration uses a simplified version of Equation 4.17 that takes the form

$$E = B' ETp \qquad\qquad (4.26)$$

where the variables are defined for Equation 4.17. This approach is most effective at scales smaller than 10^4 m and for areas of relatively homogeneous terrain (Szilagyi and Parlange, 1999).

Scintillometry is an indirect method of estimating evapotranspiration in that variables are measured that do not directly yield evapotranspiration, but they provide a theoretical basis from which evapotranspiration can be estimated. The most widely used applications of scintillometry are to produce an areal average sensible heat flux estimate that is neither a point measurement nor remote sensing. A latent heat flux estimate is derived using the sensible heat flux and other observed net radiation values. The technique is employed in research and experimental work and not for routine hydroclimatological observations.

A scintillometer is an instrument consisting of a transmitter and a receiver. The transmitter emits an electromagnetic wave, and the receiver measures the intensity fluctuations of the electromagnetic wave caused by refractive scattering in the scintillometer path caused by changes in temperature and humidity due to heat and moisture eddies (Kite and Droogers, 2000; De Bruin, 2002). Temperature fluctuations dominate the refractivity fluctuations at visible to IR wavelengths, and water vapor fluctuations account for most of the scintillation signal at radio wavelengths greater than 1 mm. Therefore, the turbulent intensity of the refraction index of air is related to the structure parameters of temperature, humidity, and a covariance term (Meijninger and De Bruin, 2000). An important feature of the scintillometer technique is that while the measurement is along the path of the light beam it is actually an estimate of sensible heat over an area because of wind fetch (Kite and Droogers, 2000). Once the structure parameters of temperature and humidity are known the fluxes of sensible heat and water vapor can be derived by applying the Monin–Obukhov Similarity Theory as described by Meijninger et al. (2002).

The scintillation method is considered a form of low-pass filtering in which small spatial and temporal changes are smoothed by the average signal (Green et al., 2000). The measured refraction of laser light in small aperture scintillometers (SAS) makes possible determination of turbulence, temperature, wind velocity, and humidity parameters. Large aperture scintillometers (LAS) commonly have an aperture diameter of 0.15 m or more, which is larger than the Fresnel length defined as a measure of the size of the most active eddies in the scintillometer signal (De Bruin, 2002; Kohsiek et al., 2002). The attraction of LAS for hydroclimatological applications over heterogeneous terrain at the watershed scale is its path length range of up to several kilometers, and its easy alignment of transmitter and receiver units. Laser SAS are usually operated over maximum distances of a few hundreds of meters (Beyrich et al., 2002).

A major advantage of scintillometry is its ability to determine spatially averaged sensible and latent heat fluxes over heterogeneous terrain at spatial scales up to 10 km (Green *et al.*, 2000; De Bruin, 2002). Spatial averaging at this scale requires scintillometers to be mounted at greater heights than conventional instruments to obtain an unobstructed propagation of their beams. Meijninger *et al.* (2002) report success in deriving water vapor fluxes with a combined LAS and radio-wave scintillometer system over a heterogeneous watershed area in the Netherlands. Sensible heat flux from a stand-alone LAS and available energy balance observations also produced reasonable area-averaged water vapor flux estimates at this site. A LAS, operated continuously for more than one year over heterogeneous terrain in Germany, successfully measured the refractive index structure used to estimate the sensible heat flux (Beyrich *et al.*, 2002).

Review questions

4.1   Why is it commonly thought that measuring precipitation is easier than measuring streamflow?

4.2   What is accomplished during a snow survey?

4.3   How does an anemometer provide data that are significant for hydroclimatology?

4.4   What is the hydroclimatic role of soil texture?

4.5   Why is soil water measurement a problem?

4.6   How does water move through the soil?

4.7   How is the evaporation process from a water surface different from the evaporation process from a dry soil surface?

4.8   How is the moisture flux estimated for water and land surfaces?

4.9   What is the significance of the Bowen ratio for evaporation studies?

4.10   Why is evapotranspiration more difficult to measure than streamflow?

5

Remote sensing
and hydroclimate data

5.1   Remote sensing data

Technology has provided an expanding array of alternative approaches for acquiring hydroclimatic data using instruments at a distance from the location being measured. The practice of distant measurement known as remote sensing is accomplished using instruments on satellites, aircraft, or the ground. In most cases, the remote sensing technique involves the interaction of electromagnetic radiation with the Earth's surface or the atmosphere. Aerial photography utilizing visible wavelengths of the electromagnetic spectrum was the earliest use of remote sensing (Engman, 1993). A wider array of the electromagnetic spectrum is utilized currently by multispectral scanners, thermal sensors, microwave sensors, lasers, radar (radio detection and ranging), and lidar (light detection and ranging) to measure or infer the fluxes of energy, precipitation, and evapotranspiration and to infer soil moisture and runoff over an area. In addition, sodar (sound detection and ranging) uses short sound pulses to determine wind characteristics.

Radar, lidar, and sodar are active remote sensors in that they emit a pulse of electromagnetic or acoustic radiation to probe the atmosphere and surface characteristics. They detect with a receiver a portion of the radiation that is reflected or backscattered by the target (Angevine *et al.*, 2003). Passive sensors detect radiation emitted from the Earth's surface or from the atmosphere. A radiometer is a passive instrument that measures the emitted radiation in a specific region of the electromagnetic spectrum.

Each instrument array has its advantages and disadvantages, and a mix of different observational techniques is needed to provide the necessary data for a specific application (Andrews, 2000; Angevine *et al.*, 2003). Some remote sensing instruments are capable of making time-resolved vertical profiles or multidimensional measurements of atmospheric properties while others provide the data for techniques to measure the fluxes of heat, momentum, or water vapor over an area using detectable quantities. One limitation on remote sensing data is that the data may not be available for past events important in framing the hydroclimatic setting (Parlange *et al.*, 1995). Nevertheless, Legates (2000b) supports remote sensing technology as broadening understanding of hydroclimatic processes and providing improved ability to analyze and model these processes by enhancing the spatial and temporal resolution of hydroclimatic data. The papers in Owe *et al.* (2001) illustrate the diversity of remote sensing data applications to hydroclimatology.

5.1.1   Atmospheric hydroclimate and remote sensing

The atmosphere changes physically on widely varying time scales that are a challenge to measure over the entire globe. Satellites are the dominant observation platform for remote sensing of atmospheric components of hydroclimate. Data provided by satellites have made a significant contribution to understanding the dynamics of Earth's climatic and hydrologic system. Earth-orbiting remote sensors provide the best means of collecting data that reveal time and space variations in atmospheric circulation on many scales that are impossible with surface-based instruments and radiosonde data discussed in Chapter 3. Cloud tracking techniques using satellite images provide the basis for inferring atmospheric motion (Menzel, 2001). The data and images provided by satellites depict the character of atmospheric motion and water vapor transport that are the ultimate expression of climate of the first kind and the atmospheric branch of the hydrologic cycle.

5.1.2   Terrestrial hydroclimate and remote sensing

Hydroclimate processes at the land surface occur at spatial scales difficult to represent by point observations. Remote sensing of terrestrial hydroclimatic variables involves a broad array of satellite remote sensing methods and ground-based instruments that provide a basis for quantifying the spatial variability of these processes. The data derived from satellite microwave remote sensors is different from the active remote sensing data from ground-based radar, lidar, and sodar (Angevine *et al.*, 2003), but significant success has been achieved in combining these data with *in situ* data discussed in Chapter 4. Each type of data has advanced understanding of climate of the second kind and the terrestrial branch of the hydrologic cycle.

A brief overview of satellite deployments and sensors is presented before discussing the nature of remote sensing data available for atmospheric and terrestrial hydroclimate components.

5.2   Satellites

Meteorological satellites provide images and quantitative information about surface features and the lowest 20 km of the atmosphere (WMO, 1996). The contribution of satellite data to hydroclimatology has expanded greatly since the United States launched the Television and Infrared Observation Satellite (TIROS-1) in April 1960 to open the weather satellite era. Early satellites were limited to providing basically cloud images, but by the 1980s reliable measurements of net incoming and outgoing radiation fluxes at the top of the atmosphere became available. More recently, satellites provide information from numerous spectral bands that contribute to understanding of the precipitable water and liquid water content of the atmosphere, winds at cloud level and at the ocean surface, atmospheric temperature and humidity profiles, cloud cover, a cloud's surface temperature, global sea surface temperatures, land surface temperatures, determination of fog, water/ice boundaries, and daily measurements for remote areas where conventional data are lacking. However, it is important to remember that satellite observing systems are viewing upwelling radiances composed of emissions from a range of heights in the atmosphere (WMO, 1996). Consequently, satellite and surface observations should be regarded as complementary and not competing data sources.

Meteorological satellites provide imagery that is digitized data even though the images may look like photographs. All meteorological satellites are equipped with a radiometer. The radiometer produces an image composed of a series of discrete point values in rows and columns called picture elements or pixels. The radiometer measures the intensity of the radiant energy coming from the Earth's surface and the atmosphere in a selected wavelength band identified as a channel. When the radiometer collects a predetermined amount of radiant energy it registers a count, and the number of counts is proportional to the radiation intensity. The area viewed by the radiometer at any given time is known as its footprint, and the total radiation from the footprint is assigned to a pixel located in the footprint's center. Some detail is lost in averaging characteristics within the footprint to produce a single value.

To enlarge the area viewed by the radiometer, a scanning system is employed that physically changes the direction in which the radiometer is pointing. A complete image of a large area of the Earth is constructed by the radiometer when all the pixels in the image have been assigned values. The width of the area scanned by the radiometer defines the swath or the path covered on the Earth's surface for a particular orbit of the satellite.

Satellites can operate in several types of Earth orbit, but the most common orbits are geostationary and polar. A circular orbit is the goal for most

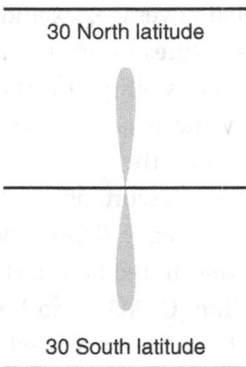

30 North latitude

30 South latitude

Fig. 5.1. Sketch of the orbital pattern of a geostationary satellite.

meteorological satellites, but satellites in general do not travel in perfect circles. The exact form of a satellite's orbit is derived from Newton's laws of motion and his law of universal gravitation, which have conceptual roots in Kepler's laws concerning planetary motion. A comprehensive summary of satellite orbits is provided by Kidder (2003).

A geostationary orbit is one in which the satellite is always in the same position with respect to the rotating Earth. The satellite orbits at an elevation of approximately 35 790 km above the Earth because that produces an orbital period equal to the Earth's period of rotation. The satellite appears stationary relative to the Earth because it is orbiting at the same rate and in the same direction as Earth (WMO, 1996), but in reality it is moving through the sky along a path resembling a narrow figure eight (Fig. 5.1). Geostationary satellites were first launched in the 1970s.

A polar-orbiting satellite circles the Earth at near-polar inclination. A true polar orbit has an inclination of 90°, which is the angle between the equatorial plane and the satellite orbital plane. Polar-orbit satellites circle the Earth at an altitude of 700 to 800 km, and the inclination of the satellite can be chosen so it is synchronized with the Sun in what is called a sun-synchronous orbit. This means that the satellite passes over a reference position on Earth at roughly the same local time during each orbital pass. The first weather satellite launched in 1960 was a polar-orbiting satellite.

5.2.1 *Geostationary satellites*

A network of geostationary satellites provides complete global coverage of all but the extreme north and south polar regions. This global satellite network is coordinated by the Coordination Group for Meteorological Satellites (CGMS) but is actually a collection of different satellites operated by independent agencies around the world. Full global coverage, excluding the polar regions, requires at least five geostationary satellites in orbit at any one time. The CGMS has agreed that operational responsibility for these five geostationary meteorological satellites is nominally the responsibility of four satellite operators. The European Organization for the Exploitation of Meteorological Satellites (EUMETSAT) inherited the Meteorological Satellite (Meteosat) program started by the European Space Agency (ESA). The other three programs are operated by Japan, Russia, and the United States. However, the Geostationary Operational Meteorological Satellite (GOMS), also known as Elektro, deployed by Russia and positioned at 76° E has not produced reliable operational imagery.

In addition to the four operators recognized by the CGMS, at least two other nations have satellite programs. An experimental geostationary meteorological satellite (Fengyun-2) launched by China is positioned at 105° E, but it has not achieved operational status. The multipurpose Indian National Satellite (INSAT-2) deployed by India and positioned at 93.5° E has meteorological capabilities but is reserved mainly for national use.

The three programs that have successfully established operational satellites for CGMS use a variety of instrument configurations on the spacecraft. The European region served by EUMETSAT has a Meteosat satellite positioned on the Greenwich Meridian and the equator above the Gulf of Guinea (Fig. 5.2). Meteosat provides imagery in one visible and eleven IR channels, including bands for ozone and carbon dioxide. A new satellite with advanced application features called Meteosat Second Generation (MSG) became operational in January 2004 and was renamed Meteosat-8. The Japanese Geostationary Meteorological Satellite (GMS), also known as Himawari or sunflower, is in a geostationary orbit at 140° E. The GMS has one scanning channel in the visible spectrum and three in the IR. In May 2003, GMS-5 handed over its observation operation to a United States satellite moved from storage to 155° E. This became necessary when GMS-5 exceeded its life expectancy and two satellite launches to replace GMS-5 failed.

The United States maintains and operates two Geostationary Operational Environmental Satellites (GOES) known as GOES-East and GOES-West. GOES satellites carry multispectral instruments with flexible scan modes for observing the atmosphere and inferring atmospheric vertical temperature and moisture structure. GOES-East is positioned at 75° W and the equator where it views North and South America and most of the Atlantic Ocean. GOES-West is positioned at 135° W and the equator, which permits it to view North America and the Pacific Ocean basin (Fig. 5.3). The two satellites together provide day and night imagery of the Earth extending from 20° W to 165° E.

Fig. 5.2. Visible image from Meteosat-7 for 2 March 2007 at 1200 UTC. (Image courtesy of NOAA from their website at http://weather.msfc.nasa.gov/GOES/.)

Operation of environmental satellites in the United States is the responsibility of the National Oceanic and Atmospheric Administration. NOAA's National Environmental Satellite, Data, and Information Service (NESDIS) operates the satellites and manages the processing and distribution of data and images these satellites produce daily. The prime customer is NOAA's National Weather Service, which uses satellite data to create forecasts and weather advisories. Satellite information is also shared with various Federal agencies, other countries, and the private sector.

Most geostationary satellites are geosynchronous because the Earth rotates at a rate slightly less than once every 24 hours. The satellite is allowed to drift within a predetermined area before corrections are made by on-board thrusters. The satellite's drift arises from anomalies in the Earth's gravitational field. The corrections maintain the satellite in a fixed position in the sky relative to the Earth's surface. Each satellite has a lifetime of about 5 years.

The scanning system of a geostationary satellite allows about 42° of the Earth's surface to be viewed from a single satellite. The geostationary satellite

Fig. 5.3. GOES-West visible image for 19 April 2006 at 2100 UTC. (Image courtesy of NOAA and the National Environmental Satellite, Data, and Information Service from their website at http://www.goes.noaa.gov/goesfull.html.)

can monitor developments in the field of view continuously and in almost real-time. However, a geostationary satellite's altitude of 36 000 km above the Earth's surface means the imagery resolution is lower than with polar-orbiting satellites deployed at lower altitudes. Also, geostationary satellites are limited to approximately 60 degrees of latitude at a fixed point over the Earth, and they provide distorted images of polar regions with poor spatial resolution. Even with these limitations, geostationary satellites provide continuous monitoring necessary for intensive data analysis. These satellites circle the Earth in a geosynchronous orbit that is high enough to allow the satellites a full-disk view of the Earth.

5.2.2   GOES instruments

Scanning and transmission systems on geostationary satellites vary with the country of origin, but they are broadly similar. In general, the basic instrumentation on these satellites measures Earth-emitted and reflected

radiation. These measurements are then used to derive atmospheric temperature, winds, moisture, and cloud cover. Some recent satellites are equipped with
instruments capable of measuring a broader range of conditions.

Satellite sensors observe the atmosphere with electromagnetic radiation
either passively or actively. Most operational systems on meteorological satellites are passive sensors receiving scattered, reflected, or emitted radiation from
the atmosphere or the Earth's surface. Sensors used on GOES satellites are
representative of the most widely used sensor systems on geostationary satellites (WMO, 1996). Therefore, a brief description of GOES instruments is provided to illustrate the general nature of these instruments.

The present GOES satellites have a sensor array that includes both imager and
sounder instruments. An imager is a radiometer designed to sense radiant and
solar reflected energy from sampled areas of the Earth. It is capable of calculating cloud cover and surface temperature remotely from space. The surface
temperature can be at the Earth's surface, a cloud's surface, the ocean's surface,
or any other surface whose temperature is required. A sounder is a radiometer
designed to provide data from which atmospheric temperature and moisture
profiles, surface and cloud-top temperatures, pressure, and ozone distribution
can be deduced by mathematical analysis. The atmospheric profiles produced by
the sounder are similar to those achieved using radiosondes. However, the
sounder system is capable of producing profiles for many more locations than
is possible with radiosondes.

The GOES imager (Fig. 5.4) has a five-band multispectral capability to detect
different wavelengths of energy in narrow wavelength bands called channels.
This allows the imager to identify visible light, emitted longwave radiation, and
other radiation wavelengths. The GOES imager has five channels that monitor

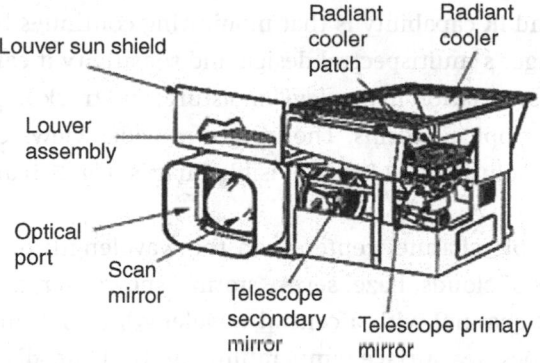

Fig. 5.4. GOES Imager diagram. (Drawing courtesy of NASA and the Goddard Space
Flight Center from their website at http://goespoes.gsfc.nasa.gov/.)

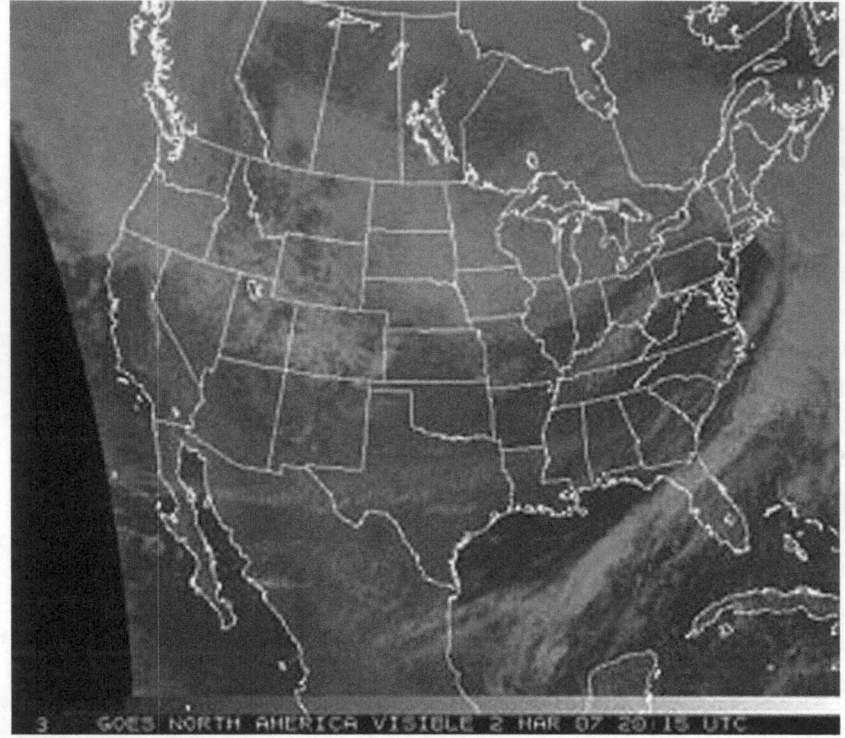

Fig. 5.5. GOES Imager visible image of North America for 2 March 2007 at 2015 UTC. Clouds cover most of Canada and the northern United States, and a band of clouds parallels the Atlantic coastline and extends southward into the Gulf of Mexico. (Image courtesy of NASA and the Global Hydrology and Climate Center from their website at http://weather.msfc.nasa.gov/GOES/.)

radiation at the specific wavelengths. Visible and IR images are acquired independently and continuously with a flexible scan system. The greatest advantage of having both visible and IR capability is that monitoring continues both day and night. With the imager's multispectral design and sensitivity it can detect temperature fluctuations, variation in low-level moisture, and track hurricanes from their inception as tropical storms. The imager provides views every 15 minutes with a spatial resolution of 1 km for visible images (Fig. 5.5) and 4 km for IR data.

The GOES imager visible channel centered on the wavelength of 0.65 μm supplies daylight images of clouds, haze, severe storms, snowcover, and volcanic activity. The near IR channel with a central wavelength of 3.9 μm reveals ground fog, fires, volcanoes, sea surface temperatures, and permits discrimination between water clouds and snow or ice crystal clouds. An IR channel centered at 6.7 μm provides a window on upper-level water vapor and is used for

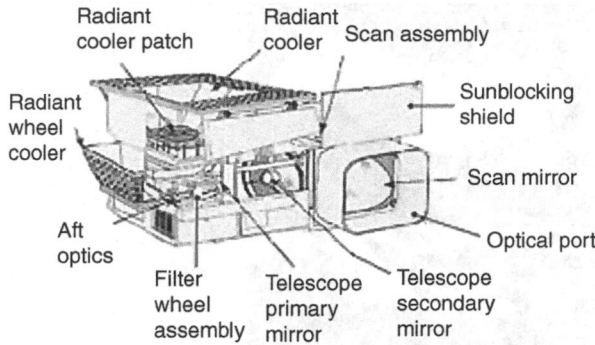

Fig. 5.6. GOES Sounder diagram. (Drawing courtesy of NASA and the Goddard Space Flight Center from their website at http://goespoes.gsfc.nasa.gov/.)

identifying upper-level moisture sources, the presence of atmospheric humidity, mid-level moisture content and advection, and tracking of mid-level atmospheric motions. Information on jet stream features, surface temperatures, and the location of heavy rainfall is provided by a longwave IR channel centered at 10.7 µm. A second longwave IR channel centered at 12.0 µm supplies images used to observe daily temperature changes and cold cloud tops, the detection of airborne dust and volcanic ash, and identification of low-level moisture. GOES-12 launched in July 2003 replaced the 12.0 µm channel with a 13.3 µm centered channel to provide improved data for estimating cloud amount and cloud heights.

The GOES sounder (Fig. 5.6) is a 19-channel discrete-filter radiometer covering the spectral range from the visible wavelengths to 15 µm. The sounder utilizes four sets of detectors to collect and identify variations within the Earth's atmosphere using a scanning broad infrared spectrum. It operates independently of the imager and makes simultaneous observations using a flexible scanning system similar to the one used by the imager. The sounder is equipped with a search and rescue transponder, and it serves as a space weather monitor.

The sounder's multi-element detector array permits it to simultaneously sample four separate fields or atmospheric columns. A rotating filter wheel brings spectral filters into the optical path of the detector array and provides the IR channel definition. It measures one visible channel and emitted radiation in 18 thermal IR bands divided into three detector groups that support several derived products (Fig. 5.7). The spatial resolution of sounder data is 1 km for the visible channel and 2 km for the IR channels.

The three-axis, body-stabilized spacecraft design of GOES enables the imager and sounder to constantly view a specified area of the Earth and frequently image clouds, monitor Earth's surface temperature and water vapor fields, and

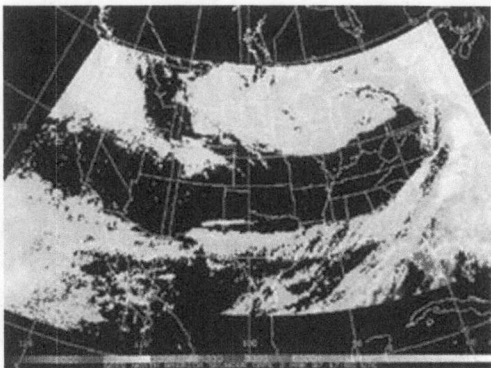

Fig. 5.7. GOES Sounder derived convective available potential energy (CAPE) for North America on 2 March 2007 at 1700 UTC. This energy relates to the energy available for thunderstorm development. (Image courtesy of NOAA and the National Environmental Satellite Data and Information Service from their website at http://www.orbit.nesdis.noaa.gov/smcd/opbd/goes/soundings/.)

sound the atmosphere of its vertical thermal and vapor structures. In addition, GOES satellites have flexible scanning that allows small-area imaging as well as simultaneous and independent imaging and sounding. These two features allow continuous data gathering from both instruments. While the imager collects digitized data, the sounder profiles temperature and moisture levels down through the atmosphere. Together the imager and sounder provide a stream of information on the movement of winds and moisture and temperature in the clouds. The viewing capability of current GOES satellites provides imagery useful beyond the previous north/south limits of geostationary satellites. This capability permits tracking of icebergs and monitoring snow and ice cover up to the Arctic and Antarctic circles.

5.2.3   *Polar-orbiting satellites*

A polar-orbiting satellite provides an observational platform for the entire Earth's surface. Polar-orbiting meteorological satellites complement the data provided by geostationary satellites. The orbit of a polar-orbiting satellite crosses close to both poles, and it follows an almost north–south direction. These satellites are inserted at altitudes of 800 to 1000 km, which imparts a relatively high speed to the satellite. Polar-orbiting satellites complete about 14 orbits per day with an orbital period of approximately 100 minutes. This results in the satellite passing over the same region on the Earth's surface twice a day at 12-hour intervals. The lower altitude of polar-orbiting satellites makes possible a finer ground resolution image than geostationary satellites, and it produces a ground swath of about 3300 km.

Numerous meteorological, commercial geophysical remote sensing (e.g. Landsat and the French Système Probatoire d'Observation de la Terre or SPOT), and communications satellites are in polar orbits. Polar-orbiting meteorological satellites have been deployed by China, Russia, and the United States. The Polar Operational Environmental Satellite (POES) Program is a cooperative effort between NASA and NOAA, the United Kingdom (UK), and France. In the United States, POES includes spacecraft developed and launched for the NOAA Polar-orbiting Operational Environmental Satellites (NPOES) program and the Defense Meteorological Satellite (DMS) program. Real-time data transmissions from the DMS satellites are encrypted, but DMS data are sent daily to the National Geophysical Data Center and the Solar Terrestrial Physics Division for archiving and can be available for civil use.

POES spacecraft have meteorological and geophysical importance because of their high-resolution global coverage and well-calibrated channels. Similar benefits are realized from the ESA European Remote Sensing (ERS) satellites equipped with both radiometers and synthetic aperture radar (SAR). The Canadian satellite RADARSAT provides SAR images for both scientific and commercial applications, and the research emphasis is on polar regions. The NASA-centered international Earth Observing System (EOS) includes the Terra and Aqua satellites launched in 1999 and 2002, respectively. These satellites are intended to monitor the health of the planet with Terra emphasizing land and Aqua emphasizing the many forms of water (Parkinson, 2003).

Most polar-orbiting satellites operate in a sun-synchronous orbit in which the orbit is tilted slightly towards the northwest and does not actually go over the poles. A sun-synchronous polar-orbiting satellite passes over a reference position on Earth at roughly the same local time during each orbital pass and is identified by the time it crosses the equator. An early morning satellite will make its ascending pass over the equator in the early morning, independent of Earth's east to west rotation. With each subsequent pass, the satellite will cross the equator southbound about 11 degrees westward due to the Earth's rotation. The time the satellite passes over the equator is usually between mid-morning and mid-afternoon on the sunlight side of the orbit. The actual orbital track is due to a combination of the orbital plane of the satellite coupled with the rotation of the Earth beneath the satellite. The orbital plane of a sun-synchronous orbit must also rotate approximately one degree each day to keep pace with the Earth's surface. In this way, it images its swath at about the same sun time during each pass, so that lighting remains roughly uniform. The repeated coverage by polar-orbiting satellites enables regular data collection at consistent times as well as long-term comparisons. Sun-synchronism produces time-constant illumination conditions of the observed surfaces, except for seasonal variations in sunlight duration.

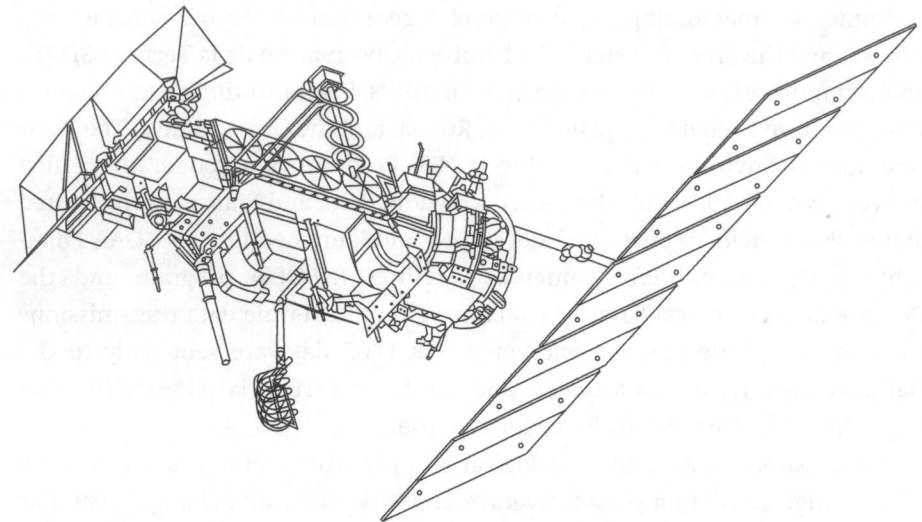

Fig. 5.8. Sketch of a POES satellite. (Drawing provided by NASA and the Goddard Space Flight Center from their website at http://goespoes.gsfc.nasa.gov/poes/.)

NOAA class satellites and Russian Meteor class satellites have orbits that cross very close to the poles on each revolution of the Earth. NOAA has two polar-orbiting satellites in the Advanced Television Infrared Observation Satellite (TIROS-N or ATN) series (Fig. 5.8). The orbits are circular and sun synchronous. The morning orbit crosses the equator at 7:30 a.m. local time and has an altitude of 830 km. The afternoon orbit crosses the equator at 1:40 p.m. local time and has an altitude of 870 km. The circular orbit permits uniform data acquisition by the satellite and efficient control of the satellite by the NOAA Command and Data Acquisition (CDA) stations located near Fairbanks, Alaska, and Wallops Island, Virginia. Operating as a pair, these satellites ensure that data for any region of the Earth are no more than six hours old.

5.2.4   *Polar-orbiting satellite instrumentation*

Instrumentation on NOAA's polar-orbiting satellites serves as a representative case due to the accessibility of the data from these instruments. The TIROS-N series satellites are three-axis-stabilized spacecraft that carry seven scientific instruments and two for Search and Rescue. The primary instrument is an Advanced Very High Resolution Radiometer (AVHRR), which is a five-channel multispectral scanner covering visible to thermal IR wavelengths in the range from 0.58 to 12.5 µm. One channel observes in the visible band, one in

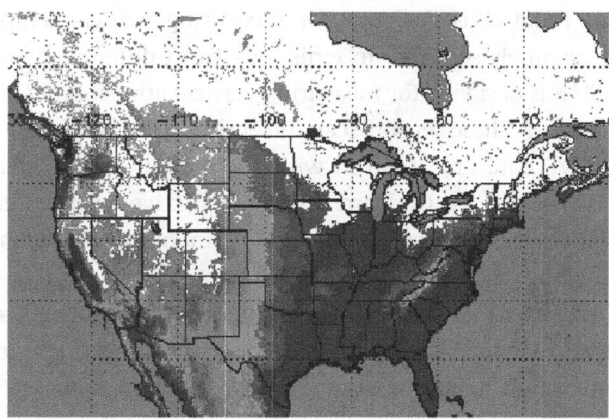

Fig. 5.9. NOAA-18 Advanced Very High Resolution Radiometer (AVHRR) 4 km resolution composite image of snowcover across Canada and the United States on 15 February 2007. (Image courtesy of NOAA and the National Environmental Satellite Data and Information Service from their website at http://www.orbit.nesdis.noaa.gov/smcd/emb/snow/.)

the near-IR, and three in the thermal-IR portion of the spectrum. The AVHRR scans the Earth's surface from side to side perpendicular to the satellite's ground track. Each scan covers an area about 2 km high and 3000 km wide.

AVHRR data are transmitted to the ground via a High Resolution Picture Transmission (HRPT) broadcast that contains AVHRR data in three formats. The HRPT provides unprocessed real-time, digital imagery at 1.1 km resolution. The onboard processor records full resolution Local Area Coverage (LAC) data at 1.1 km resolution for subsequent transmission during a station overpass. Global Area Coverage (GAC) data are produced by sampling and averaging onboard the satellite the 1.1 km AVHRR measurements and portraying them as 4 km data (Gutman *et al.*, 2000). An archive of the AVHRR data is provided by NOAA's National Environmental Satellite, Data, and Information Service (Gutman *et al.*, 2000; Jacobowitz *et al.*, 2003).

A second data transmission consists of only image data from two of the AVHRR channels, called Automatic Picture Transmission (APT). The information transmitted in the APT format provides users with imagery at 4 km resolution (Fig. 5.9). For users who want to establish their own direct-readout receiving station, low-resolution imagery data in the APT service can be received with inexpensive equipment, while the highest-resolution data transmitted in the HRPT service utilizes a more complex receiver.

Since 1998, the NOAA polar-orbiting satellites instrumentation has included the Advanced TIROS Operational Vertical Sounder (ATOVS). This instrument suite includes a 20-channel High Resolution Infrared Sounder (HIRS), a 15-channel Advanced Microwave Sounding Unit-A (AMSU-A), and a 5-channel Advanced Microwave Sounding Unit-B (AMSU-B). HIRS provides atmospheric sounding data in cloud-free regions. AMSU-A is designed for measuring vertical temperature profiles by selecting channels sensitive to emissions from different depths into the atmosphere. AMSU-B is designed for measuring vertical atmospheric humidity profiles in a similar manner. These sensors scan the entire Earth's surface over a 24-hour period. The ATOVS data are archived with the earlier TIROS Operational Vertical Sounder (TOVS) data at the Goddard Distributed Active Archive Center.

Instruments on the Terra and Aqua satellites complement the data acquired on earlier satellites, extend the record of these data, and improve the spatial resolution of some data. Aqua carries six instruments that take multiple readings across a wide span of the electromagnetic spectrum. Three instruments work together to form Aqua's sounding suite, which is composed of the Atmospheric Infrared Sounder (AIRS), the AMSU, and the Humidity Sounder for Brazil (HSB). AIRS is a 2382-channel high-spectral-resolution sounder whose primary purpose is to obtain atmospheric temperature and humidity profiles from the surface up to 40 km. The AMSU and HSB are similar to instruments on NOAA satellites, and they provide complementary measurements to the AIRS that are important under overcast sky conditions. The Advanced Microwave Scanning Radiometer for EOS (AMSR-E) provided by Japan is a 12-channel conically scanning passive-microwave radiometer measuring vertically and horizontally polarized radiation at frequencies from 6.9 to 89 GHz. The inclusion on AMSR-E of several channels measuring at wavelengths where there is little atmospheric interference permits the instrument to obtain surface data even in the presence of substantial cloud cover. This feature and the passive-microwave measurements mean that the AMSR-E has an all-weather and day-or-night capability for surface variables. The Moderate Resolution Imaging Spectroradiometer (MODIS) and the Clouds and the Earth's Radiant Energy System (CERES) instruments on Aqua are similar to instruments on Terra. MODIS is a 36-channel cross-track scanning radiometer utilizing wavelengths between 0.4 and 14.5 µm. MODIS has the finest spatial resolution for data from any of the Aqua instruments. CERES is a 3-channel broadband scanning radiometer that measures both the shortwave and longwave components of Earth's radiation balance. Two CERES instruments on Aqua operating in fixed and rotating scanning modes provide energy flux measurements that produce highly accurate radiation balance measurements (Parkinson, 2003).

5.3   Radiation data from satellites

Satellite-based measurements of solar and thermal radiation are necessary for determining the Earth-atmosphere radiation budget and for understanding the hydrologic cycle at the time and space scales of interest to hydroclimatology. The Explorer-7 satellite launched in 1959 carried an omnidirectional radiometer that provided the first satellite-based measurements of the Earth's albedo and IR fluxes. The first generation TIROS and METEOR satellites launched by the mid 1960s had scanning and non-scanning radiometers capable of measuring visible and IR wavelengths that supported early efforts to quantify the global radiation balance. The High Resolution Infrared Radiometer (HRIR) and the ultraviolet (UV) sensor on the early Nimbus series satellites provided day and night coverage and produced the first satellite data for quantifying the average Earth-atmosphere radiation balance at global, hemispheric, and zonal scales (Raschke et al., 1973). The Earth Radiation Budget (ERB) instrument package on the Nimbus-6 and Nimbus-7 satellites produced a nearly 20-year time-series of global albedo, outgoing longwave radiation, net radiation, and solar irradiance measurements. The Visible Infrared Spin Scan Radiometer (VISSR) on the first GOES satellites launched between 1968 and 1977 provided continuous spatial and temporal radiation sensing not possible from polar-orbiting satellites (Menzel and Purdom, 1994).

Outgoing longwave radiation (OLR) estimates from satellite observations have been used extensively as one component of the atmospheric radiation balance and to infer changes in cloud height and amount (Xie and Arkin, 1998). The AVHRR and the HIRS2 on NOAA POES satellites are a source of OLR estimates beginning around 1979. Neither the AVHRR nor the HIRS2 directly measure OLR, but radiance measurements by these instruments are converted to fluxes using algorithms based on theoretical model calculations. Introduction of the sounder on GOES-8 launched in 1988 made possible HIRS-like OLR estimates from sounder radiance measurements and at the high temporal resolution needed to quantify the OLR diurnal cycle (Ba et al., 2003).

The three-satellite Earth Radiation Budget Experiment (ERBE) initiated in 1984 was undertaken to provide accurate measurements of the Earth's radiation budget and to establish a long-term baseline dataset for studying climate. The Earth Radiation Budget Satellite (ERBS) launched by the Space Shuttle Challenger and the NOAA-9 and NOAA-10 weather satellites all carried the ERBE scanner and non-scanner instruments. The ERBS provided coverage between 67.5° N and 67.5° S, and the NOAA satellites provided global coverage. The ERBE scanner was a set of three co-planar detectors for longwave, shortwave, and total energy. The non-scanner was a set of five detectors designed to measure total solar energy, shortwave and total energy for the entire Earth disk, and shortwave and total energy from a medium resolution area beneath the satellite. The ERBE scanners provided over five years of data, and the non-scanners were still operational in 2002. Data products ranging from instantaneous time-sequenced instruments measurements to monthly-averaged regional, zonal, and global radiation balance parameter estimates are available from the Langley Distributed Active Archive Center (LDAAC).

The International Satellite Cloud Climatology Project (ISCCP) within the World Climate Research Program has collected IR and visible data since 1983 from a suite of weather satellites operated by several nations. ISCCP processes the data and produces datasets of global cloud cover and radiative properties to improve understanding of the Earth's hydrologic cycle and the radiation budget at the top of the atmosphere and at the surface. The data are available at 3-hour intervals and at a spatial resolution of 3 km (Rossow and Schiffer, 1999).

The CERES project is a follow-on to ERBE, and CERES space-borne radiometers measure solar, terrestrial, and total radiation at the top of the atmosphere. Cloud properties are determined using simultaneous measurements by MODIS. The Tropical Rainfall Measuring Mission (TRMM) satellite carried the first CERES instruments. The EOS Terra and Aqua satellites each carry two CERES instruments. Surface radiation balance estimates are derived from methods using combinations of radiation models, data assimilation products, and satellite measurements (Gupta et al., 2004).

The Geostationary Earth Radiation Budget (GERB) scanning radiometer on the MSG-1 (Meteosat-8) satellite is a research instrument developed by ESA to sense the Earth's total outgoing radiation and shortwave radiation. GERB began transmitting whole hemisphere high-resolution data every 15 minutes in December 2002.

5.4   Remotely sensed temperature

The equilibrium state of the surface energy balance and the efficiency of the surface emissivity determine the land surface temperature. Temperature is derived from radiance measurements by satellite radiometers rather than being observed directly. A quantitative estimate of surface temperature by remote sensors must separate the effects of temperature and emissivity in the observed radiance. Satellite-based radiometers measure a brightness temperature, T_b, derived from the radiance reaching the sensor and the antenna temperature as

$$T_b = L_\lambda(T_a)c_2 c_1^{-1}\lambda^4 \tag{5.1}$$

where $L_\lambda(T_a)$ is the blackbody radiance of an object radiating at temperature T_a, λ is the wavelength, $c_1 = 1.191 \times 10^{-16}\,\mathrm{W\,m^{-2}\,sr^{-1}}$, and $c_2 = 1.4388 \times 10^{-2}\,\mathrm{m\,K}$ (DeFelice, 1998). The brightness temperature is the apparent temperature of the imaged surface required to match the measured radiative intensity to the Planck blackbody function. Brightness temperature must be corrected for atmospheric attenuation of the surface radiance to be regarded as an estimate of land surface temperature (Schmugge et al., 2002).

Surface temperature is derived from satellite sensor radiance data and radiative transfer models. The AVHRR first launched on TIROS-N provides physically based surface temperatures representing the temperature of the emitting surface (Susskind *et al.*, 1997). TOVS measures surface temperature as part of its temperature soundings (Scott *et al.*, 1999). The AMSU on POES satellites provides enhanced temperature sounding capability that includes retrieval of mean land surface temperature (Ferraro *et al.*, 2005).

Over ocean surfaces where spectral emissivity variations are low, a "split window" method employing two thermal channels at slightly different wavelengths has been successful in estimating surface temperature. Applications of this technique to land surface temperatures have used AVHRR (Jin, 2004) and MODIS (Justice *et al.*, 2002) thermal data and land surface emissivities based on land cover databases to correct for emissivity effects. Thermal IR data from the Advanced Spaceborne Thermal Emission and Reflection Radiometer (ASTER) on the Terra satellite are used in a temperature–emissivity separation technique that permits both land surface temperature and emissivity to be retrieved (Gillespie *et al.*, 1998). AIRS data from the Aqua satellite treated with a cloud-clearing methodology permit surface temperature retrieval from radiances even under partial cloud cover conditions (Susskind *et al.*, 2003).

The three-dimensional thermal structure of the atmosphere provides important information on atmospheric stability and relative humidity. One of the primary functions of operational meteorological polar-orbiting satellites is to provide information about atmospheric temperature, and remote soundings achieve accuracies comparable with radiosondes. Real-time temperature profile data beginning with TOVS soundings are available from 1979 and are archived as global 1-day, 5-day, and monthly means (Susskind *et al.*, 1997; Scott *et al.*, 1999; Hurrell *et al.*, 2000; Seidel *et al.*, 2004). The multichannel designs of AIRS on Aqua and AMSU on POES satellites provide improved capabilities for deriving temperature profiles from the surface up to 10 hPa (Susskind *et al.*, 2003; Liu and Weng, 2005). Seidel *et al.* (2004) compare three satellite MSU datasets with five radiosonde datasets to assess the extent to which the datasets reveal similar atmospheric characteristics.

Atmospheric temperature profiles are calculated from relative density profiles measured by Rayleigh lidars. This class of lidar systems measures backscattered light by molecules from altitudes between 30 and 100 km. However, these ground-based lidars need clear skies to operate, data are available for few sites, and the data lack temporal homogeneity (Argall and Sica, 2003).

Instruments designed to make passive temperature measurements at the Earth's surface are IR temperature measuring devices, radiation thermometers, and pyrometers. IR thermometers respond to specific IR wavelengths and are used primarily for industrial and military applications (DeFelice, 1998). Radiation thermometers and pyrometers are very similar to radiometers described in Section 5.3. At the present time, these sensors are not deployed to replace mercury or alcohol thermometers for routine meteorological air temperature observations.

5.5   Derived pressure from satellite data

Atmospheric pressure is not presently measured by satellite microwave sensors. However, atmospheric pressure is estimated from information provided by satellite sensors given that the atmosphere is an ideal gas and assuming it is in hydrostatic balance. For an atmosphere at rest, the hydrostatic equation (Equation 3.3) is modified to estimate the vertical pressure profile by eliminating density (ρ) to form an alternative expression

$$\frac{\partial P}{\partial Z} = -\frac{gP}{RT} \tag{5.2}$$

where R is the specific gas constant for dry air ($287\,\mathrm{J\,K^{-1}\,kg^{-1}}$) and all other variables are defined previously. If a satellite-borne instrument provides a temperature profile, then integration over Z in Equation 5.2 from the ground upwards produces a pressure profile. The simplest case is that of an isothermal temperature profile when the pressure decays exponentially (Andrews, 2000). It is important to recognize that this is an estimated pressure profile based on hydrostatic equilibrium.

A more widely used approach is estimating cloud heights or cloud top pressure (CTP) because of the important role clouds play in the Earth's energy and water budgets. Many schemes have been proposed for estimating CTP from satellite multispectral radiance observations. Both GOES-12 and Meteosat-8 provide CTP derived using a method that correlates the cloud top temperature in an IR window channel to a thermodynamic profile. Li *et al.* (2004a) demonstrate that combining MODIS and AIRS data collected by the Aqua satellite improves CTP retrievals compared to those possible from either system alone.

Extensive research has focused on the estimations of CTP from reflectance spectra of specific gases. The basis for this approach is to use known properties of the absorption by a gas and its known distribution in the atmosphere to decipher particle scattering effects that affect line absorption (Heidinger and Stephens, 2000). Estimation of CTP from reflection spectra in the oxygen A band is described by Heidinger and Stephens (2000). The CO_2-slicing technique for calculating CTP employs radiative transfer principles and passive remote sensing data. This method takes advantage of the fact that each of the longwave

IR sounding spectral bands is sensitive to a different atmospheric layer. Li *et al.* (2004b) use combined MODIS and AIRS data to produce improved CTP retrievals using the CO_2-splicing approach.

5.6   Atmospheric humidity from satellites

Tropospheric moisture measurement by satellites is accomplished by sensing thermal emissions from water molecules in IR or microwave bands or absorption of visible/near-UV radiation. The principal difference in these techniques is that emission sensing can be made any time of the day or night, but absorption methods require sunlight. Radiation received by the satellite sensor is electronically filtered so only specified wavelengths are used to produce an electrical resistance converted into an antenna temperature. The antenna temperature is converted to a brightness temperature, T_b, based on a simplified form of Equation 5.1 expressed as

$$T_b = LT_a \tag{5.3}$$

where T_a is the antenna temperature and L is the radiance of water vapor molecules in the path from the satellite to the Earth's surface. The radiance received depends on the density of water vapor molecules (DeFelice, 1998).

The most familiar satellite water vapor products are near real-time visible and IR images provided by geostationary satellites. These images are derived from radiances measured by the satellite's sensors. MSG satellites have an imaging-repeat cycle of 15 minutes and estimate layer-mean relative humidity for two tropospheric layers (Schmetz *et al.*, 2002). The GOES imager responds to radiances at 6.7 µm to produce regional and hemispheric water vapor coverage by overlapping products from the two satellites, and the images are produced at a spatial resolution of 4 km (Menzel and Purdom, 1994). In addition, data from the GOES sounder are used to produce a derived total column water vapor image every hour (Menzel *et al.*, 1998).

Atmospheric water vapor profiles are a critical element in understanding the atmospheric component of the hydrologic cycle, and quantitative measures of atmospheric water vapor are used in a variety of operational applications. MSG and GOES satellites provide atmospheric moisture profile data with higher spatial and temporal resolution than polar-orbiting satellites (Schmetz *et al.*, 2002; Schmit *et al.*, 2002). Nevertheless, the principal satellite instrument for measuring tropospheric water vapor is the HIRS in the TOVS sensor suite. Upwelling IR radiation detected by the HIRS comes from a restricted range of altitudes defined by the strength of water vapor absorption at each wavelength (Harries, 2003). The TOVS data provide global information on the distribution and variability of tropospheric water vapor, and they form a long-term record for both land and water surfaces available in two different datasets designed for climatic analysis (Susskind *et al.*, 1997; Scott *et al.*, 1999).

The Special Sensor Microwave/Imager (SSM/I) on DMS satellites supplies water vapor measurements over oceans but not over land. However, these data are attractive for developing a long time-series because of the instrument's high stability and the ability for radiation at its sensed frequencies to penetrate cirrus clouds and atmospheric dust (Ferraro *et al.*, 1996). AIRS on the EOS Aqua satellite provides water vapor retrievals in roughly 1 km layers from the surface to 200 hPa (Susskind *et al.*, 2003). The AIRS IR measurements are complemented by the AMSU and HSB microwave sounder that allow accurate humidity profiles to be obtained under overcast conditions (Parkinson, 2003). AMSU on three POES satellites supply 4-hour global sampling and when combined with similar products from SSM/I and AMSR-E provide an excellent source of global water vapor profile products over ocean surfaces (Ferraro *et al.*, 2005). AMSU water vapor profile data have been expanded to land surfaces using an algorithm developed by Liu and Weng (2005). MODIS on both the Terra and Aqua satellites provides water vapor profiles with spatial resolution down to 250 m, which is the finest resolution for data from any of the instruments on these satellites (Parkinson, 2003).

A global water vapor dataset compiled by Randel *et al.* (1996) uses blended TOVS, SSM/I, and radiosonde data and provides gridded values for analysis of global, hemispheric, and regional water vapor variations.

5.7   Rainfall remote sensing

Practical limitations of *in situ* observation networks for measuring spatially averaged precipitation over large and inaccessible areas have promoted the use of remote sensing to quantify precipitation. Both radar and satellites are important contributors to providing real-time precipitation measurements for large areas. Meteorological radars have a spatial resolution of 1–2 km and temporal revisit times of 15–30 minutes. Satellite remote sensing products have a spatial resolution of 10–20 km and temporal repeats of 1–2 times daily. However, the quantitative estimation of errors associated with radar and satellite precipitation data is severely hampered often by the lack of an independent, precise, and accurate rain gauge value with which to compare the remotely sensed data. Aggregating the data over longer periods and increasing the number of rain gauges used in the comparison reduces uncertainties (Yuter, 2003), and calibrated estimates from weather radars provide improved spatial representation of short-term precipitation patterns (Legates, 2000a).

5.7.1   Rainfall and radar

The weather services of many industrialized nations have networks of land-based operational precipitation radars used to determine the location, size, intensity, and motion of precipitation events and to identify the type of precipitation. Ground-based radars in Europe, Japan, Canada, and the United States are used for short-term weather and flood forecasting, to estimate the distribution and amount of cumulative precipitation for a specified region, to map the three-dimensional structure of storms, and to produce precipitation estimates for hydrologic models (Yuter, 2003; Neary *et al.*, 2004). Precipitation measurement by radar is an attractive alternative because it provides coverage of a large area with high spatial and temporal resolution from a single observing point. In addition, the area can be extended by compositing data from several radars, and in this way ground-based radars provide better spatial and temporal coverage than rain gauges.

Europe and the United States have the largest number of weather radars. The Next Generation Weather Radar (NEXRAD) program in the United States supports 166 Weather Surveillance Radar-1988 Doppler (WSR-88D) systems and is managed by a joint agreement of three Federal agencies. The network of WSR-88D systems integrates advanced radar capabilities, real-time signal processing techniques, meteorological and hydrological algorithms, and automated product processing and analysis that are continually updated (Klazura and Imy, 1993; Crum and Alberty, 1993). An overview of the basic characteristics of radar-based rainfall observations is presented here, but detailed treatments are found in WMO (1996) and Yuter (2003).

The physical foundation for precipitation radars is the relationship between range-corrected, backscattered returned power and the size and number of reflecting targets in the radar beam (Fig. 5.10). Precipitation radars have a

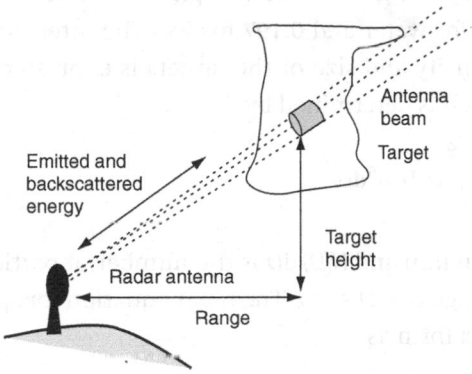

Fig. 5.10. Simplified diagram of a pulse weather radar and selected characteristics that determine the radar's effectiveness in detecting precipitation.

transmitter that switches on and off to transmit a pulse of electromagnetic energy via an antenna. The pulses are typically spaced on the order of a millisecond apart. Radar frequencies are divided into several bands, but practical considerations favor longer S-band and C-band wavelengths of 4 to 15 cm for stationary ground-based radars. These radars require larger diameter antennas that are more costly to operate than the antennas required for shorter wavelength radars deployed on satellites, aircraft, and ships (Yuter, 2003). Most precipitation radars use a circular parabolic antenna for both transmission and reception.

When the transmitted energy encounters a raindrop, some of the energy is absorbed and some is scattered in all directions. Although the raindrop serves as a reflector for the radar transmitted energy, only a small fraction of the incident radiation is scattered back toward the antenna where it is received and amplified. The backscattered power measured by the radar relates to the radar characteristics and the precipitation target characteristics. The total energy backscattered is the sum of the energy backscattered by each scattering particle. The return signal is quantitatively expressed by the radar equation, which has a radar term and a target term. The radar constant, c_r, groups numerical constants and radar hardware parameters and is expressed as

$$c_r = \frac{P_t G^2 \lambda^2 \theta_H \theta_v c \tau_p \pi^3}{1024 \ln(2\lambda^2)} \tag{5.4}$$

where P_t is the peak power of the pulse transmitted by the radar in W, G is the antenna gain and is dimensionless, λ is the wavelength of the transmitted wave in m, θ_H is the horizontal bandwidth in radians, θ_v is the vertical bandwidth in radians, c is the speed of light ($3 \times 10^8 \, \mathrm{m\,s^{-1}}$), and τ_p is the pulse duration in seconds. The physical characteristics of the precipitation particles within the radar resolution volume are represented by the complex index of refraction, $|K|^2$, which has a value of 0.93 for water and 0.197 for ice. The intensity of the return signal related to the density and size of the targets is expressed by the radar reflectivity factor, Z, which is determined by

$$Z = \sum_{\mathrm{vol}} D_i^6 = \sum n_i D_i^6 = \int_0^\infty N(D) D^6 \, dD \tag{5.5}$$

where D is the drop diameter in mm and $N(D) \, dD$ is the number of particles per unit volume in the diameter range D to $D + dD$. The radar equation for spherical drops is expressed in simplified form as

$$\overline{P_r} = c_r \frac{|K|^2 Z}{r_r^2} \tag{5.6}$$

where $\overline{P_r}$ is the average backscattered power received by the radar over several pulses in W, r_r is the range to the target relative to the radar, and the other terms are defined previously. The time delay between the original pulse transmission and receipt of the backscattered energy by the antenna is used to determine the distance to the raindrop. The relationship between the backscattered, returned energy and the size and number of the reflecting targets is the physical foundation for interpreting precipitation radar data (Yuter, 2003). Radar is capable of detecting precipitation and variations of the atmospheric refractive index generated by local variations of temperature or humidity. There are a number of basic assumptions inherent in these equations, but they serve as the basis for reasonable estimates of precipitation amounts from radar measurements (WMO, 1996).

The return signal from a radar transmitted pulse encountering a target is called an echo, and the most widely adopted approach for measuring rainfall using radar is based on the radar echo or reflectivity (Ward and Robinson, 2000). The radar echo has amplitude, a phase, and a polarization. Most operational radars are limited to analysis of the amplitude related to the size distribution and numbers of particles in the volume illuminated by the radar beam. The amplitude is used with empirical relations to determine the reflective factor to estimate the mass of precipitation per unit volume or the intensity of precipitation. Doppler radar has the capability of determining the phase difference between the transmitted and received pulse that is a measure of the mean Doppler velocity of hydrometeors (i.e. raindrops, snowflakes, or hail stones) or their motion (WMO, 1996). Radio waves reflected by objects moving away from the Doppler antenna change to a lower frequency, and waves reflected from an object moving toward the antenna change to a higher frequency. The frequency shift relative to the transmitted signal is expressed as

$$f_d = -2\frac{V_d}{\lambda_r} \qquad\qquad (5.7)$$

where f_d is the Doppler frequency or shift, V_d is the velocity, and λ_r is the radar wavelength. The change of the backscatter phase from pulse to pulse provides a measure of the change in range from the radar to the hydrometeor and the Doppler shift (Doviak and Doviak, 2003). The velocity component of a target relative to the radar beam is referred to as the radial velocity. The centimetric waves used by Doppler radar permit them to penetrate extensive fields of precipitation to identify the morphology of weather systems. This characteristic is a distinct advantage over optical and infrared waves which do not penetrate far into clouds and precipitation.

Indications of hydrometeor size serve as the basis for estimating precipitation intensity and amount by employing the reflectivity of targets or the power returned from a pulse volume. Rainfall rates are proportional to the volume of the raindrops, but the reflectivity is proportional to their surface area. The relation between radar reflectivity and rainfall rate is not constant, but extensive experimental results suggest it has the form

$$Z = aR^b \tag{5.8}$$

where Z is radar reflectivity in $mm^6 \, m^{-3}$ or dBZ, R is the rainfall rate in $mm \, hr^{-1}$, and a and b are coefficients. The most common value for a is 200, but it ranges from 70 to 500. The most common value for b is 1.6 with a range of 1.0 to 2.0. Variations in the Z–R relationship are related to physical differences in the form and size of the precipitation, radar clutter produced by ground echoes, the presence of an enhanced "bright band" related to a melting snow layer, and radar signal attenuation due to heavy rainfall. An equivalent radar reflectivity factor Z_e may be used in the Z–R relationship when precipitation aloft is measured by the radar and compared to R measured at the ground (WMO, 1996).

Reflectivity is commonly expressed in log scale decibel units (dBZ) with higher dBZ values indicating more power reflected and received by the radar. The radar computer system determines the rainfall rate and produces an estimate of the rainfall amount using a series of empirically derived equations. Light rainfall produces a reflectivity of 20–30 dBZ, moderate rainfall 30–45 dBZ, and intense rainfall 60–70 dBZ.

The NEXRAD WSR-88D radar systems deployed in the United States are active S-band Doppler radar systems operating at a wavelength of about 10 cm. WSR-88D antennas continually scan their environment in a sequence of preprogrammed 360° azimuthal sweeps at various elevations that make up a volume scan. Two common volume scans represent basic operational modes that accommodate the sensitivity range of the radar. The clear air mode is the normal operational mode in which the radar rotates slower and is sensitive to the smallest echoes. The radar rotates faster in the precipitation mode and provides more rapid data updates, but it sacrifices sensitivity at lower reflective values. Processing the returned power spectral density provides the data necessary to estimate reflectivity, mean radial velocity, and velocity spectral width, which is a measure of the variability of the radial velocities in the sample volume (Crum et al., 1993; Klazura and Imy, 1993).

Validation of radar-rainfall products is a major challenge for broad utilization of these products in hydroclimatic applications, and understanding the error structure of radar-rainfall estimates is especially critical in utilizing this information to improve quantitative precipitation forecasts and to estimate extreme rainfall and flooding (Krajewski and Smith, 2002; Yuter, 2003; Fritsch and Carbone, 2004). Experimental results indicate that improvements in quantitative precipitation forecasts are achieved by combining satellite real-time rainfall estimates with mesoscale model generated relative humidity and precipitable water and surface-radar derived instantaneous rainfall estimates (Vicente *et al.*, 1998).

5.7.2   Rainfall and satellites

Satellites are used increasingly to provide precipitation estimates over oceans and continents and especially for data-sparse regions such as deserts, mountainous regions, and humid tropical regions. Unfortunately, direct measurement of rainfall from satellites is hindered by the presence of clouds that prevents observation of precipitation with visible, IR, or microwave sensors, and this requires reliance on passive methods for satellite remote sensing of precipitation. Passive methods determine precipitation indirectly using algorithms that transform satellite-sensed radiance from clouds or raindrops into precipitation. One widely used method is to employ the flux of outgoing longwave radiation estimated from satellite observations as a basis for estimating precipitation at a variety of temporal and spatial scales (Xie and Arkin, 1998).

Passive methods are based on the radiative intensities emitted or reflected by cloud and precipitation hydrometeors using visible, IR, and microwave portions of the electromagnetic spectrum. IR and visible methods are physically indirect because precipitation is derived from the radiative properties near the cloud top. Visible methods are less widely applied with the advent of IR and microwave measurements and the lack of nighttime visible observations (Greene and Morrissey, 2000). Microwave techniques use more direct information on the vertical distribution of hydrometeors in a column of the atmosphere because the measured microwave radiation is directly related to the actual raindrops.

Active sensing of precipitation by satellite radar is accomplished by the TRMM precipitation radar, which was the first radar designed specifically for rainfall monitoring from space. The joint Japanese and United States TRMM is a low-orbit satellite stationed between 35° N and 35° S that has produced a wealth of detailed information on tropical rainfall (Kummerow *et al.*, 2000). In the microwave region, precipitation is estimated by the MSU of the TOVS on TIROS polar-orbiting satellites, the SSM/I on DMS series polar-orbiting satellites, the AMSR-E on NASA's EOS satellites, and by the TRMM Microwave Imager (TMI). Satellite measurements of precipitation by TOVS, SSM/I, and AMSR-E provide spatially uniform global coverage over both land and water for support of hydroclimatological analysis (Ferraro, 1997; Susskind *et al.*, 1997; Wilheit *et al.*, 2003).

Radiative intensity for IR and microwave wavelengths is expressed in terms of brightness temperature, which is the temperature required to match the measured intensity to the Planck blackbody function. IR brightness temperature commonly represents the physical temperature of the cloud top because most clouds are optically thick for IR radiation. Colder IR brightness temperatures often indicate higher cloud heights and higher rainfall rates at the surface. The relationship between IR brightness and precipitation is strongest for deep convection in the tropics and is less consistent in the mid-latitudes where most precipitation is produced by frontal stratiform clouds.

The radiative intensity for microwave radiation is the integrated contribution by all water drops and ice particles in the atmospheric column because microwave radiation can penetrate through cloud and precipitation layers. Microwave brightness temperatures may increase or decrease with increasing rainfall rate depending on the microwave frequency and the cloud microphysical properties. At frequencies below 20 GHz, scattering by ice particles becomes negligible and rainfall cannot be detected over land because of high surface emissivity. Ocean surface temperature and emissivity do not vary dramatically and changes in brightness temperature can be attributed to the change in the optical depth of raindrops, which is approximately proportional to integrated total rainwater amount. However, the brightness temperature increase with rainfall rate reaches a maximum that indicates saturation of microwave radiation and further increases in rainfall beyond this point have decreased brightness temperatures.

Scattering by ice particles is the dominant signature of rain clouds at microwave radiation frequencies above 80 GHz. For high-frequency microwave radiation, the brightness temperature decreases with increasing optical depth of ice particles. The lower brightness temperature indicates more large ice particles aloft, which are commonly an indication of heavier rainfall at the surface.

Measured radiative intensity in the visible spectrum is due to sunlight reflection by clouds and surface features and is limited to daylight hours. Reflectivity in the visible spectrum increases with cloud optical depth, which is proportional to the vertically integrated liquid water path if the effective droplet size remains constant. Clouds with high optical depth are more reflective and more likely to be associated with precipitation. However, the sensitivity of visible reflectivity to the liquid water path decreases with increasing optical depth and becomes virtually insensitive to optical depth at values representing a cloud producing rainfall. Consequently, reflected visible radiation indicates the physical properties near the top portion of the cloud and the relation to the rainfall rate at the surface is rather indirect (Liu, 2003).

Visible and IR images from polar-orbiting and geostationary satellites provide information on cloud tops only, but the frequent observations by these satellites permit identification of characteristics that can be related to rainfall rates and cumulative rainfall. One widely used technique is the GOES precipitation index, which is based on the fraction of cloud colder than 235 K in the IR and a fixed rainfall rate. The main disadvantage of such approaches is that they infer surface rainfall from cloud-top characteristics. Rainfall estimates are improved by using passive microwave calibrated IR estimates of precipitation at a high spatial and temporal resolution (Kidd *et al.*, 2003). An adaptation of the GOES precipitation index provides globally complete 3-hour precipitation estimates using a combination of microwave data from multisatellite observations (Huffman *et al.*, 2001). Georgakakos *et al.* (2001) use visible, IR, and water vapor data from Meteosat to estimate small-scale daily rainfall over the Nile River Basin. Chen and Staelin (2003) describe an algorithm for the Aqua AIRS for estimating precipitation using an opaque-channel approach at 15 km resolution that has potential for global precipitation applications.

A comparison of 25 satellite-based monthly global precipitation estimates is described by Adler *et al.* (2001). The satellite-based products were compared with four model and two climatological products, and the evaluation concluded that merged data products provided the best overall results. Combined precipitation estimates from low-orbit satellite microwave data, geosynchronous-orbit satellite IR data, and rain gauge data are used by the Global Precipitation Climatology Project (GPCP) established by the World Climate Research Program (WCRP) to provide monthly mean data for 1979–2001 on a global $2.5° \times 2.5°$ latitude–longitude grid (Huffman *et al.*, 1997; Adler *et al.*, 2003). These monthly data have served as the basis for 5-day global analysis (Xie *et al.*, 2003), and a daily $1° \times 1°$ latitude–longitude analysis for 1997–2001 (Huffman *et al.*, 2001). New *et al.* (2001) address how satellite precipitation estimates can be incorporated in assessing trends as multidecadal merged precipitation datasets become available.

5.8   Snow remote sensing

Remote sensing of snow involves detection and quantification of falling snow, assessment of snowcover on the land, and the snow water equivalent (SWE) of the snowpack. Detection of falling snow and assessment of the snowpack require different approaches and remote sensing strategies. Quantification of falling snow has real-time applications for transportation, construction, agriculture, and commerce, and snowcover data are used for flood forecasting, water resources management and planning, and to test hydroclimatic models.

5.8.1   *Snowfall remote sensing*

Falling snow absorbs and scatters electromagnetic energy in the same manner as rainfall and other forms of precipitation, and snow is detected by meteorological radars along with other forms of precipitation. However, snow-fall is more effective at scattering than reflecting radiation. The reflectivity of light snow is in the range of 5–20 dBZ, which is much less than the reflectivity of other forms of precipitation. The low radar reflectivity of falling snow results from the small radar cross-section of an ice sphere relative to that of a water sphere of the same size. The greater reflectivity of a water sphere is attributed to differences in the composition of the two hydrometeors. Weather radars must be sensitive to low-energy echoes to detect snowfall. WSR-88D radars in the United States operating in the clear air mode provide the high-resolution retrieval necessary to detect snowfall. However, accurate estimates of snowfall rates and snow accumulations continue to present challenges.

Snowfall rates are estimated using meteorological radars and the relationship between precipitation rate and radar reflectivity shown in Equation 5.8. However, for snow there are many natural causes that alter the nature of the Z–R relationship. An equivalent or effective reflectivity factor Z_e is defined for snowfall that assumes the backscattering particles are all spherical drops (WMO, 1996). This assumption is necessary because deviations in the Z–R relationship are due to the assortment of shapes and densities of snow particles, the alterations in snow crystal structure that change the snow particle size distributions and fall speeds, and changes in attenuation of the radar with changes in snow crystal temperature (Boucher and Wieler, 1985). Another characteristic that can be a significant problem in determining the Z for snow is the condition known as the "bright band" which results from reflectivities within a melting layer. Precipitation in the form of snow exists above the bright band, melting snow occurs in the band, and rain is present below the band. Radar detection of the bright band is a function of both the physical properties of the melting layer and the radar sensitivity and spatial resolution (Yuter, 2003). The bright band phenomenon can produce an appreciable increase in Z indicating a greater snowfall rate than actually exists. The occurrence of the bright band may vary considerably during a storm and from storm to storm and increases the difficulties in interpreting radar measurements. The bright band problem is a particular concern for estimating snowfall from stratiform clouds in which reflectivity usually decreases with height (Boucher and Wieler, 1985; WMO, 1996).

Different polarization approaches developed to improve radar measurements of rainfall are of limited use for snowfall measurement because larger snowflakes exhibit weak polarization signatures. Reflectivity measurements taken simultaneously at two wavelengths by dual-wavelength radar methods provide additional information about characteristic snowflake size, and experimental data indicate this approach produces more accurate snowfall measurements than traditional single-beam radar measurements (Matrosov, 1998).

Snow accumulation is the product of the snowfall rate and the physical processes controlling the depth of snowfall via the snow density. Snowfall density is related to the ice-crystal structure and expresses the relative proportion of the crystal volume composed of air. Numerous cloud microphysical processes contribute to the final structure of an ice crystal. The WSR-88D radar uses snow density as an input parameter in the snow accumulation algorithm that was added to the original radar product generator in response to users' needs for snowfall detection and accumulation estimates (Crum *et al.*, 1998). The concept of a fixed snow ratio is typically applied, often with some local empirical adjustment, even though the traditional 10-to-1 rule is recognized as an inadequate characterization of the true range of snow densities. Roebber *et al.* (2003) propose a 10-member ensemble of artificial neural networks utilizing surface and radiosonde data to provide more accurate estimates of snowfall density. In mountainous terrain, snowfall density issues are exacerbated by radar beam degradation by topography. Wetzel *et al.* (2004) describe a system of mesoscale models, surface observations, and multispectral satellite data used coincidentally with radar data to improve snowfall estimates in the northern Colorado Rocky Mountains.

Satellite measurements of snowfall within the atmosphere have been accomplished over oceans, but snowfall retrievals over land are elusive due to the presence of snowcover on the land. A number of studies recommend the use of radiometers operating at frequencies of 150 GHz or higher to improve ability to discriminate precipitation from snowcover. Radiation at these frequencies can measure snowfall over land because water vapor screening obscures underlying snow-covered surfaces. High-frequency AMSU microwave data from NOAA polar-orbiting satellites have been analyzed extensively as a potential means for discriminating the scattering features over land surfaces, especially snowcover, from those of ice crystals in the atmosphere. A snowfall retrieval algorithm based on AMSU measurements at 150 GHz and higher and sounding channel measurements near 54 GHz developed by Kongoli *et al.* (2003) detects rainfall and snowfall over cold and snow-covered surfaces, and the algorithm is operational for NOAA satellites.

5.8.2 *Snowpack remote sensing*

Snow on the ground responds to a number of regions of the electromagnetic spectrum that offer opportunities for remote sensing applications by both active and passive sensors. The SAR on ESA's ERS satellites provides spatial resolution suitable for snowcover assessment in medium to small watersheds. However, SAR data have complex radiometry and geometry that complicate use of the data in high-relief areas. Nevertheless, SAR data are used for snow mapping in mountainous areas and wet snow conditions (Seidel and Martinec, 2004).

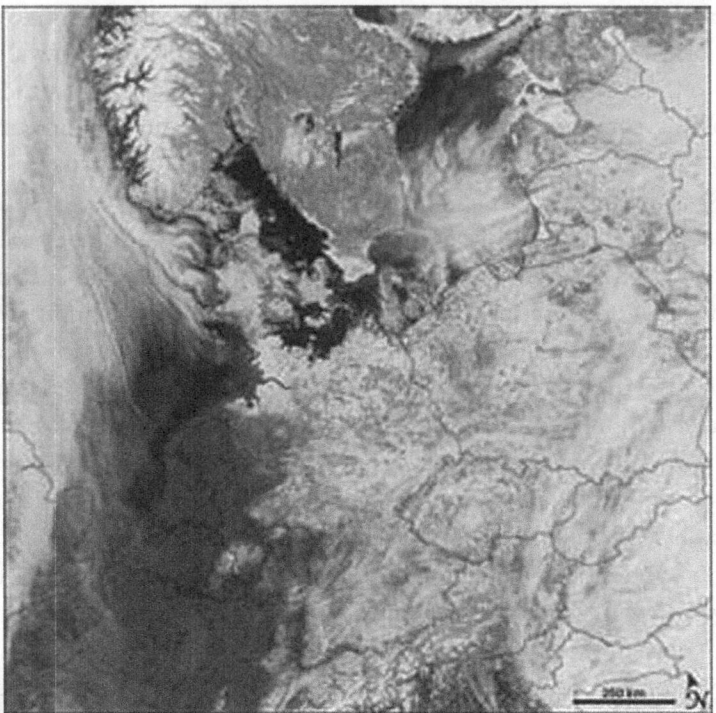

Fig. 5.11. NASA Terra satellite Moderate Resolution Imaging Spectroradiometer (MODIS) image for 13 March 2006 showing snow covering most of Europe east of Germany and Denmark. (Image courtesy of NASA and the Goddard Space Flight Center from their website at http://modis-snow-ice.gfs.nasa.gov/.)

Passive microwave applications are more commonly used for examining both snow depth and SWE. The high albedo of the snow surface compared to non-snow areas is a property easily identified using the visible bands of the GOES, Meteosat, Terra, and Aqua satellite data (Fig. 5.11), but these visible sensors are unable to see the snowcover at night or when clouds are present. Visible-wavelength satellite maps of Northern Hemisphere snow extent produced by NOAA since the late 1960s are the longest consistently derived satellite record of any environmental variable (Robinson and Frei, 2000; Hall *et al.*, 2001). IR sensors are able to make both day and night observations, but clouds interfere with observing the snowcover. Microwave sensors on satellites provide an all-weather snowcover observation capability day or night, and they have been used to map snow extent and snow depth since the early 1970s (Fig. 5.12).

The physical characteristics of the snowpack described in Section 4.5 determine its microwave properties. Snow depth and water equivalent, liquid water

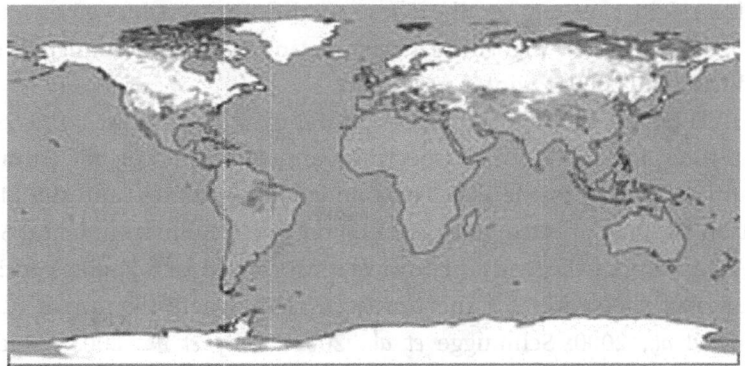

Fig. 5.12. Composite image of global snowcover for February 2004 from the
Terra satellite Moderate Resolution Imaging Spectroradiometer (MODIS). (Image
courtesy of NASA and the Goddard Space Flight Center from their website at
http://modis-snow-ice.gfs.nasa.gov/.)

content, density, grain size and shape, temperature, stratification, and soil
roughness and dielectric properties of the surface beneath the snowpack all
influence emitted microwaves from the snow surface. The extent of microwave
scattering within a snowpack is proportional to the snowpack thickness and
density, and the brightness temperature of the snow surface is related to these
properties. The naturally emitted microwave brightness temperature of a snowpack is primarily related to the number, size, and packing of snow grains along
the emission path. Consequently, deeper snowpacks generally result in lower
brightness temperatures. Liquid water within a snowpack alters the emissivity
of snow and produces a brightness temperature significantly higher than a dry
snowpack. Early morning satellite overpasses of snow-covered areas are preferred for retrieving snowcover information that minimizes wet snow influences (Schmugge *et al.*, 2002).

Brightness temperatures from different channels of satellite sensors are used
to estimate the snow depth. A commonly used algorithm developed by Chang
et al. (1987) for estimating snow depth (SD) using microwave observations is

$$SD = c_0(T_{b18} - T_{b36}) \tag{5.9}$$

where c_0 is a coefficient determined from radiative transfer model experiments
of snow and has a value of $1.59\,\mathrm{cm\,K^{-1}}$, T_{b18} is the brightness temperature in K at
18 GHz, T_{b36} is the brightness temperature in K at 36 GHz, and SD is expressed in
units of cm. Coefficients are usually developed for specific regions and snowcover conditions, but Kelly *et al.* (2003) modified Equation 5.9 by using a brightness temperature difference between 19 and 37 GHz to minimize the snow
temperature effect and by adjusting the coefficient to reflect how the snow

grain size might vary temporally. These modifications achieved encouraging results in applying the algorithm to estimating Northern Hemisphere snow depth during the 2000–1 winter season.

The AVHRR and the AMSU on NOAA polar-orbiting satellites, the SSM/I on DMS satellites, the Landsat multispectral scanner system, the SPOT multispectral scanner, MODIS on the EOS Terra and Aqua satellites, and the Medium-Resolution Imaging Spectrometer (MERIS) on the ESA Environmental Satellite (ENVISAT) are among the many microwave sensors used for mapping continental-scale seasonal snowcover for the Northern Hemisphere (Ferraro et al., 1996; Romanov et al., 2000; Schmugge et al., 2002; Kelly et al., 2003). Seidel and Martinec (2004) list the satellites, the sensors, and the image characteristics of snowcover products. Spatial resolution of snowcover mapping is an area of particular concern for hydroclimatic applications and a variety of spatial resolutions are available for mapping small watersheds to regional-scale applications.

Passive microwave sensors on satellites provide global SWE observations to complement snowcover observations. Brightness temperatures are used to determine SWE using a generalized relation similar to Equation 5.9. Snow density must be known from in situ observations or estimated. A representative value for mature mid-winter snow packs in North America is 300 $kg\,m^{-3}$, which produces a coefficient of 4.8 $mm\,K^{-1}$ for Equation 5.9 (Foster et al., 2005). Vegetation cover and snow grain size variability are the main sources of error in remote sensing of SWE, and Foster et al. (2005) suggest an algorithm based on Equation 5.9 that employs two time and space varying coefficients to account for the effects of vegetation cover and snow morphology in North America. This algorithm captures the accumulation and ablation phases of the snow season over a variety of snow surfaces. Microwave data from the Scanning Multichannel Microwave Radiometer (SMMR) on the Nimbus-7 satellite, the SSM/I, and the AMSR on the EOS Aqua satellite are used to estimate SWE of the snowpack and the presence of liquid water in the snowpack (Kelly et al., 2003; Foster et al., 2005). The spatial resolution of 25 km for the SMMR and SMM/I is best suited for regional and large basin studies, but the spatial resolution of 500 m for AMSR supports analysis of smaller watersheds. However, evaluation of SWE at spatial scales required to characterize the spatial variability typically found in mountain areas is seldom strong enough to be used in a predictive capacity for mountainous terrain (Anderton et al., 2004).

Aircraft equipped with gamma-radiation detectors provide an alternative method for estimating the average SWE of a site. Gamma radiation from the soil is attenuated by a snow layer. Aircraft flights along a prearranged line before and after snow occurs determine the attenuation due to the snow layer. An empirical equation relates the attenuation to the SWE. This technique is effective in open and relatively flat terrain, but it is less effective in hilly or forested terrain. Nevertheless, this technique is used operationally in a number of countries (Engman, 1993; Ward and Robinson, 2000).

5.9   Surface wind remote sensing

Surface wind measurement by remote sensors is accomplished using satellite microwave scatterometers and land-based lidar, radar, and sodar. Satellite scatterometers provide wind speed and direction over the oceans, under clear or cloudy skies, and both day and night. Scatterometers send microwave pulses to the Earth's surface and measure the backscattered power related to surface roughness. Roughness over land is due to terrain and vegetation differences, but over the oceans the backscatter is due largely to small waves assumed to be in equilibrium with the local wind stress. Backscatter over water increases with wind speed, and the magnitude is influenced further by the wind direction relative to the direction of the radar beam. This gives the scatterometer the capability of measuring both wind speed and direction over the ocean (Liu, 2003).

Scatterometers have been launched on ESA and NASA polar-orbiting satellites since 1978. These instruments provide data on local and regional winds and moisture advection, and they provide a nearly synoptic-scale view of global surface winds.

Lidar, radar, and sodar are employed to remotely measure wind speed for research and experimental purposes more commonly than for routine wind observations. Each of these active remote sensors is based on measuring the Doppler shift of the light, radio waves, or sound waves emitted by the instrument.

Lidar transmits radiation in the UV, visible, and IR wavelengths, and modern lidar systems use a pulsed laser to generate the radiation (Angevine et al., 2003). The transmitted energy interacts with air molecules which have thermal motion or motion due to wind. The energy is changed by its encounter with air molecules, and some energy is reflected or scattered back to the instrument. A Doppler lidar measures the shift in wavelength frequency of the backscattered energy and this information is converted into remotely measured wind velocity (Argall and Sica, 2003).

Radar measurement of wind speed usually requires the presence of an atmospheric target to produce an echo, but modern systems have sufficient sensitivity to sense clear-air returns (Angevine et al., 2003). Doppler radars provide radial velocity or wind speed toward or away from the radar when monitoring precipitation events. Doppler radar radial wind data are used in numerous analytical formulations to produce three-dimensional wind fields at various scales and serving a variety of applications (e.g. Caillault and Lemaitre, 1999; Nissen *et al.*, 2001; Liou, 2002). Specialized wind profiling radars are employed in research and experimental work studying winds within the atmospheric boundary layer (Angevine *et al.*, 2003).

Sodar operates on the principle of acoustic backscattering. It transmits a short pulse of sound which is refracted by the small-scale atmospheric turbulence structure of temperature and velocity in the boundary layer (Angevine *et al.*, 2003). The radial velocity of air is determined by measuring the intensity and the Doppler shift of the sound refracted from the turbulence. Sodar has a maximum range of several hundred meters. It primarily provides measurement of mean wind speed and direction because it samples atmospheric volume, and it samples at multiple points in space and time (Crescenti, 1997). However, expanded applications of sodar are being tested. Contini *et al.* (2004) propose a method for using a sodar system to measure mean vertical wind velocities averaged over three hours. Sodar systems are employed with other meteorological instruments studying local winds at an expanding number of monitoring sites around the world (e.g. Peters *et al.*, 1998; Sturman *et al.*, 2003; Pérez *et al.*, 2004).

5.10   Soil moisture remote sensing

Passive and active microwave remote sensors have the potential for detecting soil moisture in regions without snow or tall vegetation. Water has a very high dielectric constant for most microwave wavelengths, but the effects of variations in the water content of soils on radar backscatter are complex as shown by Mancini *et al.* (1999). In general, the longer wavelengths are better for increased sampling depth and reduced effects due to vegetation and surface roughness. Shorter microwave wavelengths cannot be used to determine soil moisture in many forested situations (Schmugge *et al.*, 2002). Nevertheless, remote sensing methods have an advantage over *in situ* observations in representing spatial variability, and they offer repeated temporal coverage. Visible and IR radiation can be used for indirect soil moisture monitoring by satellites during clear sky conditions (Robock, 2003), but the resolution of satellite data is not sufficient for all applications (Miller *et al.*, 2005).

Aircraft observations have been used for over 30 years to determine near-surface (0–5 cm) soil moisture at the field scale, but less success has been realized at the watershed scale due to the heterogeneity of land cover features. However, aircraft observations were important in showing the greater effectiveness of longer wavelength sensors, the limited sampling depth, and the effect of soil texture. This information was used to develop satellite microwave sensors to measure the natural thermal emission of the land surface using radiometers operating at the 21 cm wavelength. The intensity of the surface emission is expressed as a brightness temperature similar to thermal IR observations, and algorithms are employed to convert the brightness temperature observations to quantitative soil moisture information over large heterogeneous areas (Schmugge *et al.*, 2002). Lakshmi *et al.* (1997a, b) demonstrate how the data from the SSM/I on a DMS polar-orbiting satellite are used to estimate daily to monthly watershed-scale soil moisture. Vinnikov *et al.* (1999) use the SMMR on a polar-orbiting satellite and SSM/I data to produce a regional soil moisture record, and Jackson *et al.* (2002) use SSM/I data and multipolarization observations to develop regional soil moisture maps. Njoku *et al.* (2003) show that the AMSR-E on the NASA Aqua satellite has improved sensitivity to soil moisture compared to the SSM/I data. Cashion *et al.* (2005) use the TMI and Normalized Difference Vegetation Index (NDVI) from MODIS to characterize soil moisture conditions, but they find vegetation levels mask the soil moisture signal from TMI for several months each year. GOES imager data and AVHRR data from polar-orbiting satellites are used to model soil moisture at scales ranging from the watershed to the continent (e.g. LeMone *et al.*, 2000; Song *et al.*, 2000a; Liu *et al.*, 2003). An experimental monthly global soil moisture map based on data beginning in 1988 is provided by NASA using SMM/I data, but it is more sensitive to soil wetness than dryness. A global multiyear soil moisture map is derived from the scatterometer on the ERS-1 and ERS-2 satellites (Dirmeyer *et al.*, 2004).

Careful choice of the radar configuration is necessary to minimize vegetation and surface roughness effects for estimating watershed soil moisture with active sensors (Leconte *et al.*, 2004). SAR has proven most successful in routine remote sensing measurements of soil moisture. SAR is provided by European, Japanese, and Canadian Earth resource satellites and by aircraft. SAR images are produced by an active system that sends a microwave signal from the sensor platform to the ground and detects backscattered waves that the ground reflects back to the receiver on the same platform. The SAR data are used with empirical and theoretical models to convert the trend of the radar backscatter response to changes in soil moisture. The presence of complex terrain associated with forest canopies and agricultural crops creates a need for multifrequency, multipolarization radar to separate their effects from those of backscatter due to soil moisture (Vincent, 2003). Quesney *et al.* (2000) have shown that ERS SAR remote sensing data at the basin scale can be obtained with an accuracy of about 5%, but results degrade seasonally as vegetation cover increases. Francois *et al.* (2003) test how the ERS SAR watershed-scale soil moisture estimates can be used to improve flood forecasts. Leconte *et al.* (2004) use the Canadian RADARSAT-1 SAR to map soil moisture at the watershed scale for surfaces dominated by agriculture and herbaceous plants.

5.11   Evapotranspiration remote sensing

Satellites, aircraft, and ground-based remote sensors do not measure routinely the water vapor flux that quantifies evaporation or evapotranspiration. However, indirect evapotranspiration estimates using remotely sensed measurements of other energy and moisture flux variables are possible. Indirect estimates are useful in extending point evapotranspiration measurements to larger areas and for quantifying variables employed in energy and moisture balance models (Liu *et al.*, 2003).

Ground-based remote measurements of surface reflectance and temperature by multispectral radiometers provide an initial step in mapping evapotranspiration over diverse watershed-scale landscapes (Moran *et al.*, 1994). Remotely sensed data from aircraft-based radiometers support quantification of spatially distributed surface energy fluxes, including latent heat, for various scales (Humes *et al.*, 1994).

Satellite observations essentially depict instantaneous surface conditions and for many practical applications ETa estimates at longer times scales are needed (Schmugge *et al.*, 2002). This contrast in temporal scale has promoted exploration of numerous experimental or research approaches using one or more satellite remote sensors as a basis for estimating evapotranspiration using algorithms of various complexity in energy balance models. Temperature data from the GOES sounder are used in linear regression models to estimate surface temperature for extrapolating energy fluxes from a point to a regional scale. Incoming solar radiation is estimated from satellite observations of cloud cover, and incoming shortwave fluxes and surface albedo provided by GOES are used in radiation balance equations. Surface temperature estimates from thermal IR wavelengths provided by sensors on GOES and polar-orbiting satellites are used to estimate the outgoing longwave radiation term in the net radiation equation.

A common approach for using remote sensing data to estimate ETa is to rearrange the energy balance equation for the land surface showing the latent heat flux (LE) as a residual. This takes the form

$$LE = R_n - G - H \tag{5.10}$$

where the terms are defined for Equation 2.12. Remote sensing estimates of R_n and G are highly reliable, but estimates of H have great uncertainty because they rely on converting satellite brightness temperatures to an estimate of the land surface temperature using surface resistance and roughness variables

(Schmugge *et al.*, 2002). However, research by Kustas and Norman (1996) indicates that no universal relation exists for the behavior of surface resistance and roughness for different surfaces. A major factor may be the need to distinguish between soil and vegetation canopy components to adequately define surface roughness parameters. Kustas *et al.* (2001) show that a two-source modeling approach that considers both soil and vegetation contributions to the total heat flux provides improved estimates of ETa over heterogeneous landscapes.

Most remote sensing approaches for estimating ETa require *in situ* meteorological observations that raise concerns about the spatial representativeness of the observed data. Nevertheless, *in situ* data must be used until remotely sensed data for estimating the variations of these quantities are available. The nature of the data required for estimating instantaneous ETa using a combination of remote sensing and *in situ* observational data is illustrated by Feddes (1995). The relationship is expressed as

$$\text{LE} = (1 - r_o)K{\downarrow} + \varepsilon'\sigma T_a^4 - \varepsilon_o\sigma T_o^4 - \frac{\rho_a c_p}{r_{\text{ah}}}(T_o - T_a) - \frac{\rho_s c_s}{r_{\text{sh}}}(T_o - T_s) \qquad (5.11)$$

where LE is the latent heat flux in $\text{W}\,\text{m}^{-2}$, r_o is the remotely sensed surface reflectance, $K{\downarrow}$ is the global solar radiation in $\text{W}\,\text{m}^{-2}$, $\varepsilon'\sigma T_a^4$ is the longwave sky emittance in $\text{W}\,\text{m}^{-2}$ with ε' being the apparent emissivity of the atmosphere which is dimensionless and σ being the Stefan–Boltzmann constant in $\text{W}\,\text{m}^{-2}\,\text{K}^{-4}$, T_a is the screen-height air temperature in K, $\varepsilon_o\sigma T_o^4$ is the longwave surface emittance in $\text{W}\,\text{m}^{-2}$ with T_o the surface temperature in K remotely sensed and ε_o the land surface emissivity, ρ_a is air density in $\text{kg}\,\text{m}^{-3}$, c_p is the specific heat of air at constant pressure in $\text{J}\,\text{kg}^{-1}\,\text{K}^{-1}$, r_{ah} is the mean turbulent resistance in $\text{s}\,\text{m}^{-1}$, ρ_s is the soil density in $\text{kg}\,\text{m}^{-3}$, c_s is the soil specific heat in $\text{J}\,\text{kg}^{-1}\,\text{K}^{-1}$, r_{sh} is the soil resistance to heat transfer in $\text{s}\,\text{m}^{-1}$, and T_s is the soil temperature in K. ETa is estimated using Equation 4.18.

Coupling process-based evapotranspiration models with remote sensing data has received considerable attention in hydroclimatological applications because of interest in processes at the watershed scale. GOES imager IR data are used to quantify surface temperature in many watershed models (e.g. LeMone *et al.*, 2000; Norman *et al.*, 2003). AVHRR data from polar-orbiting satellites are used to define surface temperature and other essential parameters in algorithms for individual surface types in models for large areas and for long periods in Turkey (Granger, 2000; Kite and Droogers, 2000), in southeastern Kansas in the United States (Song *et al.*, 2000a, b), and in Canada (Liu *et al.*, 2003). Crow *et al.* (2003) demonstrate the use of radiometric temperature retrievals from both the GOES sounder and polar-orbiting TOVS to estimate monthly ETa in a regional-scale model applied to the Southern Great Plains of the United

States. Mo *et al.* (2004) use AVHRR data to estimate the leaf area index used to model ETa for the Lushi Basin in China. MODIS on the Terra and Aqua satellites have more channels and higher spectral and spatial resolution than AVHRR to support estimates of the NDVI for watershed studies of ETa (Justice *et al.*, 2002).

5.12   Runoff remote sensing

Measuring runoff involves continuous observation of a flow volume that existing remote sensing techniques are unable to achieve, but remote sensing data are used to estimate runoff. Aerial photographs are used for quantitative analysis of drainage basin networks from which selected runoff characteristics are estimated. Satellite image enlargements permit identification of channel networks by vegetation differences that support streamflow estimates for hydrologically data scarce regions (Herschy, 1997). However, the most extensive use of remote sensing data to estimate runoff is in hydroclimatic models. The broad spatial coverage and repeat temporal coverage of satellite remote sensing data contribute to determination of physical characteristics included in models and quantification of model parameters. Such data are especially important for remote areas and/or ungauged watersheds. Remote sensing of radiation, precipitation, temperature, soil moisture, evapotranspiration, and other variables are incorporated in watershed and regional models to estimate runoff. Watershed models employing a variety of approaches are discussed in Chapter 6.

Review questions

5.1   Why do the GOES satellites carry both an imager and a sounder?

5.2   Why do polar-orbiting satellites provide greater spatial resolution than geosynchronous satellites?

5.3   What are the characteristics of the primary instrument carried on NOAA's polar-orbiting satellites?

5.4   How do satellite observations of the atmosphere contrast with ground-based monitoring techniques?

5.5   What are the similarities and differences in radar and satellite monitoring of rainfall?

5.6   Why are satellites more effective than radar in determining snow depth?

5.7   Why are soil moisture and evapotranspiration especially difficult to measure with remote sensing techniques?

5.8   What are the issues regarding the use of satellites as a replacement for standard stream gauges?

6

The runoff process and streamflow

6.1   Transforming precipitation into runoff

Runoff is the water leaving a drainage basin, and it is the signature component of the terrestrial branch of the hydrologic cycle. Although runoff represents a residual portion of precipitation remaining after evapotranspiration depletes the moisture supply, runoff is significant because it is the water available for human use. The runoff process traces what happens to precipitation after it arrives at the land surface. In fact, runoff can be thought of as the integration of complex hydroclimatic processes acting on precipitation after it reaches the surface.

The allocation of precipitation at the land surface and the residual character of runoff are emphasized by rearranging Equation 2.17 as

$$P - \text{ETa} = \Delta \text{St} + \text{qr} + \text{qs} + \text{qd} \qquad (6.1)$$

where all variables are defined for Equation 2.17. This arrangement of the hydrologic equation accentuates alternative allocations for precipitation beyond evapotranspiration. The pathway followed in the case of a specific precipitation event depends on a variety of local conditions that display significant time and space variations. Consequently, the following discussion employs representative conditions as the basic framework for addressing the runoff process.

6.2   Factors affecting runoff

It is generally recognized that runoff from a given watershed is influenced by two major groups of factors categorized as climatic and physiographic. These factors are evident in the form of Equation 6.1 with the climatic factors, precipitation and evapotranspiration, on the left side and the variables on the right side being related to physiographic factors. However, this is not an exclusive listing of the variables, but rather it is a synthesis of the influence of climatic and physiographic factors on the allocation of precipitation. A number of interactive processes are involved, and the identifying climatic and physiographic factors are largely a descriptive convenience.

Seasonal variability is a distinctive characteristic of the climatic factors influencing runoff. While the quantity of precipitation has an obvious dominant role, other precipitation characteristics are important in determining the proportion of precipitation that becomes runoff. The form, intensity, duration, space and time distribution, frequency of occurrence, and the direction of storm movement have identifiable influences on runoff.

Evapotranspiration is clearly a climatic factor because ETp is driven by the daily and seasonal regime of radiant energy. However, in the runoff process the quantity of interest is the actual moisture vaporized. Quantifying ETa requires consideration of numerous related characteristics that are indirectly climatic expressions. The presence or absence of vegetation is an obvious related influence on ETa, but when vegetation is present the scope of considerations expands significantly as seasonal changes in ETa influence the moisture allocated to runoff (Czikowsky and Fitzjarrald, 2004). Furthermore, vegetation intercepts precipitation, but interception in turn depends on the species, composition, age, and density of the vegetation, and interception varies with the season of the year and the intensity of the rainfall event. Since plant transpiration relies on water drawn from the soil, soil moisture storage and the antecedent condition of soil moisture influence the quantity of ETa.

Physiographic factors are expressions of the physical characteristics of the watershed and the stream channel. Compared to the climatic factors, these influences are relatively consistent over time but some display variations over time. One group of watershed characteristics that influence runoff are related to watershed geometry and include the size, shape, slope, orientation, elevation, and stream density of the basin (Ward, 1995). Other physical basin characteristics are land use and cover, surface infiltration, soil type, permeability and capacity of groundwater formations, and the presence of lakes and swamps. Stream channel characteristics that affect runoff are related to channel hydraulic properties including its size, shape, roughness, and length. These channel features determine the channel storage capacity and largely influence the timing of runoff rather than the quantity (Mosley and McKerchar, 1993; Ward and Robinson, 2000).

Interaction of the climatic and physiographic factors produces a runoff pattern that is characteristic for a particular watershed. In general, large

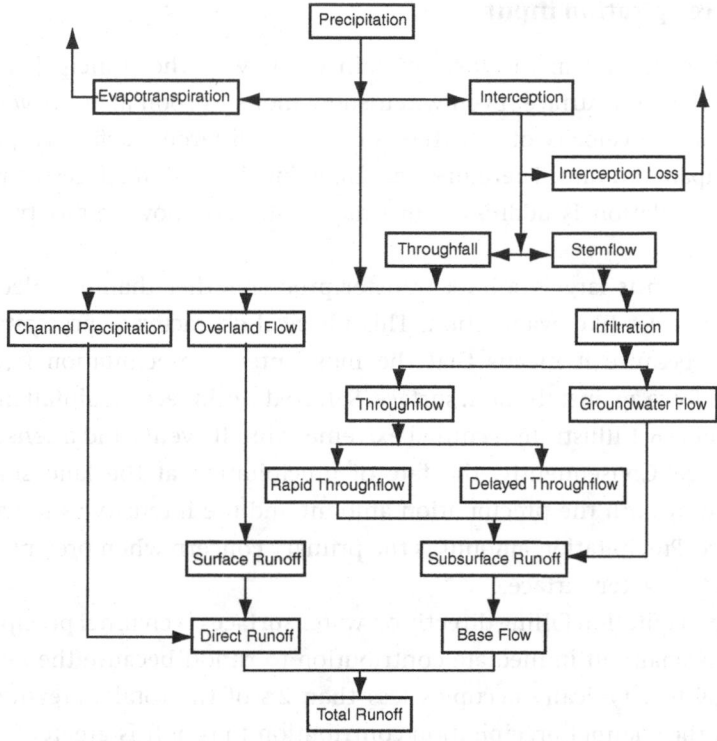

Fig. 6.1. Diagram of the watershed runoff process excluding storages.

watersheds behave differently from small watersheds, but the dominant factor is not necessarily the size of the watersheds. Two watersheds of the same size commonly display a different runoff response. The reality is that one factor may be dominant in one watershed and may be inconsequential in a nearby watershed. The interplay of influences is complex and responsive to both climatic and physiographic changes that may occur in a watershed.

Figure 6.1 portrays the components of the watershed runoff process. Basin precipitation is partitioned following Equation 6.1, and precipitation exceeding ETa arrives at the stream channel by various flow paths. Storages are omitted in Figure 6.1 to emphasize the water flow paths. Space and time variations in the precipitation–streamflow response reflect the contrasting flow paths precipitation may follow in arriving at the stream channel. In the most general sense, when precipitation reaches the land surface to initiate the runoff process it is retained there, it infiltrates into the soil, or it moves to the stream channel. A sequence of intermediate steps underlies the actual runoff process.

6.3   Precipitation input

Precipitation in the form of rain or snow is the principal moisture source to the land surface. Even when snow melts as soon as it arrives at the surface, the low velocity of snowflakes creates a different surface response to snow compared to rain. Therefore, the following discussion addresses rainfall. Snow accumulation is addressed in Chapter 4, and snowmelt is treated in Section 6.11.

Precipitation is largely a bulk transfer process rather than a molecular or diffusion process like evaporation. This physical character of precipitation is important because it means that the most intense precipitation falls with enormous force against the land surface. Selected world record rainfall intensity rates in Table 6.1 illustrate recorded extreme rainfall events and a sense of the related force. Consequently, the fate of precipitation at the land surface is influenced by both the precipitation amount and the intensity as it arrives at the surface. Precipitation amount is the primary concern when precipitation is delivered to a water surface.

Basin precipitation falling directly on water surfaces is channel precipitation. It makes a small but immediate contribution to runoff because the perennial channel system typically occupies less than 2% of the total watershed area. However, the channel precipitation contribution to runoff is greater in watersheds with large areas of lakes or swamps and when precipitation is associated with a prolonged storm (Ward and Robinson, 2000).

The presence or absence of vegetation at the land surface is a factor that influences precipitation disposition across many spatial scales. Peel *et al.* (2001) assert that the distribution of vegetation is evident even in annual streamflow variability at the continental scale. Intense rainfall on a bare soil transforms the

Table 6.1. *World extreme point rainfall intensities for selected periods*

Duration	Intensity (mm)	Location	Date
1 minute	38.1	Barot, Guadeloupe, Mexico	26 November 1970
8 minute	124.5	Füssen, Bavaria, Germany	25 May 1920
15 minute	198.1	Plumb Point, Jamaica	12 May 1916
42 minute	304.8	Holt, Missouri, USA	22 June 1947
165 minute	558.8	D'Hanis, Texas, USA	31 May 1935
1 day	1870	Cilaos, Réunion Island	15 March 1952
1 month	9300	Cherrapunji, India	July 1861
1 year	26 451	Cherrapunji, India	August 1860–July 1861

Compiled from WMO, 1986b, and Barcelo *et al.*, 1997.

surface into a muddy milieu that seals many of the openings available for water to move into the soil. Vegetation helps to absorb the impact of raindrops and diminishes the soil sealing effect of intense rain. Nevertheless, intense rain on either a bare soil or vegetation surface is likely to produce water that begins moving horizontally over the surface even during a brief storm.

The intensity, duration, and space and time distribution of rainfall have the most immediate influence on the runoff process. These characteristics commonly vary within the area covered by a given storm. Winter rainfall often tends to be relatively low intensity with little intensity variation throughout the storm. However, along the west and southeast coasts of the United States winter storms reach a maximum intensity at some intermediate time. In general, large storm systems display a more uniform rainfall distribution over space and are longer duration storms. Storms covering a small area, such as thunderstorms, tend to produce greater rainfall amounts near the center with decreasing amounts outward toward the storm margin, and they are shorter duration storms. Also, small storms have a tendency to produce the most intense rainfall as the storm begins (Elliot, 1995).

The spatial and temporal context of the upper boundary on the rainfall amount is incorporated in the concept of probable maximum precipitation (PMP). The WMO (1986b) defines PMP as the theoretically greatest precipitation depth for a given storm duration that is physically possible for a given storm area at a particular geographical location and time of year. There is no known way to develop PMP from first principles, and several estimation methodologies are employed. The methodology used by the National Weather Service in the United States derives PMP from the maximization and transposition of real storm events combined with statistical analysis of extreme rainfalls. Although PMP is a theoretical quantity, it has become the standard for dam design in many parts of the world (Douglas and Barros, 2003).

The most intense rainfall may produce runoff almost immediately. Runoff from low-intensity rainfall is commonly delayed or may not occur. The major characteristic of water delivery to the land surface is rainfall variability, and this variability is transferred to the fluxes that occur following rainfall. The reasons for the differences in the runoff response become evident when other influences on the runoff process are considered.

6.4   Interception

All of the rainfall on a surface covered by vegetation may not reach the ground beneath the plants. Depending on the nature of the rainfall and the character of the plant canopy, some water may be retained on the plant's surface. The retention of incident rainfall on the leaves and branches of plants is called interception (see Fig. 6.1). Most of the interception occurs at the beginning of storms. Evaporation of intercepted water results in the interception loss, which can be a significant proportion of total regional evapotranspiration.

Seeking shelter under a tree during a rain storm is recognition of the interception effect. Much of the early rain from a low-intensity storm is caught and held by trees and other plants. In many cases, a light rain may never reach the ground beneath a mature forest canopy. For tall vegetation, the type of tree is an important factor in determining interception because interception varies with the type of leaf, leaf size and shape, leaf density and arrangement, whether it is deciduous or evergreen, the branching form, and bark roughness (Luce, 1995). Over an area, the density of the forest stand introduces an additional dimension to interception. Rain intensity and duration and wind are meteorological factors that play a role in determining the effectiveness of leaves and branches in holding water. A larger proportion of low-intensity than high-intensity precipitation is intercepted, and a larger proportion of short-duration than long-duration precipitation is intercepted. Wind plays a complex role that inhibits or increases interception depending on wind speed and precipitation characteristics. In general, interception is greater for denser vegetation, for taller vegetation, and for wetter climates (Ward and Robinson, 2000). Local conditions, such as the daytime or nighttime dominance of precipitation occurrence, may be additional factors influencing the general magnitude of interception losses.

6.4.1   Interception storage

The water-holding capacity of a plant's surface is limited, and the amount captured during a particular storm is a function of the season, the antecedent moisture on the plant, and whether the interception capacity is exceeded by the present event. The general pattern is that interception increases exponentially during a storm until the interception capacity is achieved and the weight of additional water overcomes the surface tension holding the water on the plant. The convention is to express interception storage capacity as an equivalent depth per unit ground area and not as a physical thickness of water film on the vegetation (Ward and Robinson, 2000). The interception storage capacities of fully developed forest canopies vary by species from 1 to 6 mm with the greatest interception capacities occurring in coniferous species (Luce, 1995; Ward and Robinson, 2000). Understory vegetation and grasses have interception storage capacities as great as or greater than some forest species. Seasonal reductions in interception storage for deciduous trees and plants and grasses occur during leafless or dormant periods.

Interception loss is regulated by the same physics as evaporation, but it only occurs when the vegetation canopy is wet. Consequently, interception loss is more dependent on rainfall duration and the intervals between rainfall events than is the case for ETa (Ward and Robinson, 2000). Considerable research has examined whether the interception loss is in addition to or replaces plant transpiration. A growing body of evidence supports the position that the interception loss in most circumstances occurs in addition to plant transpiration (Ward and Robinson, 2000). One simple example is that water intercepted by dormant or dead vegetation and evaporated would represent a water loss greater than transpiration.

A combination of theoretical analysis and field data indicates that intercepted water evaporates at a greater rate than transpiration from the same vegetation in the same environmental conditions. The higher evaporation rate from wetted vegetation surfaces relates to the roles of surface resistance and aerodynamic resistance in the evaporation process as discussed in Chapter 4. Surface resistance is a physiological and environmental factor imposed by the vegetation canopy, and wet vegetation surfaces of all types reduce surface resistance to zero (Calder et al., 1984). The effect is that the stomatal resistance of the vegetation is by-passed and evaporation rates increase (Shuttleworth, 1995). Aerodynamic resistance is related to the roughness of the vegetation surface, and it tends to be greater for trees than for grass. Consequently, a difference in the interception loss among trees, short vegetation, and grass is likely due to differences in aerodynamic resistance and its relationship to energy delivery to the vegetation. The energy required to support the higher evaporation rates of intercepted water in forests may be provided by advected energy from surrounding areas, from heat stored in the canopy, and by radiation balance modifications within the forest (Lundberg et al., 1997; Ward and Robinson, 2000). The interception loss in temperate forests typically ranges between 9% and 48% of annual precipitation (Pypker et al., 2005), but interception losses can be nearly 60% for wind-exposed forests (Mossin and Ladekarl, 2004). The interception loss for grasses is between 7% and 36% of precipitation during the growing season, but the interception loss may be as little as 3% during the dormant season (Lull, 1964).

6.4.2   Throughfall and stemflow

During prolonged rainfall, the interception storage capacity of the vegetation can be exceeded and water accumulates at the lowest points on leaf edges as increasingly larger water droplets. When the surface tension between the water and the leaf film is exceeded, the water falls from the leaf. In addition, water droplets may be shaken from the vegetation by the wind or by the impact of rain on the leaves. Throughfall is the rainfall that reaches the ground either directly through the vegetation canopy or as drip from leaves, twigs, and stems

(Ward and Robinson, 2000). Rainfall intensity beneath a forest canopy may be more uniform due to interception and throughfall. The small drops during a light rain are combined into larger drops that drip from the leaves, while large raindrops are broken by the foliage into smaller drops. Water droplets falling from canopy heights of more than 7.5 m exceed the force of rainfall when they reach the ground (Lull, 1964).

Stemflow is another pathway intercepted rain may follow in reaching the ground. Intercepted rain can move down the twigs and stems of plants when the interception storage capacity is exceeded. The roughness of the plants' bark is largely responsible for determining the amount of stemflow that occurs compared to throughfall. Stemflow can deliver a substantial amount of moisture to the area around the base of a tree that results in high infiltration rates (Ward and Robinson, 2000), while the ground a few meters away is only dampened.

6.5   Infiltration

The first raindrops that reach the soil surface either directly or as throughfall enter the soil profile through the pores and openings in the surface by infiltration. This process is driven by capillary forces and by gravity as discussed in Section 4.7. Capillary forces draw the water into the smaller pores of the soil matrix, but this movement of water is relatively slow and the quantity is small. Gravity moves water through the larger openings at a faster rate and in greater quantities. Infiltration displays great variability at both temporal and spatial scales in response to topography, soil physical properties, land use, and season of the year. Table 6.2 illustrates infiltration rate differences based on selected surface types.

Infiltration plays a fundamental role in the hydrologic cycle because it directly influences percolation, groundwater, and surface runoff contributions. Changing infiltration rates respond to surface ponding and recovery periods between storm events permitting water to percolate deeper into the soil profile. Even during a given storm, changing infiltration rates alter the proportion of rainfall retained on the surface for runoff.

6.5.1   Depression storage

When precipitation arriving at the soil surface exceeds the infiltration rate, ponding occurs on the surface in the many shallow depressions of varying size and depth that are present on practically all surfaces. This retention of water in microdepressions constitutes depression storage, which is a function of both surface geometric irregularities and the overall slope of the surface.

The specific magnitude of depression storage is difficult to determine, but values for natural surfaces ranging from 3 mm for flat areas to 13 mm for

Table 6.2. *Steady-state infiltration rates for various soil types*

Soil type	Infiltration rate (mm h^{-1})
Clay	1
Clay loam	5–10
Loam	10–20
Sandy loam	20–30
Sand	> 30

moderate slopes are suggested (Jens and McPherson, 1964). Values for urban surfaces are typically 1.5 mm (Miller, 1977). The bounded capacity of depression storage means that it is effective at the beginning of a rainfall event and its influence ceases rapidly as the depressions are filled with water. Where depression storage is important, evaporation and infiltration usually empty surface depressions between rainfall events (Dingman, 1994).

6.5.2   Soil moisture

The greatest proportion of rainfall infiltrates into the soil except during the most intense and longest duration storms as presented in Section 4.7. Retention and movement of water in the soil matrix is a function of the shape and size of the pores and related tension forces. When water infiltrates into the soil, air is displaced from the voids and soil water increases. If the amount of water entering the soil from the surface is great enough, the process continues until all of the pores are filled with water except for a small amount of trapped air. During this wetting process, the small pores fill first because they exert the largest surface tension forces. Following cessation of water inflow from the ground surface, water in the pores is removed by capillary and gravity flow processes. During this drainage and drying phase the large pores empty first for the same reason they were filled last. The larger surface tension forces exerted by the small pores are more effective in retaining water during the drainage phase. When gravity flow of water from the soil matrix becomes negligible, the soil water content of the soil profile approximates field capacity. For well-drained soils, field capacity is regarded as an effective upper limit to the amount of water that is available to vegetation for uptake by its root system. For a given soil and plant species, the lower limit of the soil water content below which the plant cannot extract water is known as the wilting point (Rose, 2004). Redistribution of soil water due to capillary flow continues after gravity flow ceases.

Soil moisture performs multiple roles significant to the runoff process by providing opportunities for partitioning water among several modes of subsequent movement. It contributes to the partitioning of precipitation to evapotranspiration by serving as a storage that supports moisture returns to the atmosphere by evaporation directly from the soil and transpiration by vegetation (Robock, 2003). Soil evaporation decreases once the top few millimeters of soil have dried because water diffuses slowly upward through the soil. Water stored in the upper few tens of centimeters of soils is available for vegetation to utilize in transpiration due to the penetration of plant roots to variable depths. Soil moisture is perhaps the single most climatically important soil variable because of its substantial influence on energy fluxes.

Soil moisture is a factor in partitioning precipitation between surface and subsurface flow routes. Soil moisture influences the surface infiltration rate and water ponding on the surface which initiates surface runoff. Water that infiltrates the surface moves through the soil profile vertically and horizontally due to gravity and the soil water content of various layers.

6.6   Overland flow

Water that exceeds depression storage capacity moves horizontally across the ground surface as quasi-laminar sheet flow or as small trickles and minor rivulets. The shallow film of sheet flow acts as a small storage in the sequence of alternating fluxes and storages water negotiates at the land surface. The depth of the water film on the soil depends on the land surface slope, the obstructions to water movement in a film, the rainfall rate, and the infiltration rate. Water is continuously infiltrating from this film into the soil matrix throughout a rain storm, and rain must occur at a higher rate than the infiltration rate for sheet flow to accumulate to any depth. On natural hillslopes under typical rainfall conditions, overland flow or sheet flow has a depth of 2 mm and a speed of less than $0.1 \, \mathrm{m \, s}^{-1}$ (Dingman, 1994). Sheet flow is an especially important runoff component in arid and semiarid regions, but significant overland flow is not common in humid regions unless soil saturation occurs (Rose, 2004). In urban areas, sheet flow is observed as the thin film of water running down sidewalks during rain. Overland flow tends to concentrate rapidly into microdrainage patterns that join to form rills, gullies, and channels for runoff.

6.7   Throughflow or interflow

Water that infiltrates into the soil may move laterally through the upper soil horizons as throughflow in the form of unsaturated flow or as shallow perched saturated flow. This condition is most likely to occur during prolonged rainfall on a hillslope when the lateral hydraulic conductivity of the surface soil's horizons greatly exceeds the vertical hydraulic conductivity through the soil profile (Ward and Robinson, 2000). Downslope water movement is considered as throughflow if it returns to the surface or a stream without first reaching the regional saturated zone.

Mechanisms supporting throughflow include unsaturated Darcian flow in the soil matrix, preferential flow in macropores that form routes that largely bypass the soil matrix, and flow in saturated zones of very limited vertical extent caused by soil horizons that impede vertical percolation. The relative importance of specific subsurface mechanisms is determined by watershed geology, soils, topography, and the amount, intensity, and spatial and temporal distribution of rainfall (Dingman, 1994). The rate of water movement to stream channels by throughflow varies with the different flow mechanisms.

A saturated zone must exist adjacent to a stream to provide a streamward hydraulic gradient, so water cannot travel to a stream entirely as unsaturated flow. Nevertheless, the development of a moist layer near the surface soon after infiltration begins leads to a pronounced downslope flow component in the unsaturated soil matrix governed by Darcy's law (Dingman, 1994). Downslope movement in unsaturated near-surface layers occurs soon after infiltration begins and produces a pronounced downslope flow rate. During intense rainfall, the downslope flow in unsaturated near-surface soil layers is at rates approaching those for saturated flow (Ward and Robinson, 2000).

Macropores are preexisting pathways in the soil matrix through which water can move in response to gravity (Hillel, 2004). Cracks and fissures between crusted areas of dry soil, worm holes, coarse sands and gravels, plant roots, and dry organic matter extending into the soil all serve as macropores. These openings are 0.3 to 3 mm in diameter (Ward and Robinson, 2000). Water movement through macropores results in more rapid wetting at deeper depths, and macropores can conduct water downslope considerable distances at relatively high velocities (Hillel, 2004). Unsaturated flow of this type is especially common in forested areas with shallow soils and on steep slopes (Luce, 1995).

Hydraulic conductivity in many soil profiles tends to be greater in surface layers than deeper in the profile. During prolonged rainfall on hillslopes, water enters the upper soil layers more rapidly than it can drain vertically through the lower layers. Several perched saturated layers can occur in the soil profile corresponding to textural changes. The accumulated water in perched saturated layers moves laterally in the direction of greater hydraulic conductivity (Ward and Robinson, 2000; Hillel, 2004). The flow rate in perched saturated layers is relatively slow, but the presence of macropores can produce much faster flow rates (Ward and Robinson, 2000). Drainage from perched layers can persist for many days and helps sustain streams between rainfall events in humid regions (Dingman, 1994).

6.8   Groundwater

Water drains vertically below the root zone under the force of gravity until it reaches impermeable or non-conductive strata. This deep percolating water accumulates as perched groundwater in a temporarily saturated zone above less permeable material or it enters the saturated porous rock or unconsolidated materials that form the regional groundwater system. A geologic unit that stores and transmits enough water to be hydrologically significant is known as an aquifer (Hillel, 2004). The fluctuating upper boundary of the saturated zone is the water table. In many parts of the world, groundwater recharge is mainly a seasonal phenomenon during the rainy season or the snowmelt season. In either case, groundwater recharge is a residual quantity remaining after precipitation has been allocated to surface runoff, soil moisture, and evapotranspiration. Under most natural conditions, groundwater discharges into rivers, lakes, or directly into the ocean, but in some instances groundwater is drawn upward by capillary forces (Dingman, 1994). Water at depth commonly moves very slowly, and the outflow of groundwater into streams may lag the occurrence of precipitation by days to years. The slow movement of groundwater accounts for it being considered water in storage in the same way a lake or reservoir is viewed as a storage mechanism.

Groundwater flow tends to be very regular, but groundwater flow patterns are controlled by the elevation and location of recharge and discharge areas, the heterogeneity of the geologic materials, the thickness of the strata, and the configuration of the water table (Bair, 1995). Two major types of idealized groundwater flow are unconfined and confined flow. In unconfined flow, the pressure at the water table is atmospheric and the hydraulic head is equal to the water table elevation above some specified datum. Recharge to unconfined aquifers typically occurs from water percolating vertically from the surface above the aquifer. The elevation of the water table varies as the flow through the aquifer changes. Consequently, flow in unconfined aquifers is analogous to surface flow in streams.

A confined aquifer is bounded by strata with significantly lower hydraulic conductivity than the strata forming the aquifer. Recharge for a confined aquifer typically occurs from water infiltrating at the highest elevation end of the strata where the aquifer is not confined and a water table is present. The flow in confined aquifers is analogous to flow in pipes because the boundary of the flow for a confined aquifer does not change (Hillel, 2004).

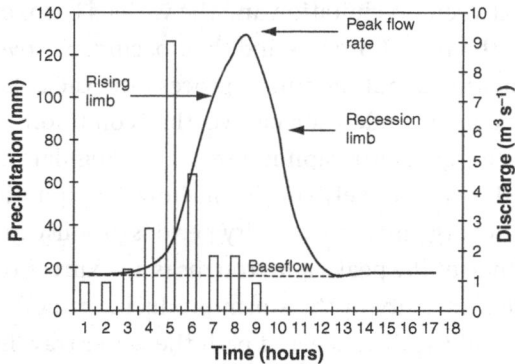

Fig. 6.2. Hypothetical runoff hydrograph for a drainage area of about 4 km^2 showing common hydrograph components and the hyetograph.

The direction and rates of groundwater movement are controlled by the geology of a watershed that may not be readily apparent from the surface. The groundwater storage volume is a function of the specific yield and depth of the saturated layer. Specific yield is the volume of water that will drain from an aquifer relative to the total volume of the aquifer or the water released from storage per unit surface area of aquifer per unit change in the water table. Average specific yield values for unconfined aquifers typically range from 0.01 to 0.40 (Bair, 1995). Differences in the extent of the aquifer system are important in determining the volume of water actually retained in the regional groundwater system. Groundwater discharge into the surface stream system contributes water identified as base flow that sustains streamflow during dry periods and represents the main long-term component of total runoff (Ward and Robinson, 2000).

6.9   The hydrograph

Streamflow is a valuable source of hydroclimatic information. It conveys the complex details of how the climate system and the hydrologic cycle are interacting. The runoff process can be viewed directly for relatively small areas, but streamflow for a large watershed results from distant events and relationships not readily apparent. It is necessary to understand how water arrives at the stream channel in order to assess the coupling of climate and the hydrologic response.

Streamflow is portrayed graphically using a hydrograph that is a plot of discharge against time (Fig. 6.2). A hydrograph relates stream discharge to the water supplied by rainfall or snowmelt. It is a graphical representation of the sequence of relationships between precipitation and the various basin environmental factors important to the runoff process and their occurrence over time. Hydrographs provide information about the runoff process and the behavior of streams during drought, floods, or under normal weather conditions. In general, a long-term stream discharge hydrograph is a series of irregular increases and decreases superimposed on a relatively consistent flow. Precipitation over the watershed produces a discharge increase and dry periods produce gradually declining flows. The magnitude of the peaks and troughs is an expression of the size of the drainage area, the intensity of the storm producing runoff, antecedent watershed conditions, and the distance and path the water travels before reaching the stream gauge. Information supporting detailed analysis of stream hydrographs is found in the hydrology literature (e.g. Dingman, 1994; Ward and Robinson, 2000).

The nature of the storages, delays, and time of travel is different for each of the surface and subsurface paths water follows in becoming runoff as presented in Section 6.2. Discharge in the channel results from the integration of flow from all runoff sources. Many of the sources exert distinctive influences on the quantity and timing of streamflow, which elucidate how the runoff process works in a watershed and how the processes are different from one watershed to another.

6.9.1   Hydrograph components

Transforming precipitation into runoff by hydroclimatic processes is evident in the hydrograph form. Hydrographs also portray information on the change in runoff rates with time, the peak runoff rate, and the volume of runoff. The hydroclimatic role in the runoff process can be characterized by the path followed by water in arriving at the stream channel after it is delivered to the surface by precipitation. Tracing a rainfall event serves as the framework for assessing other combinations of precipitation.

A common convention to facilitate description of the discharge hydrograph is to recognize two streamflow components designated as event flow and base flow. Event flow, also called direct flow, surface flow, storm flow, or quick flow, is water that enters the stream channel promptly in response to individual water-input events. Event flow is dominantly water that moves over the surface to the stream or travels as throughflow.

Base flow is water that enters the stream from persistent, slowly varying sources, and it maintains streamflow between water-input events. It is usually assumed that most, if not all, base flow is supplied by groundwater. However, streamflow between water-input events can also derive from drainage of lakes or wetlands or from the slow drainage of relatively thin soils on upland hillslopes. Some surface stream baseflow comes from throughflow in the soil (Ward and Robinson, 2000). Streams that receive large proportions of flow as groundwater tend to have relatively low temporal flow variability.

Hydrograph separation is a convenient method for gaining insight into the array of runoff components represented by streamflow. A concentrated rainfall event produces a typical hydrograph with a single peak and a skewed distribution curve (see Fig. 6.2). Multiple peaks can result from variations in rainfall intensity, a succession of storms, or other causes. Therefore, the shape of the hydrograph provides an integration of the climatic and watershed characteristics responsible for runoff.

The customary method for examining hydrographs is to recognize commonly recurring features of the curves. This is a reasonable approximation approach, and it provides a perspective that is useful for developing a general understanding of the watershed response. Complete details of methods for analyzing storm hydrographs are found elsewhere (e.g. Dingman, 1994; Ward and Robinson, 2000).

For descriptive purposes, the hydrograph is composed of a rising limb, a crest, and a recession limb. Some time after the beginning of rainfall, the flow rate begins to increase relatively quickly from a preexisting level. This period of rapid discharge increase is the rising limb of the hydrograph. The slope of the rising limb is largely determined by the storm intensity influencing the proportion of rainfall allocated to surface runoff. Rainfall following surface routes causes the rising limb to be very steep because water is delivered quickly to the channel.

The peak discharge defines the hydrograph crest. This is approximately the time when surface inflow related to the rainfall event ceases. Large contributions by surface flow and throughflow contribute to high peak flows.

Declining flows following the crest form the recession limb of the hydrograph. In general, surface inflow ceases and water is provided by basin storages. The groundwater contribution strongly influences the character of the recession flow, which is described mathematically by exponential, regression, and wave transform equations that result in decreasing flow to near the pre-event value (Sujono et al., 2004). The exponential function is the most commonly used form. Base flow is conventionally identified on the hydrograph by a line extending from the foot of the rising limb to the point of intersection on the recession limb (Mosley and McKerchar, 1993).

6.9.2   Hydrograph insights

Hydroclimatic relationships are evident when the time and space scales of hydrographs are expanded. Discharge time-series for a number of years

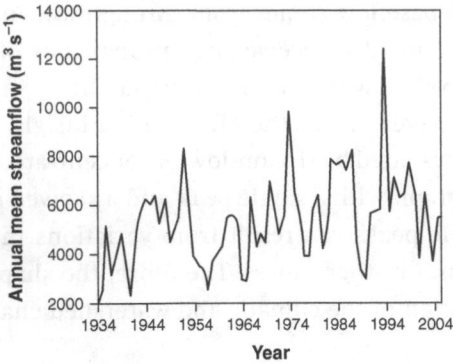

Fig. 6.3. Annual mean streamflow for the Mississippi River at St. Louis, Missouri (39° N), for 1934–2004. (Data courtesy of the U.S. Geological Survey from their website at http://waterdata.usgs.gov/nwis/.)

emphasize information on the year-to-year variations in both peak discharge and low flows. Figure 6.3 displays annual mean streamflow for the Mississippi River at St. Louis, Missouri (39° N). The drainage area above this gauge is $1\,812\,200\,km^2$, and the average annual discharge is $5386\,m^3\,s^{-1}$. The lowest annual mean streamflow is $2233\,m^3\,s^{-1}$ in 1934 and the highest is $12\,435\,m^3\,s^{-1}$ in 1993. Other prominent wet years are 1951 and 1973, and other dry years are 1940 and 1956. Eight of the nine years when streamflow is less than one standard deviation below the mean occur prior to 1967. In contrast, eight of the nine years when streamflow is greater than one standard deviation above the mean occur after 1967. Overall, the 82-year streamflow time-series indicates a slight increasing trend.

Seasonal cycles imbedded in annual streamflow data are evident in the greater variability of monthly mean streamflow shown in Figure 6.4. However, the increasing streamflow trend suggested in Figure 6.3 is more difficult to identify in the mean monthly data. Extreme wet events in 1951, 1973, and 1993 are amplified in Figure 6.4 by the magnitude of individual high stream-flow months. The maximum monthly value is $22\,905\,m^3\,s^{-1}$ for July 1993, but six additional wet years are evident when monthly streamflow exceeds $15\,000\,m^3\,s^{-1}$. Dry events are seen in the monthly data in terms of the duration of low flows and the absolute streamflow values. The 1934 dry event results from persistently low monthly streamflow and not from a single month of extremely low flow. The lowest monthly flow ($888\,m^3\,s^{-1}$) in the record occurs in January 1940 (see Fig. 6.4) which is the second driest year (see Fig. 6.3). The second lowest monthly value ($1360\,m^3\,s^{-1}$) occurs in December 1989, but 1989 is only a moderately dry year as shown by the annual mean streamflow. A final

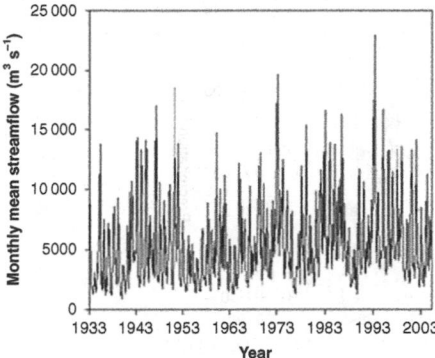

Fig. 6.4. Monthly mean streamflow for the Mississippi River at St. Louis, Missouri (39° N), for April 1933 to June 2005. (Data courtesy of the U.S. Geological Survey from their website at http://waterdata.usgs.gov/nwis/.)

feature of Figure 6.4 is that the slightly undulating pattern of minimum monthly discharge delineates a base flow component not evident in the annual mean streamflow.

6.10   Rainfall runoff

The streamflow responses to rainfall and snowmelt produce identifiable differences in discharge hydrographs. Rainfall events produce streamflow dominated by surface runoff and or near-surface flow. These conditions can produce abrupt increases in streamflow, especially for small watersheds or urbanized watersheds. Land use and geology can delay runoff and produce less abrupt streamflow increases and more gradual decreases.

The climatic role in the runoff process is characterized by the path followed by water in arriving at the stream channel after it has been delivered to the surface by precipitation. This perspective provides a basis for distinguishing the climatic influence as distinct from the geomorphologic influence imposed by terrain characteristics related to the watershed's size, shape, and relief.

6.10.1   Similar basin area

Examining watersheds with similar size areas reduces some of the physical complexity, but does not eliminate all of the non-climatic influences. However, many watershed physical characteristics are recognized as being highly correlated with the size of the drainage basin, so comparing similar size watersheds reduces a good portion of the expected variability among watersheds. Three small streams emphasize watershed response differences in

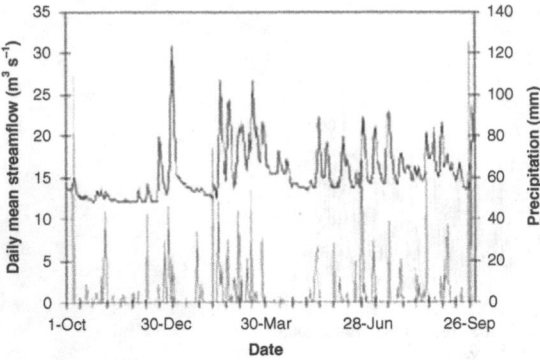

Fig. 6.5. Daily mean streamflow (bold line) for Econfina Creek near Bennett, Florida (30° N, 90° W), and daily precipitation (gray line) for Apalachicola, Florida, for 1 October 1941 to 30 September 1942. (Streamflow data courtesy of the U.S. Geological Survey from their website at http://waterdata.usgs.gov/nwis/. Precipitation data courtesy of NOAA's National Climate Data Center and the Oak Ridge National Laboratory, Carbon Dioxide Information Analysis Center from their website at http://cdiac.ornl.gov/epubs/ndp/ushcn/usa_daily.html.)

selected areas of the United States. Daily streamflow for the period 1 October 1941 to 30 September 1942 is used to minimize human influences on runoff for all three watersheds. Precipitation for these months at a representative station near each watershed is used to define the moisture input. Runoff from small watersheds facilitates comparisons by maintaining daily values within ranges that focus on the outcome of the runoff process.

Econfina Creek (Fig. 6.5) in the south central Florida panhandle drains a $317\,\text{km}^2$ watershed of relatively low relief underlain by limestone. Monthly precipitation for the period ranges from a low in April to a high in September, but 49% of the annual total is received from May to September. However, little evidence of this regime is seen in the streamflow response, which displays low flows from October to late December and the most consistent high flows in February and March. Most daily precipitation pulses are muted in the stream-flow response by the domination of the low relief and subsurface routes in limestone delivering water to the stream channel. Average daily streamflow is $15.6\,\text{m}^3\,\text{s}^{-1}$, and the dominant role of base flow is apparent in the consistent flow exceeding $12\,\text{m}^3\,\text{s}^{-1}$. Even during the summer months with a high energy demand for moisture, a major portion of precipitation appears to be allocated to subsurface routes that reduce evapotranspiration losses. Discharge peaks produced by storm events in all months are less than 3-times greater than the low flows during October to late December.

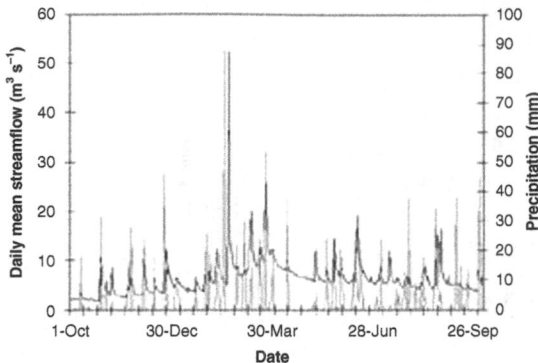

Fig. 6.6. Daily mean streamflow (bold line) for Cartecay River near Ellijay, Georgia (34° N, 85° W), and daily precipitation (gray line) for Rome, Georgia, for 1 October 1941 to 30 September 1942. (Streamflow data courtesy of the U.S. Geological Survey from their website at http://waterdata.usgs.gov/nwis/. Precipitation data courtesy of NOAA's National Climate Data Center and the Oak Ridge National Laboratory, Carbon Dioxide Information Analysis Center from their website at http://cdiac.ornl.gov/epubs/ndp/ushcn/usa_daily.html.)

The Cartecay River (Fig. 6.6) drains 348 km^2 of the southern tip of the Blue Ridge Mountains in northwestern Georgia. The watershed has relatively steep slopes, and the variable flow throughout the year indicates dominant surface runoff superimposed on a seasonally replenished variable base flow component. Monthly precipitation is greatest in March and least in April. The average daily streamflow is 6.5 m^3 s^{-1}. An increasing energy demand for moisture during the summer is evident in the gradually declining streamflow during the summer and fall even though precipitation is 48% of the annual total during these months. Although there is close agreement between daily precipitation pulses and streamflow peaks throughout the year, peak flows in the winter and early spring are especially concordant with precipitation. During this period, soil moisture is maximized, most precipitation is allocated to runoff, and surface runoff is a high proportion of total streamflow. The peak flow in late February is 29-times greater than the minimum flow in late October.

Antelope Creek (Fig. 6.7) drains 320 km^2 along the west slope of the southern Cascade Range in northern California. The watershed is characterized by relatively steep terrain and highly variable streamflow. Cool season precipitation is dominant, and January and February account for 49% of the annual total. Average daily streamflow is 6.5 m^3 s^{-1}. June through September receives little precipitation and this is evident in the absence of streamflow peaks in the hydrograph during this period. A summer and fall low-flow period without sharp runoff pulses contrasts markedly with the hydrographs for Econfina

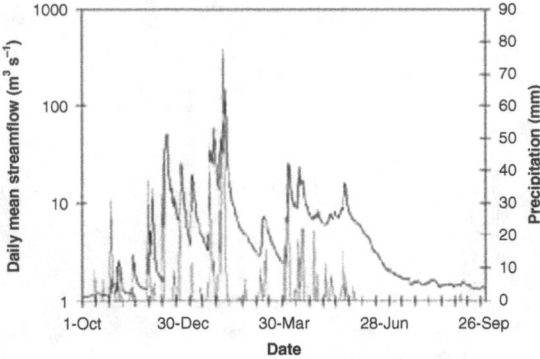

Fig. 6.7. Daily mean streamflow (bold line) for Antelope Creek near Red Bluff, California (40° N, 122° W), and daily precipitation (gray line) for Redding, California, for 1 October 1941 to 30 September 1942. (Streamflow data courtesy of the U.S. Geological Survey from their website at http://waterdata.usgs.gov/nwis/. Precipitation data courtesy of NOAA's National Climate Data Center and the Oak Ridge National Laboratory, Carbon Dioxide Information Analysis Center from their website at http://cdiac.ornl.gov/epubs/ndp/ushcn/usa_daily.html.)

Creek and the Cartecay River, which both display a streamflow response to storm events during these months. An additional contrasting feature is that the Antelope Creek peak flow in February is 2.8-times greater than the March peak flow for the Cartecay River. Antelope Creek received 178 mm of precipitation during a 3-day storm, and peak streamflow followed in one day. The 88 mm of precipitation for the Cartecay River was a single-day event and peak streamflow occurred four days later. These differences indicate variations in the runoff process within each watershed that are not related to drainage area. Furthermore, a limited groundwater role in the Antelope Creek runoff process is indicated by the 10-times larger mean daily flows in October for Econfina Creek and the 1.6-times larger low flows for the Cartecay River.

6.10.2   Similar basin climate

Holding climate constant permits watershed physical characteristics responsible for water arriving at the stream channel to be emphasized in comparing stream hydrographs. A selection of streams in Northern California simplifies assessment of watershed conditions by maintaining a similar general climate setting in terms of precipitation seasonality. However, precipitation amounts, the energy-driven moisture demand, and watershed physical characteristics are different among the watersheds. The monthly data portrayed in Figure 6.8 are expressed as percentages of annual runoff, which has the advantage of facilitating comparisons among watersheds with different areas.

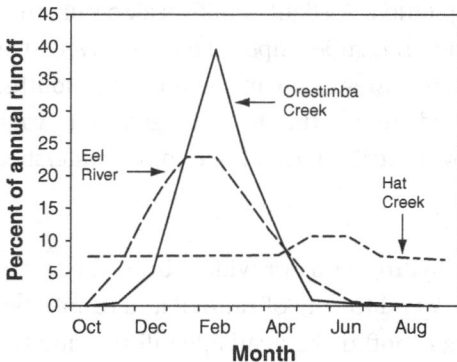

Fig. 6.8. Monthly percentage of annual runoff for 1971–2000 for three California streams in a generally similar climate setting that emphasizes the influence of topography on the runoff process. (Data courtesy of the U.S. Geological Survey from their website at http://waterdata.usgs.gov/nwis/.)

The Eel River basin (40° N) occupies 8094 km² on the western slopes of the Coast Ranges in northwestern California. It receives average annual precipitation of 150 cm and produces 79 cm of average annual runoff (Rantz, 1972). Precipitation is concentrated in November to March which accounts for 79% of the annual total. Runoff during December to March dominates the streamflow regime for the Eel River. The coincidence of the precipitation and runoff implies that storm events are the major runoff producers for the watershed. The low runoff percentages during May to October and the abrupt increase in November runoff reflect the widespread occurrence of low permeability strata in the watershed that contributes to meager base flow.

The Orestimba Creek watershed (37° N) covers 348 km² on the eastern slopes of the Coast Ranges in west-central California. This semiarid area receives about 25% of the precipitation amount occurring in the Eel River basin. The average annual precipitation for Orestimba Creek is 41 cm, and this produces 4 cm of average annual runoff (Rantz, 1972). The Orestimba Creek runoff regime is typical of a semiarid intermittent stream. Little or no runoff occurs during July through November, and runoff increases abruptly beginning in December to reach a peak in February. Over 80% of the annual runoff occurs from January through March. The high runoff during the rainy season and the absence of runoff during the summer indicate a stream system dominated by surface runoff.

Hat Creek (41° N) drains 421 km² of the Modoc Plateau region of northeastern California. The watershed is dominated by layered basalts related to volcanic activity in the southern Cascade Range. The average annual precipitation for Hat Creek is 130 cm, and average annual runoff is 28 cm (Rantz, 1972). November through March accounts for 69% of annual precipitation. A slight runoff peak occurs in May and June, but this pulse of increased runoff is superimposed on a relatively steady base flow component indicated by the consistent flow in the other 10 months. Highly permeable basalts promote reduced surface runoff, and significant groundwater storage accounts for high base flow contributions all year in this watershed.

6.10.3   Ungauged basins

For gauged watersheds, hydrographs provide a basis for assessing the relationship between the quantity and timing of rainfall and runoff, but procedures are required for estimating runoff using available information that simulates runoff production from a measured rainfall quantity. Rainfall–runoff assessments commonly involve determination of the peak runoff rates, the depth or volume of runoff, or a storm hydrograph. Each approach employs techniques designed to incorporate specific features of the runoff process.

Estimating peak flow

The simplest way to view the rainfall–runoff process responsible for watershed discharge is to use a system-based conceptual perspective that treats the watershed as a black box. In this approach, some function transforms a given time-varying input into a time-varying output without detailed consideration of the physical processes producing the response (Dingman, 1994). The black box provides little understanding of the processes involved in the transformation. In the United States, the widely used empirical rational method for estimating peak runoff in designing ditches, channels, and storm water control systems for small areas of up to 80 hectares with no significant flood storage illustrates this approach. The rational method is expressed as

$$Q_p = 0.28 C_s I_p A \tag{6.2}$$

where Q_p is the peak flow in $m^3 s^{-1}$, 0.28 is a unit conversion factor, C_s is a dimensionless empirical coefficient based on soil type, slope, vegetation cover, soil moisture content, and land use characteristics, I_p is the average rainfall intensity in $mm\,hr^{-1}$ during the time of concentration, and A is the watershed area in km^2. The time of concentration is the time required for water to move from the most distant point in a watershed to the outlet, and it is mainly a function of the watershed size and shape (Mansell, 2003). The empirical runoff coefficient C_s is the major source of uncertainty in applying the rational method. The coefficient incorporates all the factors influencing the relation of peak flow to average rainfall intensity in the watershed other than response time and watershed area. A common feature of these coefficients is they are based on judgments rather than experimental data (Pilgrim and Cordery, 1993). Typical

Table 6.3. *Rational method runoff coefficients (C$_s$) for selected land use and soil groups and a 2–6% slope*

Land use	Well-drained soil (Hydrologic soil group A)	Poorly drained soil (Hydrologic soil group D)
Industrial	0.68^a, 0.85^b	0.69, 0.86
Commercial	0.71, 0.88	0.72, 0.89
Residential	0.23, 0.32	0.32, 0.40
Cultivated	0.13, 0.18	0.23, 0.29
Pasture	0.20, 0.25	0.40, 0.50
Forest	0.08, 0.11	0.16, 0.20

^{a} Runoff coefficients for storm recurrence intervals less than 25 years.
^{b} Runoff coefficients for storm recurrence intervals of 25 years or more.

runoff coefficients for selected conditions are shown in Table 6.3. Larger values of C_s indicate increased runoff potential as is seen by comparing the values for forested areas, residential land, and commercial property. A comprehensive list of runoff coefficients is found in McCuen (2005).

The popularity of the rational method is due to its simplicity, but it is based on assumptions that are rarely realized under actual circumstances. The methodology assumes that rainfall is uniformly distributed over the entire drainage area, that the rainfall fraction that becomes runoff is independent of rainfall intensity and volume, and that the predicted peak discharge has the same probability of occurrence as the rainfall intensity (Ward, 1995). Choosing the correct rainfall duration is one of the major challenges of the rational method because the duration must be just long enough for maximum runoff to occur. Nevertheless, this methodology is widely used in urban settings for forecasting and predicting peak flow because of the difficulty in obtaining enough information to adequately characterize the spatial and temporal variability of hydrologic processes (Mansell, 2003). A modification commonly employed in urban applications is to partition the watershed into subareas based on various surface characteristics and to sum the products of the runoff coefficient and area representing each surface type (Mays, 2005). In other applications, a storage coefficient is added to the rational method to account for a recession time longer than the time required by the hydrograph to rise.

Estimating runoff volume

The transparent box perspective provides a more process-oriented approach to understanding runoff. This approach emphasizes that streamflow is a spatially and temporally integrated response determined by varying input rates and the time required for water to travel from where it arrives on the watershed surface to the stream and then to the point of measurement. The essential aspects of this perspective are that water moves within the watershed in an infinite number of surface and subsurface flow paths. Each flow path in a watershed is an accumulation of lateral water inflows that vary in space and time.

Estimating the depth or runoff volume requires more information about the watershed than is needed for estimating the peak runoff rate. The U.S. Soil Conservation Service (SCS), now the Natural Resources Conservation Service (NRCS), curve number (CN) procedure is used globally for determining the depth of runoff (Michel *et al.*, 2005). This approach is an empirical method derived from infiltrometer tests and measured rainfall and runoff on small plots and basins (Pilgrim and Cordery, 1993). It is best used as a means to transform a rainfall frequency distribution into a runoff frequency distribution (Jacobs *et al.*, 2003), but it has been applied successfully at a range of watershed scales (Rose, 2004). The CN method combines infiltration losses, surface storage, and short-duration high-intensity rainfall events nested within larger events to estimate accumulated runoff using the relationship

$$Q = \frac{(P - I_a)^2}{(P - I_a + S)} \tag{6.3}$$

where Q is runoff in mm, P is the rainfall depth in mm, I_a is the initial abstractions due to surface storage, interception, and infiltration prior to runoff and is commonly approximated as $0.2S$ in mm. S is a parameter given in metric units by

$$S = \frac{25\,400}{CN} - 254 \tag{6.4}$$

where CN is the SCS curve number designed with a range of 0 to 100 and determined empirically to be a function of the ability of soils to infiltrate water, land use, and the antecedent soil moisture condition (AMC). In a watershed encompassing varying characteristics, an area-averaged composite CN can be computed for the entire watershed (Mays, 2005). The general form of Equation 6.4 is well established by both theory and observation, but the AMC is one of the greatest uncertainties of the CN methodology.

The SCS recognizes three AMC categories based on dormant season and growing season antecedent soil moisture amounts. AMC II is the average moisture condition and is often used as the representative value. AMC I applies to dry soil conditions, and AMC III applies to wet soil conditions. The infiltration characteristics of soils are used by the SCS to define four hydrologic soil groups. These groups range from Group A, which has low runoff potential and high infiltration rates, to Group D,

Table 6.4. *U.S. Soil Conservation Service runoff curve numbers (CN)*
for antecedent soil moisture condition II and selected hydrologic soil
groups and land uses

Land use	Hydrologic soil group A	Hydrologic soil group D
Industrial	81	93
Commercial	80	95
Residential	61	87
Cultivated	65	86
Pasture	39	80
Forest	25	77

Data from U.S. Soil Conservation Service, 1986.

which has high runoff potential and low infiltration rates. Remotely sensed microwave soil moisture measurements and other new methods are helping to define
spatial soil variations at higher resolution scales. Curve numbers represent the
averages of median site values combining the AMC, hydrologic soil group, and
numerous land use conditions, and CN tables developed by the SCS are presented
in the original documentation (U.S. SCS, 1964) and in later editions. An example of
CN values is shown in Table 6.4. Michel *et al.* (2005) revised the original CN formula
to comply more explicitly with accepted soil moisture accounting procedures
while maintaining the efficiency of the original methodology.

Storm hydrographs

Storm hydrographs require the most complex information and provide
the most comprehensive runoff estimates. The unit hydrograph is probably the
most widely applied method. It derives its name from the characteristic that the
area under the hydrograph is equal to 1 mm. The unit hydrograph assumptions
are that a rainfall excess is uniformly distributed over the watershed, the rainfall
excess rate is uniform, and the runoff rate is proportional to the runoff volume for
a rainfall excess of a given duration (Ward, 1995). The central hypothesis emerging from these assumptions is that the watershed response to the rainfall excess
is linear and a single rainfall hyetograph defines the input. In this way, the unit
hydrograph is the watershed response to a standard input for a specified time.
Consequently, the duration must be included in the name of the unit hydrograph
(Pilgrim and Cordery, 1993). For example, a 1-hr unit hydrograph is produced by
1 mm of rainfall excess occurring over a watershed in 1 hr at a rate of 1 mm hr^{-1}.

Standard hydrograph separation techniques are used to determine direct
runoff, which is taken as equal to rainfall excess. The average depth and

intensity of basin rainfall in each time increment of the storm are estimated. The rainfall excess for each period of the storm is calculated using a loss model. For a single-period storm, a unit hydrograph is derived by dividing the ordinates of the surface runoff hydrograph by the depth of surface runoff in mm. The duration of the rainfall excess determines the time period of the unit hydrograph. Several analytical techniques are available to derive unit hydrographs for multiperiod storms (Pilgrim and Cordery, 1993).

The time–area method is a process-based procedure to determine the runoff volume and peak runoff rate from small watersheds. The approach does not consider interflow and assumes that surface and channel flow velocities do not change with time. The physically based kinematics approach is theoretically the most complete. The method is based on the solution of the continuity and momentum equations and the relationship between the watershed geometric characteristics and the drainage system (Ward, 1995). The antecedent precipitation index and normalized antecedent precipitation index methods estimate runoff where the initial watershed moisture condition is probabilistic. These methods are data intensive and most applications are solved using computers.

6.11   Snow and runoff

The role of snowmelt in the hydrologic cycle is defined by the response of snow to radiant energy. Snow stores water at the Earth's surface for various periods, and snowmelt has a delayed effect on a discharge hydrograph because snowfall does not influence streamflow until the snow melts. Seasonal snowpack melting is related to the snowpack radiation balance which supplies the energy to convert snow into liquid water. However, the radiation balance depicted by Equation 2.11 requires expansion when applied to a snow surface due to the relatively complex nature of snow compared to other surface materials.

6.11.1   Radiation balance and snow

The snow surface is semitransparent to solar radiation, and this allows transmission of shortwave radiation into the snowpack. Radiation absorption then occurs within a volume rather than being limited to a plane coincident with the snow surface. This radiation transmission and absorption characteristic influences the nature of both the radiation balance and the energy balance of snow.

Shortwave radiation incident at the snow surface is greater than that at any depth. The decay of the flux with distance into the snow follows an exponential curve defined by Beer's law

$$K{\downarrow}_z = K{\downarrow}_o \, e^{-az} \tag{6.5}$$

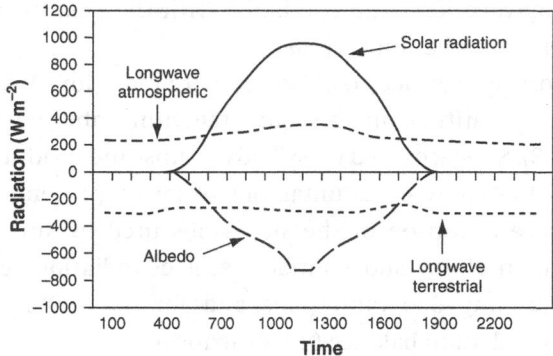

Fig. 6.9. Radiation balance for a snow-covered surface on 22 April 1954, a clear day, at the Central Sierra Snow Laboratory just west of the crest of the Sierra Nevada in California (40° N, elevation 2273 m). (Developed from data reported by the U.S. Army Corps of Engineers, 1956.)

where $K{\downarrow}_z$ is shortwave radiation at depth z, $K{\downarrow}_o$ is shortwave radiation at the surface, e is the base of natural logarithms, and a is the extinction coefficient (m^{-1}). The attenuation or depletion of shortwave radiation with depth in the snowpack is directly related to the extinction coefficient. The value of the extinction coefficient depends on the physical characteristics of the transmitting medium and the wavelength of the radiation. Consequently, the extinction coefficient of shortwave radiation is greater for snow than for ice because of the difference in the physical medium. Shortwave penetration may be 1 m in snow and 10 m in ice (Oke, 1987).

Snow albedo is another important radiative characteristic of snow. Albedo can change from 0.25 to 0.80 in a few hours with the deposition of a thin snowcover. The albedo of the snow surface may then decrease from 0.80 to 0.40 in a matter of days as the snow ages. Albedo varies with wavelength within the shortwave portion of the spectrum, but the highest albedo is for the shortest wavelengths. Albedo decreases to quite low values in the near-infrared wavelengths. The high albedo of snow at the shortest wavelengths is the reverse of the condition for most soil and vegetation surfaces (Oke, 1987).

In general, the daytime net radiation surplus for a snow surface is small compared with most other natural surfaces. Snow's high albedo results in little shortwave radiation absorption. Even a portion of the shortwave radiation transmitted into the snowpack is reflected and contributes to the high albedo. Absorption of the radiation transmitted into the snowpack does not offset the stronger influence of the shortwave loss due to albedo. In the longwave portion of the spectrum, the absolute magnitude of $L{\uparrow}$ is usually relatively small because the snow surface temperature is low, but the $L{\downarrow}$ flux is also small (Fig. 6.9). The result is a small net longwave loss, and combined with the small shortwave surplus results in a small net radiation surplus.

Equation 2.11 best portrays surface conditions where radiation fluxes are a function of solar angle and latitude and a steady-state atmospheric boundary layer (Pomeroy *et al.*, 2003). Snow-covered grasslands, plains, and tundra approximate these conditions, but snow in mountainous terrain represents a more complex problem because a portion of the sky is obscured by surrounding topography. In addition, an alpine snow surface receives radiation reflection and emission from surrounding topography (Plüss and Ohmura, 1997). In mountainous terrain the snow radiation balance takes the form

$$R_n = (I_s + D_s + D_t)(1 - \alpha) + L_a{\downarrow} + L_t{\downarrow} - L{\uparrow} \tag{6.6}$$

where I_s is direct solar radiation, D_s is diffuse solar radiation, D_t is diffuse solar radiation from the surrounding terrain, L_t is the incoming longwave radiation from the surrounding terrain, and the other variables are defined for Equation 2.11. All units are in $\mathrm{W\,m^{-2}}$ except for α which is dimensionless. In mountainous terrain, longwave radiation from surrounding topography can make an important contribution to the surface radiation balance. Since longwave radiation is present day and night, its influence on the radiation balance is significant in understanding areal variations in the energy balance and snowmelt. Slope and aspect are additional important considerations because they affect snow accumulation, snowmelt energetics, meltwater fluxes, and runoff contributing area (Pomeroy *et al.*, 2003).

6.11.2   Energy balance and snowmelt

Much of the current knowledge of the energy balance and melt characteristics of snow can be traced to the work of the U.S. Army Corps of Engineers (1956). The energy balance of snow is complicated by the penetration of shortwave radiation into the snowpack and by internal water movement and phase changes within the snowpack. Viewing the snowpack as a volume helps to understand the energy fluxes and the physical changes that occur in the snowpack as the melt process occurs.

The traditional energy balance (Equation 2.12) is typically rewritten for a snow surface to include terms describing energy sources available to melt snow. The energy for snowmelt comes from net radiation, conduction and convection transfers of sensible heat from the overlying air, condensation of water vapor from the overlying air, conduction from the underlying soil, and from rainfall. The relationship is represented by

$$SN_m = R_n + H + LE + G + D \tag{6.7}$$

where SN_m is the energy available for snowmelt, D is energy transported to the snowpack by snow or rain, and all other variables are defined for Equation 2.12. All units are in $W\,m^{-2}$. The position of R_n in Equation 6.7 acknowledges that net all-wave radiation is the dominant energy component responsible for snowmelt (Plüss and Ohmura, 1997; Suzuki and Ohta, 2003). Net radiation has a maximum absorption just below the surface during the day (Oke, 1987). Consequently, the plane of greatest energy, the highest temperature, and the greatest potential mass flux are not coincident with the snow surface. At night, with only longwave radiative exchanges, the active surface is at or near the snow surface. This produces a sharp temperature gradient in the upper layer of the snow largely due to the low thermal conductivity of snow. The snow surface becomes very cold because radiative losses are not offset by heat flows from within the snowpack.

Snow is an effective insulating cover for the ground. As little as 0.1 m of fresh snow will insulate the underlying ground. Large radiation variations are limited to the snowcover because snow has low heat conductivity. Surface radiative losses are not replaced quickly by heat fluxes from below. The strong cooling of the atmosphere near the Earth's surface stabilizes the atmosphere against convection and contributes to the occurrence of colder local temperatures. During summer daylight hours, snow quickly comes to a uniform temperature of 0 °C throughout the snowpack. This is due to penetration of the strong shortwave radiation and percolation of water to deeper layers that assist in transferring heat within the snowpack.

The melt process is especially important hydroclimatically because it represents the return of moisture to the liquid phase, and it involves energy fluxes that are climatically driven. Snow undergoes a continuous metamorphism until it melts, and the metamorphic changes are driven by the energy state of the snowpack. A centimeter of water melted in a snowpack at 0 °C requires $39\,W\,m^{-2}$ of energy.

The conversion of snow to water is known as ablation. At this time, the mass of the snowpack is reduced in response to the loss of water. Ablation can be thought of as the opposite of the accumulation stage during which the snowpack increases. Melt or ablation of the snowpack does not occur as long as the energy balance is negative and $SN_m < 0$. This condition cools the snowpack and increases the snow "cold content" or the amount of energy required to bring the entire snowpack to 0 °C. A positive energy balance results in $SN_m > 0$, and this adds energy to the snowpack and produces warming of the snow until the entire snowpack is isothermal at 0 °C. Snowmelt does not occur in significant amounts until the entire snowpack is isothermal at 0 °C, but once this condition is reached $SN_m > 0$ results in melt (Marks and Winstral, 2001). The ablation period of a seasonal snowpack is divided into three phases once a sustained positive net energy input supports $SN_m > 0$ continuously.

The snowpack first experiences warming when the average snowpack temperature increases more or less steadily until the snowpack is isothermal at $0\,°C$. It is important to note that no melting occurs, rather the snowpack temperature changes. The heat required to raise the snowpack average temperature to the melting point before melt occurs defines the cold content of the snow. The cold content (SN_{cc}) is expressed as

$$SN_{cc} = -c_i\rho_w h_m(T_s - T_m) \tag{6.8}$$

where c_i is the heat capacity of ice at $0\,°C$ ($2.05\,J\,g^{-1}\,K^{-1}$), ρ_w is the mass density of water ($1.00\,g\,cm^{-3}$), h_m is the water equivalent of the snowpack (cm), T_s is the average temperature of the snowpack (°C), and T_m is the melting point temperature ($0\,°C$) (Dingman, 1994). The cold content expresses the energy required to raise the snowpack temperature to the melting point, and it can be determined anytime before melt begins.

Snowpack ripening occurs as additional energy warms the snowpack after it is isothermal. Ripening produces meltwater that is initially retained in the pore spaces of the snow grains by surface-tension forces. No meltwater is released from the snowpack during this stage, but at the end of this phase the snowpack is isothermal at $0\,°C$. Under this condition, the snowpack is considered to be ripe and cannot retain any more liquid water within the snowpack. The energy necessary to bring a snowpack to a ripe condition (SN_r) is equal to the cold content plus the latent heat required by the amount of melt produced and is represented by

$$SN_r = \theta_{ret} h_s \rho_w \lambda_f \tag{6.9}$$

where θ_{ret} is the maximum volumetric water content that the snow can retain and is estimated from empirical studies, h_s is the depth of the snowpack, ρ_w is the mass density of water ($1.00\,g\,cm^{-3}$), and λ_f is the latent heat of fusion ($3.35 \times 10^5\,J\,kg^{-1}$) (Dingman, 1994).

After the snowpack is ripe, any further energy inputs produce water releases from the snowpack. Additional meltwater cannot be held by surface tension against the pull of gravity within the snowpack after the ripe condition is reached. Water begins to percolate downward ultimately to become water output. The net energy input required to complete the output phase is the amount of energy needed to melt the snow remaining at the end of the ripening phase (SN_o) and is computed as

$$SN_o = (h_m - h_{wret})\rho_w \lambda_f \tag{6.10}$$

where h_{wret} is the liquid water retaining capacity of the snowpack (cm), and all other variables are defined previously (Dingman, 1994).

In many situations, the snowpack does not progress steadily through this sequence. Some melting occurs at the surface of a snowpack prior to the ripening phase (Ward and Robinson, 2000). The meltwater percolates into the cold snow at depth and refreezes releasing latent heat that raises the snow temperature. Nevertheless, the three-phase structure provides a useful framework for describing the process and understanding how the energy balance of the snowpack drives the ablation process. The amounts of net energy required for each of the melt phases are readily computed. Streamflow contributed by snowmelt is thus likely to occur during clear sky conditions and produces a hydrograph with lower peak flows but sustained high flows.

6.11.3 Thermal indices and snowmelt

Snowmelt can be computed using heat conservation principles outlined in the previous section, but it is difficult and expensive to fulfill the data requirements of the energy balance of the snowpack. An empirical temperature index approach has a long history of use for estimating snowmelt largely because air temperature is often the only reliable and consistently available weather variable measured for remote areas, and it is well correlated with radiation, wind, and humidity so that residual errors are usually not a factor (Luce, 1995). However, Ohmura (2001) asserts there is a physical basis for air temperature as an effective parameter for estimating snowmelt that is evident when R_n in Equation 6.7 is expressed as individual terms. The expanded form of the energy balance is

$$SN_m = K{\downarrow}(1 - \alpha) + L{\downarrow} - \varepsilon\sigma T^4 + H + LE + G + D \tag{6.11}$$

where ε is the emissivity of snow with a value close to one, and all other terms are defined previously. The $K{\downarrow}$ and $L{\downarrow}$ terms depend on the composition and temperature of the overlying atmosphere and on the relief of the surrounding topography as discussed for Equation 6.6. The σT^4 term is fundamentally different from all the other terms on the right side of the equation in that it is determined entirely by the other terms which represent external fluxes. The σT^4 term adjusts in response to the other fluxes by altering the radiative emission rate to obtain a new equilibrium. The longwave surface emission is a function of surface temperature corresponding to the equilibrium state. There are mutual dependencies of different degrees between other terms in equation 6.11, but σT^4 is not autonomous nor can it alter itself spontaneously as is the case for the other terms. Since atmospheric longwave radiation is the dominant heat source for snowmelt, its relationship with σT^4 in forming the longwave radiation balance defines a close physical coupling between surface temperature and snowmelt that establishes air temperature as a significant index for estimating melt (Ohmura, 2001).

A common temperature index approach for estimating daily snowmelt (SNm_d) is

$$SNm_d = k'(T_a - T_b) \tag{6.12}$$

where k' is a melt coefficient or melt factor in mm $°C^{-1}$ that varies with latitude, elevation, slope inclination and aspect, forest cover, and time of the year and must be empirically estimated for a given site, T_a is the air temperature in $°C$, and T_b is the reference temperature usually taken as the snow melting point temperature, $0\,°C$. In the absence of site-specific data, k' can be estimated using various generalized expressions incorporating common site factors (Dingman, 1994). The temperature index approach has been applied with success for spring snowmelt over large basins (Luce, 1995), but temperature indices have not been successful for estimating snowmelt from open grasslands or for small basins at high elevations (Linacre, 1992).

Mathematically, Equation 6.12 implies that melt occurs when $T_a > 0\,°C$, which is not always true (Bras, 1990). The energy balance of a snowpack indicates snowmelt can occur for air temperatures below $0\,°C$, especially during clear, calm days when solar radiation dominates the energy balance. Also, no melt can occur on clear nights when outgoing longwave radiation is significant even though the air temperature is above $0\,°C$.

Forest cover is probably the most common factor used in developing a melt coefficient for Equation 6.12. Forest cover serves as a surrogate because it has a significant effect on many of the variables affecting the snowcover energy balance (Suzuki and Ohta, 2003). Climatic factors are important in accounting for differences when physiographic conditions are constant (Marks and Winstral, 2001).

The degree-day method for estimating daily snowmelt is based on empirical evidence that daily snowmelt is expressed as a linear function of average air temperature. A degree-day is a departure of one degree per day in the daily mean temperature from a reference temperature. In general, it is assumed that there is no melt for temperatures below freezing and that melt is directly proportional to the number of degrees above freezing. Degree-day snowmelt for one day (SNm_{dd}) in mm is expressed as

$$SNm_{dd} = D_f(T_a - T_b) \tag{6.13}$$

where D_f is a degree-day factor in mm $°C^{-1}$, and the temperature variables are defined for Equation 6.12. Empirical degree-day factors range from 2 to 6 mm $d^{-1}\,°C^{-1}$ depending on specific site characteristics (Seidel and Martinec, 2004). Rango and Martinec (1995) suggest gradually increasing the degree-day factor over the course of the snowmelt season, and this approach is incorporated in the widely used Snowmelt Runoff Model (Seidel and Martinec, 2004). The degree-day method is the standard tool for estimating snowmelt runoff, and the accuracy of this method is comparable to more complex energy balance for-mulations (Rango and Martinec, 1995). Clark and Vrugt (2006) successfully employ a two-parameter snow model using physically realistic parameterized temperature to estimate snow accumulation and melt in an alpine setting. The WMO (1986a) comparison of 11 snowmelt runoff models illustrates the capabil-ities of different modeling approaches.

6.11.4   Snowmelt runoff

It is difficult to quantify the fraction of runoff derived from snowmelt due to a lack of detailed knowledge of the runoff process. However, a smaller proportion of snowfall than of rainfall is evaporated and transpired, so it is clear that snowfall contributes proportionally more to runoff for those areas that receive a seasonal snowpack. It has been suggested that more than half the annual runoff in much of the Northern Hemisphere is derived from snowmelt, and the timing of winter–spring streamflow is important in these regions (Hodgkins and Dudley, 2006).

In considering a snowmelt calculation for hydroclimatology, it is important to realize that all parts of a watershed may not have a snowpack or an evenly distributed snowpack. Temperature, wind, topography, and vegetation all influ-ence the space and time variability of snow accumulation and ablation. Hill slopes in particular may contain mixed surfaces of snow, bare ground, and vegetation (Marks and Winstral, 2001; Pomeroy et al., 2003). However, Anderton et al. (2004) conclude that topographic controls on the redistribution of snow by wind are the most important influence on snow distribution at the start of the melt season because the spatial variability of melt is largely deter-mined by the snow distribution rather than the spatial variability of melt rates. This means that the contributing area for snowmelt may be very different from the area responsible for rain-related runoff in a specific watershed. At the same time, the snowmelt runoff process is very different for high-elevation water-sheds with a continuous snowcover relative to lower-elevation watersheds that have a snowpack on only a portion of their total area. A final complication is that while snow accumulates by elevation, snow melts by slope aspect. In the Northern Hemisphere, snowmelt on south-facing slopes occurs more rapidly than on valley-bottom sites or on north-facing slopes in response to energy balance differences (Pomeroy et al. 2003). These distinctions are important for snowmelt forecasting.

Meltwater is removed from the snowpack by gravity drainage. Since the majority of the melting occurs near the snow surface in response to available energy, there is a distinct daily rhythm to the production of meltwater in response to solar radiation. After being produced, the meltwater percolates through the snowpack. Consequently, there is a series of delays beginning with the transport of meltwater through the snowpack that produce a distinct lag in meltwater delivery to the stream channel (Meier, 1990).

Water arriving at the bottom of the snowpack either infiltrates into the soil or accumulates to form a saturated zone at the base of the snowpack. If the ground under the snowpack is not frozen or saturated, water percolating out of the base of the snowpack can infiltrate into the soil in the same manner as rainfall. However, snowmelt releases large quantities of water more slowly than the rate occurring in most flood-producing rainstorms, and the probability of the infiltration capacity being exceeded is low. In this case, the snowmelt water moves to the stream channel by the subsurface processes of throughflow or groundwater flow. The lag time for water reaching an upland stream channel is determined by the time required for vertical percolation in the snowpack and the travel time through the subsurface processes.

When the ground beneath the snowpack is frozen or saturated, infiltration into the soil is impeded and water percolating through the snowpack accumulates at the base of the snowpack. A saturated zone forms in the snowpack, and water moves downslope within the snowpack and just above the ground surface. When this is the dominant process, a wave of melt water arrives at the saturation zone in response to the daily energy balance, and the input wave produces a daily output wave that travels downslope in the basal saturated zone (Dingman, 1994). The lag time associated with this process is determined by the time required for water to percolate vertically through the snowpack. On slopes steep enough to promote drainage, the net storage effect of water draining through the snowpack is about 3 to 4 hours for moderately deep and fully ripened snow. In shallow snow, the saturated zone may include the entire snowpack and create a slush layer with a liquid content almost great enough to promote downslope flow. The original structure of the snowpack is the basic impediment to flow by the complete layer (Meier, 1990).

The snowpack storage effect is the primary factor in regulating the volume and time distribution of snowmelt runoff. The physical properties of the snowpack continue to change from the time of deposition to melt, and the storage effect is closely related to the snowpack energy balance. After melt occurs, watershed physical characteristics become a factor in the delivery of the meltwater to the stream channel. Small headwaters streams display the daily rhythm of the melt process with relatively short delays determined by the route the water takes in reaching the channel. After a week or more of active melt, the result is a peak flow by mid-afternoon followed by a rapid recession in the

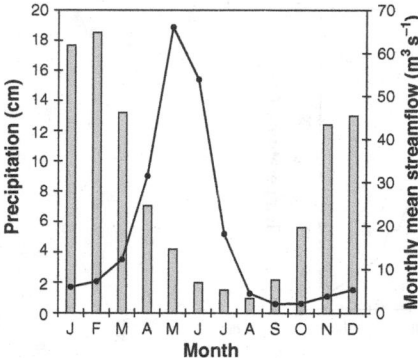

Fig. 6.10. Average monthly streamflow (1971-2000) for the Merced River at Pohono Bridge near Yosemite, California (38° N, elevation 1170 m), and average monthly precipitation (1971–2000) for Yosemite National Park, California, showing the contrasting seasonal regimes resulting from snow accumulation and melt. Shaded columns are precipitation, and the line is streamflow. (Streamflow data courtesy of the U.S. Geological Survey from their website at http://waterdata.usgs.gov/nwis/. Precipitation data courtesy of NOAA's National Climate Data Center and the Oak Ridge National Laboratory, Carbon Dioxide Information Analysis Center from their website at http://cdiac.ornl.gov/epubs/ndp/ushcn/usa_monthly.html.)

evening as temperatures fall below freezing. This daily peak and recession pattern recurs for the remainder of the melt season (U.S. Army Corps of Engineers, 1956). In larger watersheds, the melt process begins at the lower elevations and proceeds to successively higher elevations. As an increasingly larger proportion of the watershed participates in the runoff process, the daily melt sequence becomes masked by runoff contributions from other storages. For example, in forested watersheds there is no basin-wide surface runoff from snowmelt. Practically all snowmelt runoff enters the stream channel as subsurface flow, groundwater flow, or a combination of both (Seidel and Martinec, 2004). At this spatial scale, the snowmelt influence becomes more apparent in monthly flows that increase to a peak in late spring or early summer when active snowmelt and all of the other storages are contributing to streamflow.

The seasonal snowpack effect on the annual distribution of runoff is illustrated by the Merced River (Fig. 6.10) which drains 831 km² surrounding Yosemite Valley in California. This watershed on the western slopes of the Sierra Nevada receives 114 cm of average annual precipitation, 96% of the watershed area is above the annual snowline, and average annual runoff is 64 cm (Rantz, 1972). The snowmelt influence on runoff is evident in the abrupt increase in April runoff, which is two months after the February precipitation maximum. May and June runoff accounts for nearly 60% of the annual total, but

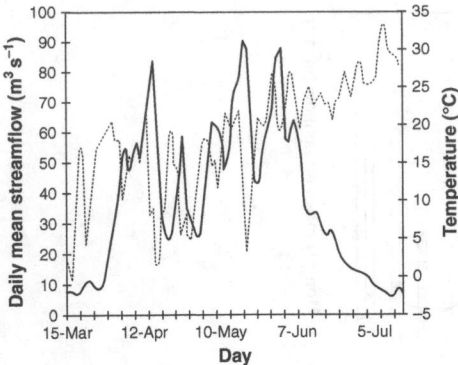

Fig. 6.11. Daily streamflow (solid line) for the Merced River at Pohono Bridge near Yosemite, California (38° N, elevation 1170 m), and daily maximum air temperature (broken line) for Yosemite National Park, California, for 15 March 2002 to 15 July 2002. The streamflow response to increasing and decreasing temperature is an implied link with snowmelt. (Streamflow data courtesy of the U.S. Geological Survey from their website at http://waterdata.usgs.gov/nwis/. Temperature data courtesy of NOAA's National Climate Data Center and the Oak Ridge National Laboratory, Carbon Dioxide Information Analysis Center from their website at http://cdiac.ornl.gov/epubs/ndp/ushcn/usa_daily.html.)

May and June precipitation is only 6% of annual amount. November through March accounts for 76% of annual precipitation and most occurs as snow that remains until the height of the snowmelt season in May and June.

Figure 6.11 shows the daily streamflow response to snowmelt for the Merced River during the 2002 melt season. The daily maximum air temperature at a nearby station is included as a general indicator of energy available to drive snowmelt (Seidel and Martinec, 2004). The mean daily streamflow follows the increasing and decreasing daily maximum temperature with a lag that persists until after 1 June when most snowmelt has occurred. It is notable that the three peak flows are each preceded by several days of gradually increasing flows, and the greatest difference in the three events is $7\,\mathrm{m^3\,s^{-1}}$.

6.12   Lakes as surface storage

Lakes, wetlands, and reservoirs constitute essential components of the terrestrial branch of the hydrologic cycle due to their ability to retain water and influence streamflow levels and water quality. At the same time, these water bodies serve as giant evaporation features that support an upward moisture flux limited only by available energy (Coe, 2000). The net hydroclimatic result is not the same for all these water features, and their regional hydroclimatic significance is highly variable.

From the perspective of the global hydrologic cycle, natural lakes are wide places in rivers. While this is a nearly true characterization for most small lakes, large lakes are more accurately characterized as a body of water surrounded by land. Lakes occur where the physical setting favors accumulation of water and the hydroclimatic environment supports persistence of the water body. Crater Lake, Oregon, occupies a collapsed volcanic crater. The Great Lakes of Canada and the United States occupy huge basins formed by isostatic adjustment, glacial excavation, and moraine and outwash deposition. Lake Tanganyika in East Africa occupies a portion of an expansive Rift Valley. Lake Eyre in central Australia occupies a shallow depression that intermittently contains water following seasonally heavy rainfall. Lakes can be either freshwater or saline depending on the hydroclimate and geomorphology of the site. Many lakes in arid regions become salty due to evaporation exceeding precipitation, which concentrates salts arriving from tributary streams. This condition is exacerbated when the lake is endorheic or does not have a natural outlet.

Two commonly recognized hydroclimatic functions of lakes are that they provide storage that reduces the time variability of flow in the rivers that drain them, and they increase evaporation by providing large evaporating surfaces. Exorheic lakes which are drained by outflowing rivers serve as natural reservoirs that reduce high streamflow and augment low streamflow to the extent that they can adjust the water volume retained in the basin. By providing an unlimited moisture source, lakes support higher evaporation rates than would occur if the lake area was land instead of water. However, the net result is not the same for all lakes. Lakes in the temperate latitudes of Europe and North America contribute to an increase in river flow because the precipitation falling on their surfaces is greater than the evaporation from them, but the increase is relatively small and varies between 5% and 15% of the inflow. Rouse et al. (2005) estimate evaporation in western Canada is 32% greater due to the presence of lakes than would occur from uplands alone. In the Southern Hemisphere, most lakes with outlets have a negative influence on river flow, lowering it considerably because of intense evaporation from the water surface. The lake-related decrease in streamflow for African rivers is 30% to 90%. In South America, the lake influence produces a lowering of river flow by 4 to 5 times (Vikulina et al., 1978).

Lakes integrate the precipitation and runoff processes operating in the lake watershed and the hydroclimatic relationship for the lake surface. The depth and areal extent of water in lakes are indictors of water gains and losses that are functions of changes in hydroclimatic variations for both the lake and the surrounding watershed. The water balance for a lake is expressed as an equation based on the conservation of mass as

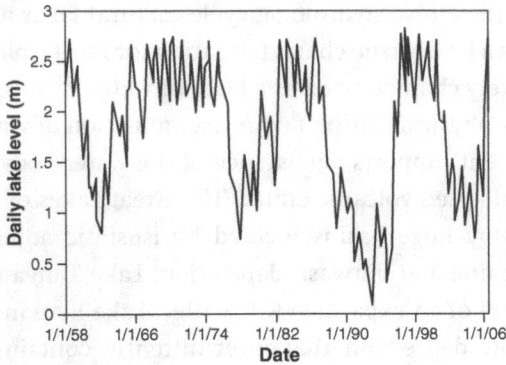

Fig. 6.12. Daily lake level relative to an elevation datum of 1885 m for Lake Tahoe, California/Nevada (39° N), for 1 January 1958 to 1 August 2006. (Data courtesy of the U.S. Geological Survey from their website at http://waterdata.usgs.gov/nwis/.)

$$\frac{dV}{dt} = Q_i + P_w - E_w - Q_o + Q_{ui} - Q_{uo} \qquad (6.14)$$

where V is the lake volume, Q_i is the surface inflow due to the relationship between P and ETa for the watershed, P_w is the precipitation on the lake surface, E_w is evaporation from the lake surface, Q_o is the outflow from the lake, Q_{ui} is the underground inflow to the lake, and Q_{uo} is the underground outflow (Street-Perrott, 1995).

Solving Equation 6.14 for different time intervals permits assessment of lake volume changes on streamflow ranging from daily to annual periods. A highly simplified measure is to use the change in lake level, which is related to lake volume and water storage, to assess the lake influence because the hydroclimatic role of lakes may change over the course of the year. For example, lakes in the western United States commonly reach high stages in the winter or spring while lakes in the eastern United States reach high stages in spring and early summer before beginning to decline. Seasonal changes in lake levels are superimposed on annual changes in lake levels responding to long-term hydroclimatic relationships which may account for abnormally high or low lake levels during periods of wet or dry years.

Figure 6.12 displays water level fluctuations for Lake Tahoe, California/Nevada (39° N). Lake Tahoe is the largest alpine lake in North America with a surface area of 487 km², a volume of 156 km³, and a drainage area of 819 km². Mountains surrounding the lake rise to 3297 m, and the lake surface elevation is 1886 m. Precipitation on the lake is greatest from November to March, but the dominant inflow to the lake is snowmelt runoff in May and June from the surrounding alpine watershed. The annual lake level time-series (Fig. 6.13) displays seasonal changes in the Lake Tahoe water level that are masked by the

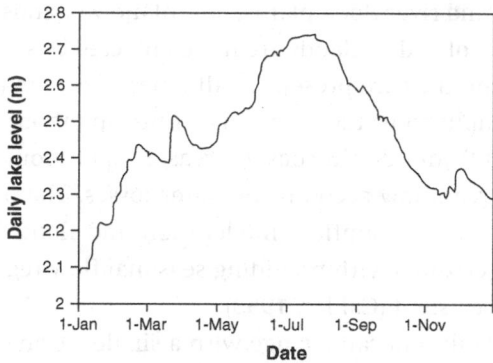

Fig. 6.13. Daily lake level relative to an elevation datum of 1885 m for Lake Tahoe, California/Nevada (39° N), for 1 January 1998 to 31 December 1998 showing the annual response to snowmelt inflow to the lake. (Data courtesy of the U.S. Geological Survey from their website at http://waterdata.usgs.gov/nwis/.)

long-term characteristics. Precipitation-related increases are evident in March, April, and December, but high sustained levels in late June, July, and August are products of snowmelt from high elevation areas of the watershed.

Figure 6.12 displays alternating periods of high and low levels of various durations for Lake Tahoe. The most prominent is the wet period from 1967 to 1975, and the dry period from 1986 to 1992. Not all years during these periods are wet or dry, but there is persistence toward wetness or dryness. The wet periods oscillate around a stationary mean, and the dry periods are characterized by abrupt declines and recoveries. Three of the four dry periods display similar minimum lake levels, but the event from 1986 to 1992 is longer in duration and reaches a lower minimum lake level than the other three periods.

6.13   Wetlands and runoff

Wetlands are characterized by the presence of standing water for some period during the growing season so that water is the dominant factor determining soil development and the types of plant and animal communities (Winter, 2000). Wetlands include swamps, marshes, bogs, tundra, and similar surface areas, and the structure of wetland ecosystems is closely linked to surface water depth and hydroclimatic fluxes (Bidlake, 2000). Wetlands occupy a greater variety of physical settings than lakes because wetlands occur on flat areas and in coastal locations that lack the depressions required by lakes. Slow surface runoff, restricted water infiltration, or groundwater discharge to or near the land surface promote wetlands (Winter, 2000). Many wetlands are associated with coasts, estuaries, deltas, and river flood plains, but inland wetlands extend from tropical to boreal areas. Isolated wetlands are not connected by streams to other surface water bodies, and they are present in all types of terrain (Winter and LaBaugh, 2003). Water input to wetlands occurs as precipitation, surface and groundwater inflow, and from tides in coastal areas. Evaporation, flow to groundwater systems, and streamflow account for water losses. The hydroclimatic influences of wetlands are streamflow moderation and flood storage. Although wetlands are often credited with providing seasonal flow regulation, this effect is marginal or non-existent (Calder, 1993).

Wetland hydrology is dynamic and can change with a single storm event or during an unusually hot day. Eddy covariance (see Section 4.8.4) is a useful technique for measuring turbulent surface fluxes in wetlands where reliable measurements of the rate of change of energy storage are difficult due to the inundated surface (Bidlake, 2000). Evapotranspiration is an especially important hydroclimatic component of wetlands, and vegetation effects on wetland evapotranspiration are the focus of numerous studies. While some studies report wetland evapotranspiration exceeds adjacent open water evaporation, other studies conclude wetland evapotranspiration is the same as or less than open water evaporation (Mao *et al.*, 2002). The highly variable composition of wetland plant communities and the variety of physical settings supporting wetlands contribute to promoting a range of energy partitioning responses that underlie varying evapotranspiration rates.

6.14   Reservoirs and streamflow

Reservoirs are artificial lakes produced by dams that impede streamflow. They function like lakes in almost all respects, but reservoirs are located by human design and are usually operated to achieve multiple goals. Reservoirs mimic natural lakes regarding their hydroclimatic influences on water storage and evaporation, and their water balance is expressed by Equation 6.14. However, flow regulation is more evident with reservoirs as the storage volume is managed to accommodate flood control, navigation, irrigation, hydroelectric production, recreation, water supply, water quality, fish and wildlife requirements, and other conservation purposes. The regulated streamflow masks much of the natural discharge variability and alters the natural discharge quantity to such an extent that observed streamflow may bear little resemblance to the natural flow.

Reservoirs are typically operated as storage or run-of-river impoundments. Storage reservoirs have sufficient capacity to offset seasonal streamflow fluctuations and provide relatively constant flow throughout the year. Run-of-river impoundments use low dams to create a pool that stores a daily or weekly

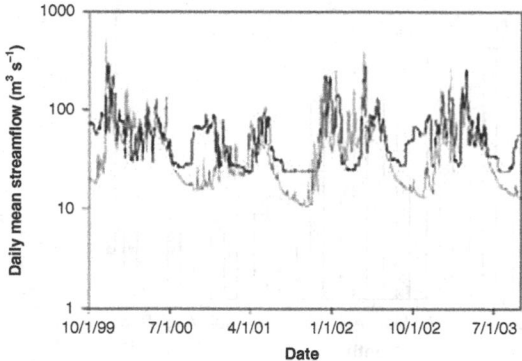

Fig. 6.14. Daily mean streamflow for the North Santiam River at Niagra, Oregon (45° N), downstream from Detroit Reservoir. The bold line is outflow due to reservoir operations, and the gray line is the gauged reservoir inflow. (Data courtesy of the U.S. Geological Survey from their website at http://waterdata.usgs.gov/nwis/.)

water volume used primarily for hydropower production or navigation. These dams cause few alterations to natural streamflow. Storage reservoirs have the greatest impact on the hydrologic cycle and streamflow, and large storage reservoirs alter streamflow the most.

The North Santiam River in western Oregon displays the impact of a storage reservoir on natural streamflow (Fig. 6.14). Detroit Dam (45° N) creates a multi-purpose reservoir with a storage of $0.396 \, km^3$. The $1132 \, km^2$ watershed above the dam produces a mean daily streamflow of $59 \, m^3 \, s^{-1}$. Reservoir operations reduce high flows as seen by the $205 \, m^3 \, s^{-1}$ difference between inflow and outflow peaks in late 1999 and the $100 \, m^3 \, s^{-1}$ difference in April 2002. Maintaining low outflows greater than inflows is consistently demonstrated during the exceedingly dry period from July 2000 to November 2001. Precipitation from October to February was as little as 25% of average within the watershed, and the 12 months from October 2000 through September 2001 were the second driest period on record for many Oregon stations.

Complex influences on streamflow are evident when reservoir storage is operated to support multiple purposes and numerous water diversions are factors. Streamflow in the Sacramento River at Freeport, California, (38° N) is altered by multiple human influences on streamflow related to the operation of 43 major reservoirs, a major water diversion into the basin, and several diversions out of the basin (Shelton, 1995). The Sacramento River watershed at Freeport drains nearly $68\,000 \, km^2$, and records of both gauged and calculated streamflow are available for this site for the years 1949 to 1994. The gauged record at Freeport is combined with the gauged record for the Yolo Bypass to represent regulated

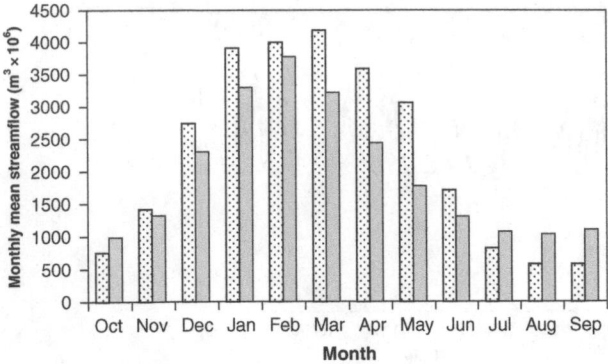

Fig. 6.15. Mean monthly streamflow for the Sacramento River at Freeport, California (38° N). Shaded columns are the observed flow regulated by reservoir operations and irrigation and flood diversions. Stippled columns are the calculated natural flow determined by adjusting observed flow for the influences of reservoir operations, evaporation, and irrigation and flood diversions. (Data courtesy of the U.S. Geological Survey from their website at http://waterdata.usgs.gov/nwis/.)

streamflow. The Yolo Bypass is a flood control feature that protects the city of Sacramento. Water is diverted from the Sacramento River above the city and flows through a levee-bounded area that rejoins the river below Freeport. This facility is dry for many months each year, and it is used only when flows exceed a potentially damaging stage which does not occur in all years. Calculated stream-flow is employed as an estimate of natural streamflow by adjusting measured streamflow for changes in reservoir storage, known or estimated channel losses, evaporation, and water diversions into or out of the watershed. These adjustments attempt to account for human intervention in the runoff process.

The average annual calculated streamflow for the Sacramento River is $27\,136\,km^3$, and 79% of this flow occurs in December through May (Fig. 6.15). Winter precipitation and spring snowmelt in the alpine regions of the watershed contribute to the March calculated streamflow being 7.7-times greater than September streamflow. The 46 years of regulated streamflow indicate an average annual value of $23\,417\,km^3$. The difference between calculated and regulated streamflow is due largely to diversions for irrigation and urban use which increase evapotranspiration. This influence is evident during July through October when regulated streamflow exceeds calculated streamflow.

Annual variations in the relationship between calculated and regulated streamflow (Fig. 6.16) reflect the complex hydroclimate of the Sacramento River watershed and the changing character of human influences during the 46 years. Both data series show 1983 is the wettest year and 1977 is the driest year. The difference between calculated and regulated streamflow is smallest in 1977 and is

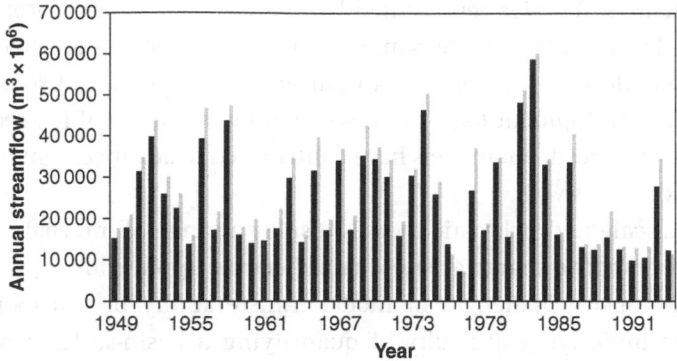

Fig. 6.16. Annual streamflow for the Sacramento River at Freeport, California (38° N). Solid columns are the observed flow regulated by reservoir operations and irrigation and flood diversions. Gray columns are the calculated natural flow determined by adjusting observed flow for the influences of reservoir operations, evaporation, and irrigation and flood diversions. (Data courtesy of the U.S. Geological Survey from their website at http://waterdata.usgs.gov/nwis/.)

greatest in 1978. Calculated streamflow is larger than regulated streamflow in all years except 1976 and 1994. These are the second and third driest years in the 46-year record, and reservoirs were drawn down with the expectation that winter precipitation would replenish storage. Meager winter precipitation in 1977 did not replenish reservoir storage, and 1977 regulated streamflow was reduced to 29% of the average annual regulated streamflow. Abundant precipitation in 1978 produced a striking contrast as regulated streamflow was 44% of the average and calculated streamflow was 136% of the average and the 11th highest annual calculated streamflow in the 46 years. The streamflow disparity in 1978 reflects efforts to refill surface reservoir storage and to increase diversions to restore depleted soil moisture for agriculture (Shelton, 1995).

Other influences of storage reservoirs include disrupting natural flood cycles, disconnecting rivers from wetlands and floodplains, disrupting fish migrations, altering riverine and terrestrial habitats, and altering downstream sediment deposition. Sediment accumulation in storage reservoirs reduces their water volume, and the reduced sediment delivery by rivers to the oceans alters coastal processes (Syvitski *et al.*, 2005).

6.15   Watershed models

The time and space variability of the physical processes that transform precipitation into runoff and the equations that express these physical processes are organized into models that simulate the watershed response. The spatial and temporal scales represented by models vary across a broad range that address flows due to a single rainfall event on an area of a few hectares to the continuous flow over a year for a basin of several thousand square kilometers. Model development has progressed as understanding of hydroclimatic processes has advanced, but models have contributed to advancement in understanding processes as well.

Watershed-scale hydroclimatic models display a range of internal complexities and time response characteristics that reflect the application for which the model is developed. Challenges of model transferability or transportability involving the underlying difficulty of quantifying a basin-scale response to small-scale spatial complexity of physical processes have resulted in development of numerous models for specific applications. The result is an array of watershed models with various levels of sophistication and complexity, but they display shared traits that facilitate categorization. One common grouping is empirical, conceptual, and physical models based on the purpose and characteristics of the model structure. Empirical models relate climate inputs to hydrological outputs using simple data requirements but without prescribing the physical processes involved. Conceptual models represent the watershed as an idealized representation of storages. They perform well at daily and monthly time steps and are responsive to changes in climatic inputs while maintaining the integrity of watershed processes. Physical models strive to incorporate relevant watershed physical processes in a linked system, and these models are founded on the fundamental water balance relationship among precipitation, evapotranspiration, and runoff. Physical models are data intensive and operate at fine temporal scales (Arnell, 1996).

Other model characterizations address specific details regarding assumptions underlying the model structure. Models can take the form of either linear or non-linear structures depending on their statistical or system theory formulation. Stochastic models simulate the conversion of watershed precipitation into runoff using probability distributions of hydroclimatic variables, and they contain a large random component. Deterministic models strive to quantify the actual physical processes involved in transforming precipitation into runoff. The runoff simulated by a deterministic model given a specific precipitation input is fixed or determined by the physical processes represented in the model. The spatial variability of hydroclimatic variables is incorporated in watershed models by employing either lumped or distributed approaches to address heterogeneity. Lumped models treat the basin as a homogeneous whole that can be represented by a single set of parameters. This approach assumes there is no variation across the watershed. Distributed models divide the basin into separate units to achieve spatial representation, but there is some level of approximation and smoothing related to the size of the grid (Mansell, 2003). Semi-distributed models employ a combination of lumped and distributed model characteristics by treating separate units as homogeneous areas to simulate runoff. Physically based distributed models are popular because they are realistic and have a theoretical advantage that makes them applicable over a wide range of physical environments (Jones, 1997; Ward and Robinson, 2000).

An important feature of watershed models is their coupling of external forcing factors and internal dynamics. They transform climatological forcings, in particular energy and moisture, into the watershed response in the form of runoff. Climate variability and change is imparted to the models through altered climatic forcings, and physical alterations of the watershed are simulated by changes in the internal dynamics of the terrestrial hydrologic system. Consequently, watershed models are an economically feasible way to assess land-use changes and the only possible way to quantitatively evaluate the impact of climate variability and change on the runoff process.

A widely used application for watershed models is to improve understanding of the runoff process at various time and space scales. The goal for this modeling endeavor is the ability to reliably estimate water quantity and quality conditions. Parameterization issues regarding watershed models are widely reported (e.g. Grayson et al., 2002; Sankarasubramanian and Vogel, 2002; Oudin et al., 2005) along with results of watershed simulations at sites around the world (e.g. Montaldo et al., 2004; Miller et al., 2005). However, one must be aware that the application of hydroclimatic models involves a high degree of uncertainty that is a consequence of insufficient data availability and quality, high spatial heterogeneity, high temporal variability, and limited knowledge of the physical processes operating at various scales incorporated in the model structure (Menzel et al., 2002).

Contemporary concerns for climate variability and climate change influences on the water supply have focused attention on coupling hydroclimatic watershed models with GCM output, regional climate model (RCM) output, and satellite imagery. However, watershed models require hydroclimatic data on scales far smaller than those generally available from higher resolution GCM, RCM, and satellite data. Techniques for downscaling coarse-resolution data to watershed-scale modeling needs have progressed with the development of complex algorithms for data downscaling and the adaptability of robust watershed models for utilizing downscaled data in simulating the runoff process. Watershed models linked with downscaled high-resolution data include SWAT (Soil Water Assessment Tool), TOPMODEL (TOPography based hydrological MODEL), and VIC (Variable Infiltration Capacity) (Beven, 1997; Lohmann et al., 1998; Jha et al., 2004).

GEWEX studies initiated by the World Climate Research Program have promoted linking watershed models and high-resolution data to improve understanding of atmospheric dynamics and thermodynamics and atmospheric interactions with the Earth's surface (Randall *et al.*, 2003). A major emphasis of this program is better understanding of the contribution of land surface processes to the sources and limits of predictability in the hydrologic cycle. GEWEX embraces seven regional studies from different climatic regions of the world. Study areas are the Asian monsoon region, Australia's semi-arid Murray–Darling Basin, the maritime modified continental climate of the Baltic Sea region in northern Europe, the cold continental climate of the Mackenzie River Basin in Canada, the variable continental climate in the Mississippi River Basin in the United States, and the tropical climate of the Amazon River Basin and the extratropical humid climate of the La Plata River Basin in South America (Lawford, *et al.*, 2004).

The GEWEX Continental-Scale International Project (GCIP) in the Mississippi River Basin was designed to develop regional-scale coupled atmosphere–land models using remote sensing data and data from existing meteorology and hydrology networks. The purpose of the coupled models was to predict seasonal, annual, and spatial variability of water resource changes in the basin and to produce a number of special datasets (Leese *et al.*, 2003). The GEWEX Americas Prediction Project (GAPP) is an extension of GCIP. GAPP uses GCIP data and techniques and expands the study area to the Pacific coast and parts of northern Mexico (Lawford *et al.*, 2004).

The Mackenzie River Basin GEWEX study focuses on the large amount of freshwater in Canada and cold region processes. High-latitude energy and water cycles are sensitive to land surface processes, and RCMs and GCMs simulate too much rainfall and evaporation in this region (Rouse *et al.*, 2003). By focusing on the water cycle of the Mackenzie River Basin, the Canadian GEWEX program strives to improve understanding and modeling of high-latitude water and energy cycles that can aid in economic and policy decisions (Stewart *et al.*, 1998). The Baltic Sea project is described by Raschke *et al.* (2001), and the other GEWEX projects are summarized by Lawford *et al.* (2004).

Review questions

6.1   What is the nature of the relationship between precipitation and interception?

6.2   What is the difference between throughfall and throughflow?

6.3   What is the role of precipitation in determining the quantity of runoff that arrives at the stream channel by each of the flow routes identified as surface runoff, throughflow or interflow, and groundwater?

6.4 What are the commonly recognized components of a stream hydrograph? What information does each component convey about the runoff process?

6.5 What factors are thought to influence the shape of a hydrograph?

6.6 Why is spatial heterogeneity the key to understanding runoff generation?

6.7 Why is the radiation balance equation for a snow surface commonly depicted in a more complex form than the radiation balance for other land surfaces?

6.8 How are the ripening of a snowpack, the output phase, and a snowmelt flood related?

6.9 What contributes to making modeling of snowmelt a challenging problem?

6.10 How are the water balance components for a lake different from those for land?

6.11 What is meant by "lumped" and "distributed" watershed models?

6.12 What is the major contribution provided by watershed models?

7

Hydroclimate spatial variations

7.1   Spatial scale

Hydroclimate displays notable variations from place to place in response to multiple influences exerted by the Earth's climate system on the hydrologic cycle. These spatial variations exist on a continuum and produce an infinite number of hydroclimatic expressions. Various constructs are employed to categorize segments of the continuum in an effort to achieve understanding of hydroclimate, and these constructs embrace scale issues that are connected fundamentally with the measurement process. Scale manifests itself in heterogeneity explicitly through the measuring instrument and implicitly through the physical environment (Cushman, 1986). In many instances, the spatial variability of one process greatly exceeds the spatial variability of another process even though both are important to hydroclimate. This is illustrated by contrasting the spatial variability in atmospheric forcing apparent in precipitation heterogeneity with the mosaic-like pattern of spatial variability in runoff due to changing boundary conditions associated with the distribution of biotic, pedologic, and geomorphic features (Shelton, 1989).

Meteorologists concerned with dissimilar scales of atmospheric behavior ranging from global to local reduce the entire range of phenomena into a small number of categories based on observational and analytical techniques needed to describe and understand the phenomena. Recognizing that short time-scale phenomena tend to occur on a small space-scale and long-term phenomena affect large areas, meteorologists focus on weather analysis at the synoptic or regional scale that embraces units the size of nations in Europe or states in the United States. Climatologists employ global, regional, and local spatial scales somewhat similar to those employed in weather analysis, and they

212

utilize many of the observational and analytical techniques used by meteorologists to understand the present state of the climate system at a given location. However, different techniques are needed to assess global and regional climates even though basic processes occur at all spatial scales. It cannot be assumed that all forms of climatic data for a specific location can be acquired with the same precision and resolution for an area the size of a state or an area the size of a continent. Climatology expands the time scale employed in weather analysis to focus on monthly, annual, and decadal states of the climate system.

Hydroclimatic processes display enormous spatial variability at the Earth's surface due to differences in soils, geology, land use, and climatic inputs. Hydrology utilizes microscale to macroscale perspectives to understand the variability of hydrological phenomena. Although all hydrological phenomena are governed by the same laws, combined microscale hydrologic behavior does not necessarily represent hydrologic behavior at the mesoscale or macroscale. Consequently, microscale observations and macroscale hydrological theory must accommodate the spatial integration of heterogeneous non-linear interacting processes as they are merged into problem-solving applications (Ward and Robinson, 2000).

Examination of hydroclimatic spatial variability at global, regional, and local scales is aided by the dual constructs of climate and the hydrologic cycle introduced in Chapter 1. These constructs provide a framework for examining basic hydroclimatic processes common at all scales.

7.2   Global atmospheric hydroclimate

Global-scale hydroclimate includes the entire climate system and the interactions among subsystems. The seasonal migration of maximum solar radiation inputs between the Northern Hemisphere and the Southern Hemisphere triggers seasonal changes in global temperature, pressure, winds, and precipitation. These rhythms within the climate system impart identifiable characteristics to the hydroclimatic signature for a specific location. However, complexity is added to the hydroclimatic signature of a place by the fluxes of energy and mass introduced into the lower atmosphere from terrestrial and oceanic surfaces. Land areas comprise 29% of the Earth's surface area, but the diverse and variable character of land areas strongly influences the rates at which heat and moisture enter and exit the lower atmosphere. The variability of these fluxes affects the distribution of heat and moisture in the atmosphere, the climate of a given place, and the nature of global weather and climate patterns (Mather and Sdasyuk, 1991). Monthly, seasonal, and annual time scales are employed to develop understanding of global hydroclimate.

Atmospheric motion is a response to unequal heating of the Earth's surface by solar radiation and thermal radiation emitted by the atmosphere and the Earth's surface. The general circulation of the atmosphere and the oceans redistribute energy to overcome the unequal distribution of the initial energy input. At the same time, atmospheric general circulation provides the transport mechanism for moisture to replenish continental precipitation. Global patterns portray the most evident interconnection of climate of the first kind state variables, and the atmospheric transport that is the foundation of the atmospheric branch of the hydrologic cycle emerges from the global patterns of the state variables.

7.3   The radiation balance

The simplest form of the global radiation balance is portrayed in Figure 2.3. Details of the spatial characteristics of the radiation balance are provided by observational and remote sensing data discussed in Chapters 3, 4, and 5. The radiation balance at the top of the Earth's atmosphere and the radiation balance at the Earth's surface are important to hydroclimatology because they emphasize the unequal distribution of energy that is the primary forcing for the climate system. Furthermore, the global condition of the radiation balance serves as a benchmark for assessing the variations in the radiation balance for regions and for individual sites. The nature of the global radiation balance is an important initial step in understanding the related regional and local components discussed later in this chapter.

The Earth's radiation balance is driven by the annual receipt of solar radiation received at the upper edge of the Earth's atmosphere minus the reflected shortwave energy that constitutes the Earth's albedo. The combined downward and upward shortwave fluxes form the net shortwave radiative flux at the top of the atmosphere (Fig. 7.1). The solar radiation input component is driven by differences in the intensity and duration of solar radiation produced by the annual cycle of Earth–Sun geometry, and the Earth's albedo modulates the solar input. Averaged globally, the incoming solar radiation is $342\,\mathrm{W\,m^{-2}}$, the reflected solar radiation is $107\,\mathrm{W\,m^{-2}}$, and the net shortwave balance is $235\,\mathrm{W\,m^{-2}}$.

Global albedo increases monotonically from equatorial to polar regions. Minimum values of 20% and less occur over the tropical oceans where clouds are sparse, while the albedo of subtropical continents is around 30%. Albedo is a maximum at polar latitudes exceeding 60% due to snowcover, clouds, and large solar zenith angles (Hartmann, 1994). Albedo variability due to land and water

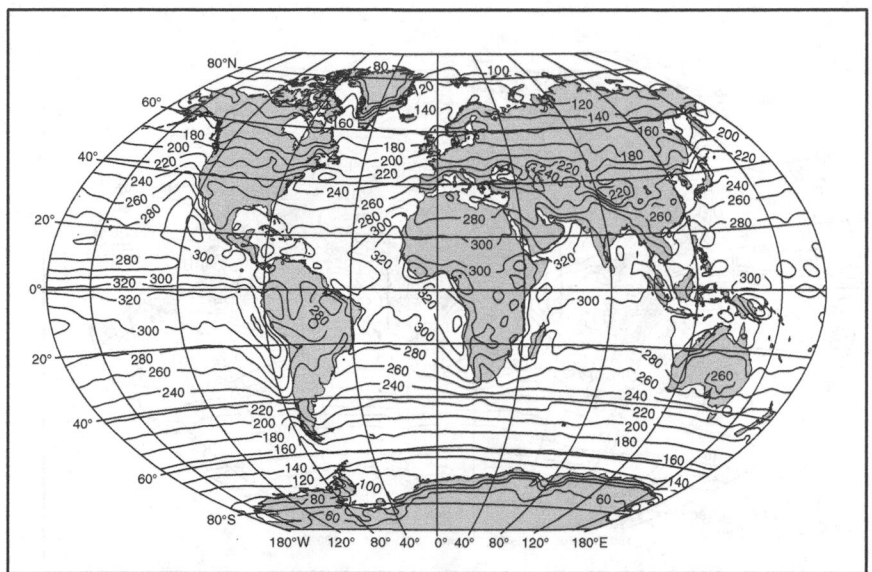

Fig. 7.1. Global annual mean net solar radiation at the top of the atmosphere. Units in Wm^{-2}. (NCEP Reanalysis data courtesy of NOAA/OAR/ESRL PSD, Boulder, Colorado, USA, from their website at http://www.cdc.noaa.gov/.)

contrasts and the presence and absence of clouds accounts for much of the spatial complexity between $40°N$ and $40°S$ evident in the net shortwave flux in Figure 7.1. Expressed as a net value, the atmosphere above the equatorial region receives significantly greater solar radiation than the atmosphere above the polar regions. The systematic solar radiation decrease with increasing latitude indicates the Earth–Sun geometry dominance on the solar radiation input before the solar radiation begins its transit through the Earth's atmosphere.

Upwelling longwave terrestrial radiation at the top of the Earth's atmosphere (Fig. 7.2) is the thermal radiation emitted by the Earth–atmosphere system. The influence of the solar radiation input on the distribution of outgoing longwave terrestrial radiation is evident in the inverse relationship between outgoing terrestrial radiation and latitude. However, the latitudinal gradient is much weaker for outgoing terrestrial radiation than for solar radiation. The highest values of $280\,Wm^{-2}$ and greater occur over the warm subtropical deserts and over tropical oceans where clouds occur infrequently. This pattern is a response to outgoing thermal radiation being related to the temperature of the emitting surface. The poles produce the lowest values of $180\,Wm^{-2}$ and less. Cold poles and cold cloud tops in tropical latitudes produce lower values while warm surfaces and relatively clear sky regions produce higher values.

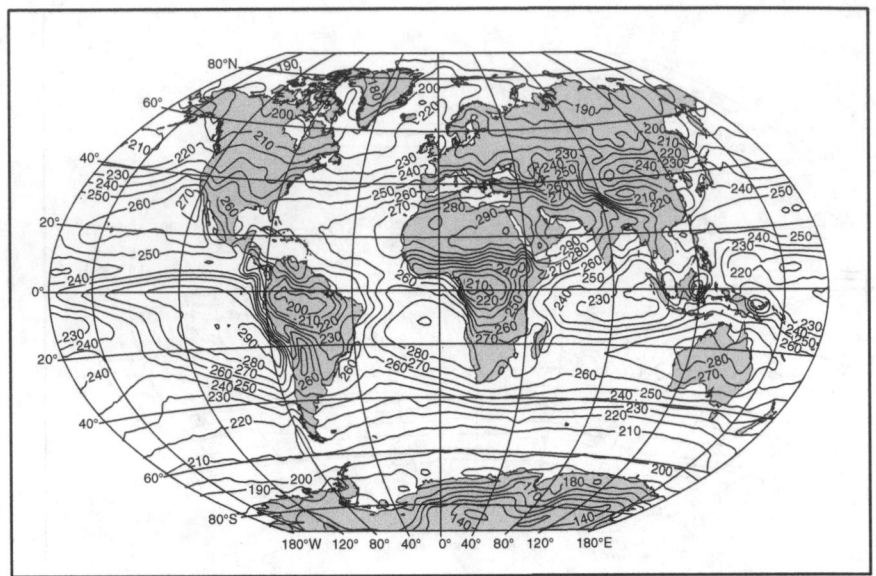

Fig. 7.2. Global annual mean outgoing longwave radiation at the top of the atmosphere. Units in $W\,m^{-2}$. (NCEP Reanalysis data courtesy of NOAA/OAR/ESRL PSD, Boulder, Colorado, USA, from their website at http://www.cdc.noaa.gov/.)

The atmospheric influence on radiation exchanges and the different thermal responses of land and water account for much of the greater complexity of the outgoing terrestrial radiation pattern. Albedo influences contribute to quantitative differences at specific latitudes and energy redistribution within the climate system supports larger outgoing terrestrial radiation fluxes at high latitudes compared to solar radiation values at similar latitudes. The global average thermal energy flux at the upper edge of the Earth's atmosphere is $235\,W\,m^{-2}$, which equals the net solar radiation flux at this level. The zonal mean top of the atmosphere radiation balance (see Fig. 2.6) highlights the radiation balance excess in the tropics and the deficit at the high latitudes.

Absorption and reflection of solar radiation by the atmosphere depletes the quantity of solar radiation reaching the Earth's surface. The net shortwave flux at the Earth's surface is represented by the first two variables on the right side of Equation 2.11. The spatial distribution of this quantity is an inverse relationship with latitude evident in conditions at the top of the atmosphere (see Fig. 7.1), but with complexity added due to solar radiation being absorbed by the atmosphere and reflected by the atmosphere and the Earth's surface. Since water has a lower albedo than land at high sun angles, the equatorial oceans have a lower albedo than adjacent continents and a larger available solar radiation flux.

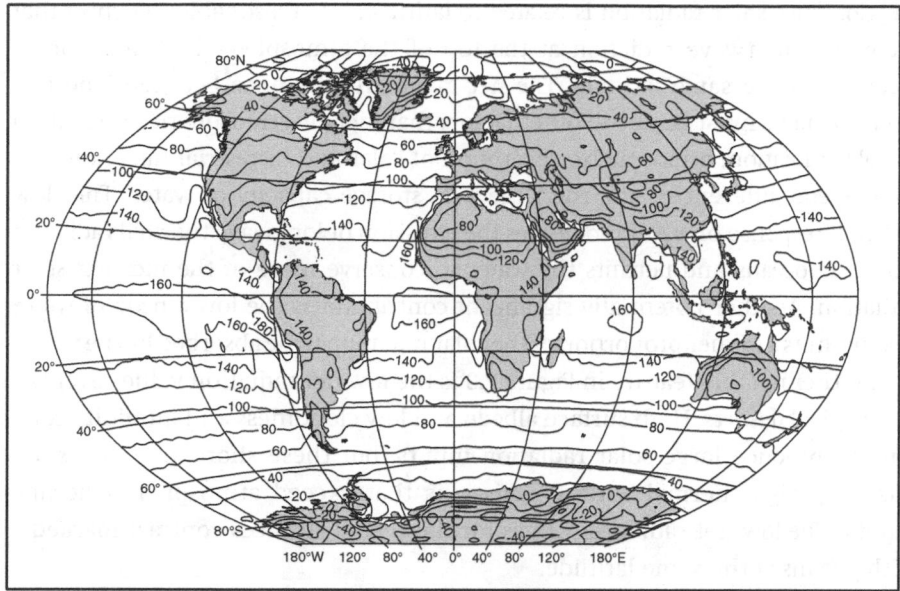

Fig. 7.3. Global annual mean net radiation at the Earth's surface. Units in $W\,m^{-2}$.
(NCEP Reanalysis data courtesy of NOAA/OAR/ESRL PSD, Boulder, Colorado, USA,
from their website at http://www.cdc.noaa.gov/.)

The concept of net radiation expressed in Equation 2.11 is a useful paradigm
at the Earth's surface for defining available energy to drive non-radiative processes. Figure 7.3 shows that over most of the globe the net surface radiation is
downward, but polar regions experience a net radiation loss. Variations in the
global pattern of net radiation are related to the spatial variation in solar
radiation, cloud cover, and the albedo and thermal differences of land and
water surfaces. Two readily identified characteristics of the spatial variation in
net radiation result from the interplay of radiation exchanges. A systematic
decrease in net radiation with increasing latitude demonstrates the dominant
role incoming solar radiation plays in quantifying annual net radiation. The
radiation flux decreases from values of $160\,W\,m^{-2}$ or greater near the equator to
values of $-40\,W\,m^{-2}$ poleward of 80° latitude. The intensity and duration of
solar radiation at the surface is related to latitude through fundamental
Earth–Sun geometry responsible for delivering the solar beam and the Earth's
rotation which distributes the energy longitudinally. The relationship between
net radiation and latitude is most evident over the world's oceans.

A second feature apparent in Figure 7.3 is that net radiation is greater over
oceans than over land at the same latitude. Net shortwave radiation varies due to
spatial differences in atmospheric attenuation of shortwave radiation and surface

albedo. Since solar radiation is related to latitude, it is reasonable to expect that incoming shortwave radiation at the top of the atmosphere is similar for all locations at the same latitude. The net radiation difference between land and water surfaces must be due to other factors that include atmospheric attenuation of solar radiation, atmospheric absorption of solar and terrestrial radiation, surface albedo differences, and the large heat storage capacity of water. The slow thermal response of water moderates the emission of longwave thermal radiation from the oceans and permits the water to conserve more of the incident solar radiation. Another potentially significant contributor is the low albedo of water that permits a higher proportion of the solar radiation to be absorbed by the water.

Another notable feature in Figure 7.3 is the low net radiation values over the subtropical deserts. High surface albedo and low cloudiness and humidity combine to produce large solar radiation inputs, but these shortwave fluxes are offset by high thermal surface emissions that exceed atmospheric thermal inputs. The low net radiation values for these land surfaces contrast markedly with oceans at the same latitude.

7.4   Temperature

The radiation balance and the energy balance are linked conceptually by net radiation. In the energy balance (Equation 2.12), positive net radiation quantities are partitioned among non-radiative exchanges, and negative net radiation is sustained by energy provided by non-radiative exchanges. When net radiation is positive, a major portion of the available energy is used for evaporating water as long as water is present. In the absence of water, the proportion of net radiation allocated to sensible heat increases and the temperature of the surface and the overlying air increases as energy is transmitted by conduction and convection. Energy transfers in the soil are limited because they occur by conduction, which is a relatively slow process. Water in the soil can accelerate soil heat flux, but the presence of water in the soil can also promote a greater latent heat transfer and reduce the proportion of net radiation available for soil heat flux.

Ambient air temperature is a measure of the rate of molecular motion, or kinetic energy, which is equated with sensible heat flux. The temperature registered by a thermometer is an indication of the quantity of sensible heat present in the substance in which the thermometer is placed. An imbalance in radiant energy fluxes leads to non-radiative energy gains or losses at the surface that result in surface temperature changes. In this way, the radiation balance and energy balance equations summarize the important link between net radiation and temperature.

Global average annual temperature patterns reflect variations in net radiation due to the link between net radiation and non-radiative fluxes. The spatial pattern of temperature is not an exact replication of net radiation, but the strong influence of net radiation is apparent in an equator to pole temperature gradient. The most obvious contrast between the pattern of net radiation and temperature is due largely to the difference in the thermal response of surface materials, and it is manifested by the more rapid heating and cooling of land areas compared to water bodies. The temperature response over continents is complex and benefits from examination using representative months and isotherms or lines joining points of equal temperature. Isotherms sacrifice some data resolution, but they facilitate identifying generalizations from large data displays.

Maps of mean January temperature (Fig. 7.4) and mean July temperature (Fig. 7.5) portray seasonal temperature variations resulting from changes in the energy endowment determined by radiant energy fluxes. The seasonal variations are superimposed on several general characteristics. The surface temperature is greatest in the equatorial zone where it exceeds 26 °C across a broad latitudinal band. Outside the equatorial region, temperatures decrease with increasing latitude and the relationship is observed best over the oceans and in the Southern Hemisphere. The lowest temperatures are confined to polar latitudes, and especially continents at high latitudes, due to the minimum annual solar radiation in polar regions.

Additional large-scale patterns are evident when the January and July patterns are compared. The latitudinal temperature gradient is greatest in the winter hemisphere because the noon solar angle and day length both decrease as latitude increases. The effects of the reduced solar radiation are amplified by the large continental expanse in the Northern Hemisphere and the temperature gradient is greatest in the Northern Hemisphere winter. The behavior of isotherms as they extend across continents in the middle and high latitudes is another significant characteristic. In the winter hemisphere, isotherms shift equatorward over continents in response to the more rapid cooling of land compared to water. In the summer hemisphere, isotherms shift poleward as the continents become warmer than adjacent oceans at the same latitude. This has the effect of making it appear that isotherms bend toward the equator during the winter and toward the poles in the summer.

7.5   Atmospheric humidity

Atmospheric humidity is constantly changing as water vapor enters and leaves the atmosphere at the Earth's surface. These water vapor fluxes are

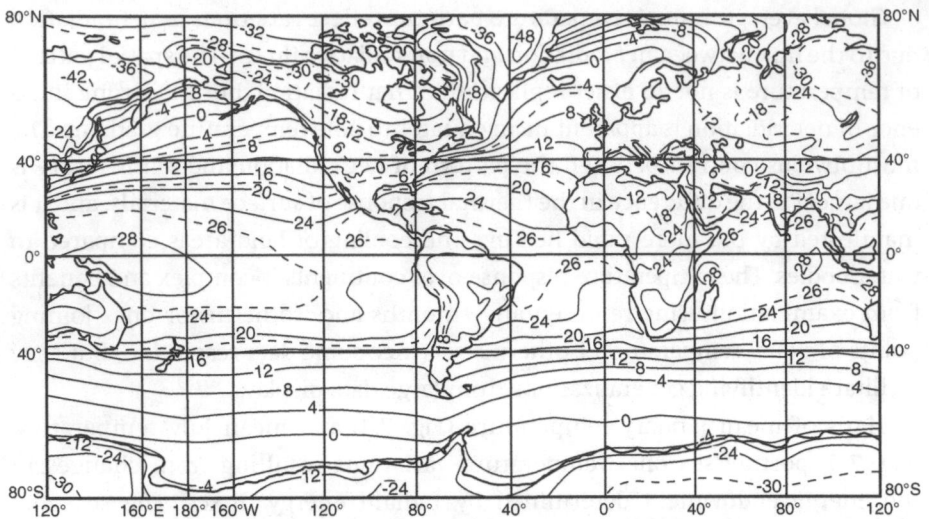

Fig. 7.4. Global January mean surface air temperature. Units in °C. (From Peixoto and Oort, 1992, Figure 7.4a. Used with kind permission of Springer Science and Business Media.)

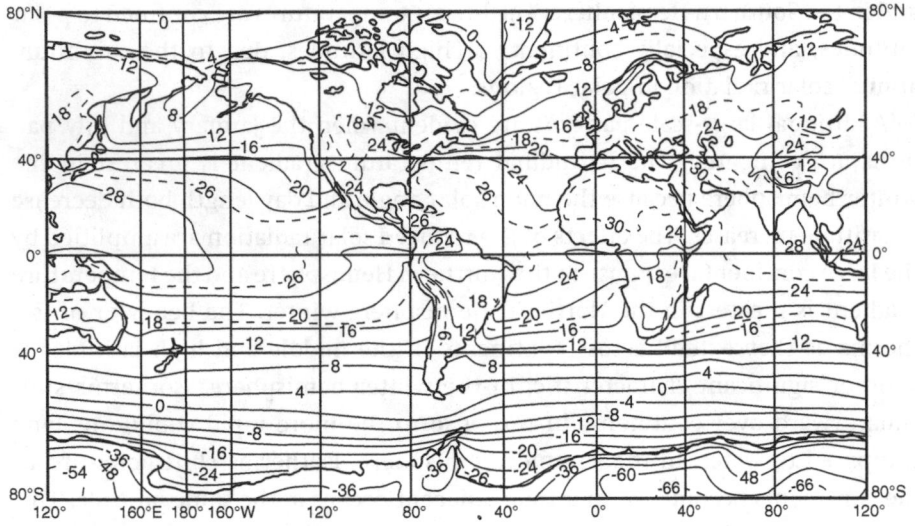

Fig. 7.5. Global July mean surface air temperature. Units in °C. (From Peixoto and Oort, 1992, Figure 7.4b. Used with kind permission of Springer Science and Business Media.)

coupled with energy exchanges that account for significant energy transfers between the Earth's surface and the atmosphere (see Fig. 2.3). The distribution of atmospheric moisture is complex, but one evident control is that moisture condenses out of saturated air (Allen and Ingram, 2002). Consequently, spatial variations in atmospheric moisture provide important insights into the atmospheric branch of the hydrologic cycle. The amount of water vapor in the atmosphere is crucial for quantifying the atmospheric water balance and for understanding the quantity of water vapor present to support the precipitation formation process. In addition, the latent heat content of atmospheric water vapor can have a profound effect on the efficiency with which heat is transported horizontally by the atmosphere (Pierrehumbert, 2002). The spatial distribution of atmospheric water vapor is a composite of surface moisture sources, atmospheric moisture sinks, and available energy to drive phase changes of water at the surface.

Methods for expressing atmospheric humidity introduced in Section 3.6 include specific humidity, which is the ratio of the mass of water vapor to the mass of humid air. Specific humidity is particularly useful for expressing the global distribution of atmospheric water vapor because its value is not influenced as the atmosphere expands or contracts in response to pressure changes. Also, specific humidity is not temperature dependent, which makes it a good indicator for comparing atmospheric water vapor at locations with different air temperatures. Specific humidity (q) is represented by

$$q = \frac{m_v}{m} = \frac{m_v}{m_v + m_d} \tag{7.1}$$

where m_v is the mass of water vapor in kg, m is the mass of the atmosphere in kg, and m_d is the mass of dry air in kg, for a specific volume. Specific humidity as the ratio of masses is dimensionless, but the convention is to express it in units of g of water vapor per kg of moist air.

Figure 7.6 shows the annual mean distribution of specific humidity at the Earth's surface. Other commonly used specific humidity expressions are the value for a layer of air between two pressure levels or the vertical-mean specific humidity for the entire atmospheric column. The quantities are different in each representation, but the general distribution patterns are similar. Several distinctive characteristics are evident in the global pattern of annual mean specific humidity at the Earth's surface. Two prominent features are the gradient of decreasing quantities from the equator to the poles and the ocean–land contrast at the same latitude.

High humidity in equatorial regions is expected due to the coincidence of abundant energy and moisture at these latitudes. High saturation vapor pressures in equatorial regions support the capacity of the atmosphere to retain large water vapor volumes. At high latitudes, the atmosphere is cool, saturation vapor pressures are lower, and the volume of atmospheric water vapor is low. A general pattern of zonal symmetry is evident in both hemispheres, but the

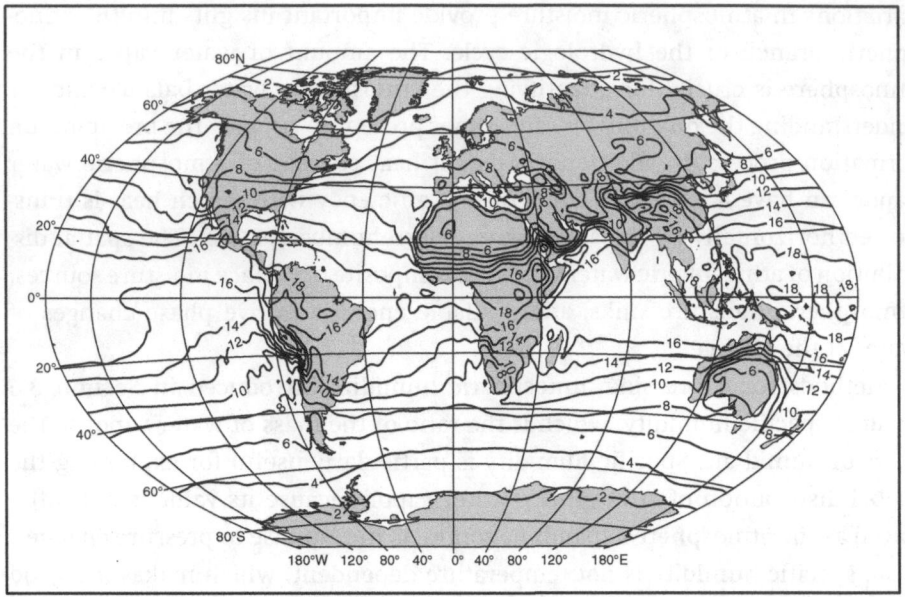

Fig. 7.6. Global annual mean specific humidity. Units in g kg^{-1}. (NCEP Reanalysis data courtesy of NOAA/OAR/ESRL PSD, Boulder, Colorado, USA, from their website at http://www.cdc.noaa.gov/.)

pattern is more regular in the Southern Hemisphere where the large ocean expanse presents a consistent surface that promotes expression of atmospheric moisture limited only by available energy to drive the upward flux. The presence of large land areas in the Northern Hemisphere disrupts the zonal pattern and produces a more complex scheme.

The most apparent influence superimposed on the pattern of decreasing specific humidity with increasing latitude is the contrasting effect of ocean and continental surfaces. As a general rule, specific humidity is greater over oceans than over continents at the same latitude. The high specific humidity values for the Amazon Basin in South America and the Congo Basin in Africa are major exceptions. The zonal Southern Hemisphere pattern due to ocean-dominated surfaces contrasts markedly with the Northern Hemisphere complex specific humidity pattern shaped by a combination of oceans and continents. Specific humidity isopleths are deflected near the western and eastern coasts of continents by topography and the presence of warm and cold ocean currents. Deserts in North Africa and Central Australia are evident by their low specific humidity relative to their respective zonal averages. The high-latitude expanse of North America and Eurasia contributes to a prominent reduction in specific humidity for the core regions of these continents (Peixoto and Oort, 1992).

Surface air generally contains more moisture during the high-sun season when available energy is most abundant, and expected seasonal changes in specific humidity driven by the energy balance are magnified by land–ocean differences. Specific humidity in January (Fig. 7.7) for a specific latitude is greater in the Southern Hemisphere than in the Northern Hemisphere. The relationship is reversed in July (Fig. 7.8). These patterns are most evident over the oceans where seasonal changes are smallest. January and July continental specific humidity changes are relatively small for areas near the equator, but seasonal differences are large at higher latitudes. Consequently, July specific humidity is 50% greater than January values over continental interior areas of the Northern Hemisphere. The Southern Hemisphere displays smaller seasonal changes over its continental area poleward of 20° S because these land areas are less expansive. Seasonal hemispheric water vapor exchanges related to seasonal atmospheric pressure changes are addressed in Section 7.6.

The vertical structure of specific humidity reveals a rapidly decreasing pattern with increasing altitude. More than 50% of water vapor is concentrated below the 850 hPa surface, and more than 90% is confined to the layer below 500 hPa (Peixoto and Oort, 1992). This pattern is expected due to the Earth's surface serving as the source of atmospheric moisture and the atmospheric drying with increasing height due to reduction in the saturation vapor pressure.

7.6   Atmospheric pressure

The spatial maldistribution of global energy drives an equator to pole temperature gradient at the surface. The spatial pattern of temperature is linked dynamically with a vertical temperature gradient in the atmosphere. These thermal patterns in the horizontal and vertical dimensions contribute to definition of variations in atmospheric pressure in both dimensions, and they trigger atmospheric motion with a variety of time and space scales.

The atmosphere conforms to the ideal gas laws as expressed in Equation 3.2 even though it is a combination of gases in an unconfined state. The downward directed force due to the weight of the overlying atmosphere is atmospheric pressure. Mass is one of the fundamental quantities of the atmospheric system and the mass distribution is closely related to the pressure field. We assume the total mass of the atmosphere taken over a year is practically constant, but it varies in time because of changing constituents. The mass of water vapor is an

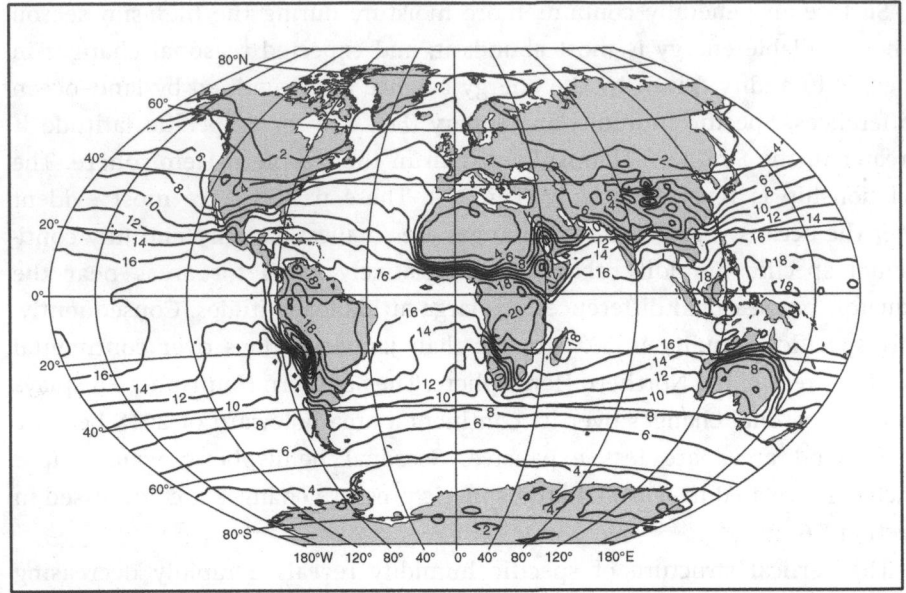

Fig. 7.7. Global January mean specific humidity. Units in $g\,kg^{-1}$. (NCEP Reanalysis data courtesy of NOAA/OAR/ESRL PSD, Boulder, Colorado, USA, from their website at http://www.cdc.noaa.gov/.)

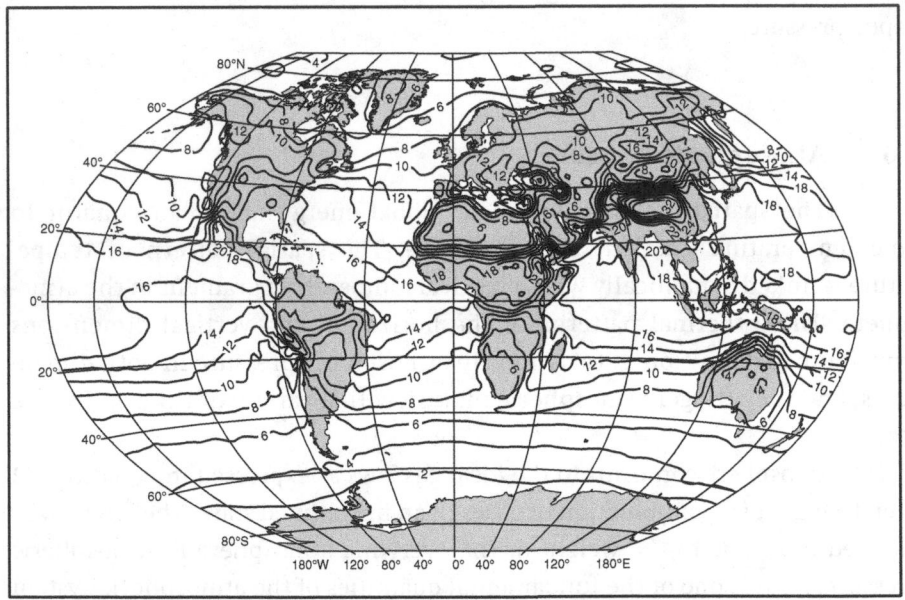

Fig. 7.8. Global July mean specific humidity. Units in $g\,kg^{-1}$. (NCEP Reanalysis data courtesy of NOAA/OAR/ESRL PSD, Boulder, Colorado, USA, from their website at http://www.cdc.noaa.gov/.)

important component of the total atmospheric mass, and water vapor is subject to short-term fluctuations (Trenberth and Smith, 2005). Under hydrostatic conditions the surface pressure is a good measure of the total atmospheric mass in a unit vertical column.

Vertical differences in pressure complement the horizontal differences depicted by the surface pressure field. Understanding vertical variations is aided by recognizing the relationship between temperature and pressure in the vertical dimension. Since air density decreases as temperature increases, a warmer layer of air must cover a greater geometrical height to embrace the same mass of gas. Consequently, pressure changes more rapidly in cold air than in warm air. The end result is that the height difference between any two given pressure levels, known as the thickness, varies directly with temperature. This accounts for the atmosphere being thicker or deeper over the equator than over the poles. Thus, the tropopause is higher over the equator (16 km) than over the poles (8 km). At any specified height, a pressure gradient exists directed from the equator to the poles. The rapid decrease of atmospheric density and pressure with height is due to the compressibility of the atmosphere, and it contrasts markedly with the oceans where density displays little vertical variation (Peixoto and Oort, 1992).

The mean sea-level pressure map is commonly used for synoptic weather analysis and is probably the most familiar depiction of horizontal pressure differences. Average monthly conditions for January and July are shown in Figures 7.9 and 7.10, respectively, to illustrate seasonal variations. Atmospheric pressure for a specific site results from the mass of the gas molecules extending from the surface to the upper edge of the atmosphere. This is commonly envisioned as a column of air above a unit area. A mixture of gases representative of the standard atmosphere exerts a downward-directed force equivalent to 1013.2 hPa.

General features evident in Figures 7.9 and 7.10 indicate the nature of pressure differences due to the global pattern of net radiation and features related to atmospheric dynamics. An initial examination of atmospheric pressure emphasizing description of the observed features rather than an explanatory analysis of the features permits a focus on space and time variations. A more comprehensive view emerges when atmospheric circulation is discussed in Section 7.7. Also, it is important to note that the depiction of pressure patterns using isobars, or lines joining points of equal pressure, commonly includes references to high pressure and low pressure. Such references are based on relative comparisons and not absolute pressure values. In addition, the common convention is to identify the lines as isobars even when the pressure is expressed in hPa.

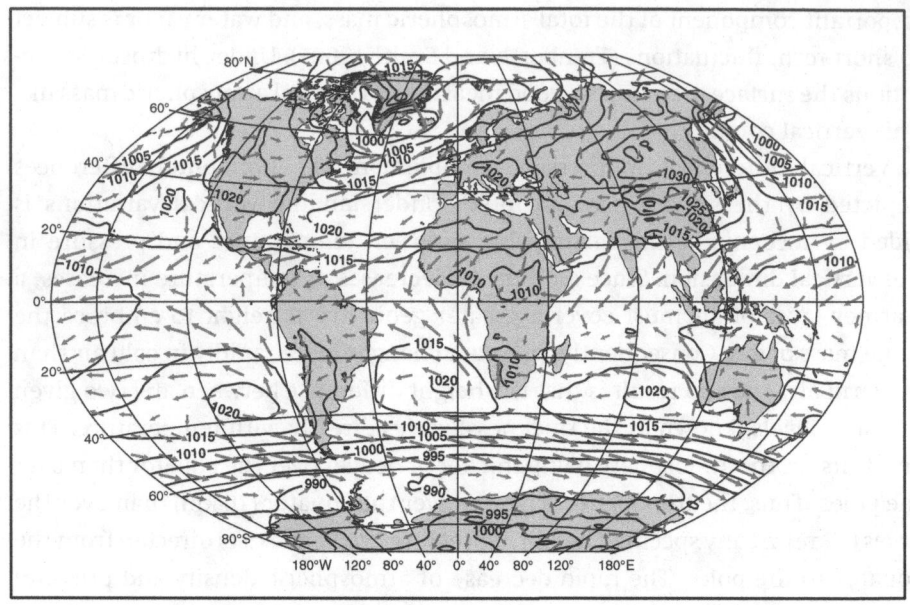

Fig. 7.9. Global January mean sea-level atmospheric pressure. Units in hPa. Wind direction shown by arrows, and wind speed by the length of the arrow shaft. (NCEP Reanalysis data courtesy of NOAA/OAR/ESRL PSD, Boulder, Colorado, USA, from their website at http://www.cdc.noaa.gov/.)

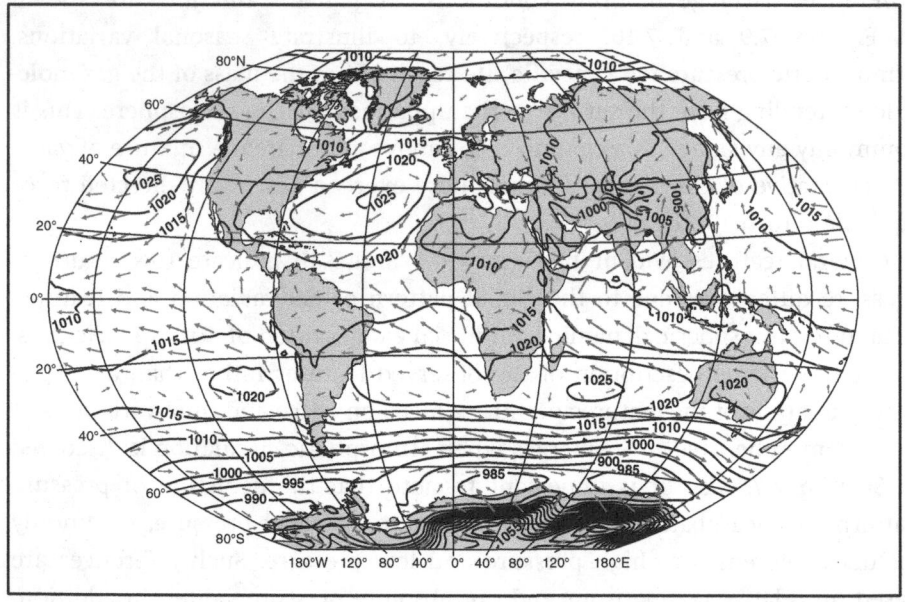

Fig. 7.10. Global July mean sea-level atmospheric pressure. Units in hPa. Wind direction shown by arrows, and wind speed by the length of the arrow shaft. (NCEP Reanalysis data courtesy of NOAA/OAR/ESRL PSD, Boulder, Colorado, USA, from their website at http://www.cdc.noaa.gov/.)

In January (see Fig. 7.9), a region of nearly continuous low surface pressure is present in the equatorial zone. The low pressure is coincident with the high net radiation and warm temperatures typically found in the low latitudes. The surface pressure near the subtropics at 30° latitude is generally higher with a definite cellular structure. These semipermanent features centered over the oceans are known as subtropical highs. High pressure is present over the continents where land temperatures are cold but at slightly higher latitudes compared to the subtropical highs. The area over Asia displays a circular form similar to the subtropical highs, but the high pressure over North America merges with the Atlantic subtropical high to form an elongated cell that extends to North Africa. The oceanic high-pressure cells in the Southern Hemisphere are clearly defined, and poleward of these cells is an extensive zone of generally low pressure that constitutes the Antarctic low-pressure trough. The Northern Hemisphere counterpart to this low pressure in January is the Aleutian and Icelandic low-pressure cells centered at the northern margins of the Pacific and Atlantic oceans, respectively. The polar regions in both hemispheres display a predominance of high pressure.

In July (see Fig. 7.10), seasonal variations in sea-level pressure are most apparent in the Northern Hemisphere. The equatorial low-pressure zone shifts predominantly into the Northern Hemisphere. The large oceanic subtropical highs move toward the pole in the summer hemisphere and are centered near 40° latitude, while in the winter hemisphere they move equatorward to about 30° latitude. The largest seasonal variations are found over the Asian continent where the winter high over Siberia is replaced by a low-pressure system north of the Indian subcontinent. A similar situation takes place over North America, but the pressure change is less intense. The annual variation of surface pressure over Siberia exceeds 25 hPa, while the change over North America does not exceed 10 hPa. At high latitudes, the low-pressure systems in the Northern Hemisphere are weaker during the summer, whereas in the Southern Hemisphere the almost continuous zonal low-pressure belt around Antarctica displays little seasonal variation.

The seasonal variation of global surface pressure shows a net positive residual during the Northern Hemisphere summer. Although this appears to contradict the basic assumption of a constant atmospheric mass, it is attributed to the additional contribution by the added water vapor associated with variations in the net evaporation minus precipitation at the Earth's surface. Observations of the precipitable water in the atmosphere show an excess of water vapor on the order of $0.2 \, \text{g kg}^{-1}$ during the Northern Hemisphere summer due to the higher values of evaporation and higher surface temperatures over the continents. This excess in water vapor mass is compatible with the observed global pressure difference of 0.2 hPa. The hemispheric values imply the existence of important shifts or exchanges of mass across the equator (Trenberth and Smith, 2005).

Global atmospheric surface pressure reveals a pattern that is sometimes disparate with the global patterns of net radiation and temperature. The deviation is especially evident with regard to the presence and latitudinal shifting of the oceanic subtropical high-pressure cells. High pressure at about 30° latitude implies that the observed pattern of surface pressure involves factors other than temperature and may include atmospheric motion as a causal component as well as a resultant. Considering pressure patterns higher in the atmosphere contributes to understanding global atmospheric circulation and its possible role as a causal agent.

The state of the middle-troposphere is indicated by the 500 hPa atmospheric pressure field. This constant pressure surface represents the height above sea level that divides the Earth's atmospheric mass into two approximately equal parts. Typical heights for the 500 hPa surface are around 5500 m, but the actual heights vary depending on thermal and dynamical factors that influence the vertical distribution of the atmospheric mass. Gravity plays a significant role in the processes determining vertical variations in atmospheric mass, and calculation of geopotential heights accounts for variations in gravity due to differences in latitude and elevation. In this way, geopotential heights approximate the actual pressure surface height above mean sea level. For most applications, geopotential heights are used interchangeably with geometric heights so that a constant pressure surface and a constant geopotential height are similar (Wallace and Gutzler, 1981). Figures 7.11 and 7.12 display 500 hPa geopotential heights, but the following discussion uses the 500 hPa geometric surface due to the similarity of the fields and to provide simplicity in referring to the fields.

Thermal influences on the characteristics of air columns are immediately apparent in Figures 7.11 and 7.12. The ideal gas law expressed in Equation 3.2 establishes that atmospheric pressure varies in response to temperature or density differences. Warm air is less dense and occupies a larger space resulting in the height to the 500 hPa surface being higher. Cool air is more dense, occupies a smaller space, and the 500 hPa surface is closer to the Earth's surface. This fundamental difference in the height of air columns underlies the observation that vertical atmospheric pressure changes more rapidly in cold air than in warm air.

Figures 7.11 and 7.12 reveal the 500 hPa surface in both seasons is highest over the equatorial region where air is warmest and lowest over polar regions occupied by cool air. The complex pattern of pressure cells at the surface is replaced in the upper atmosphere by one simple pressure gradient from the equator to the poles.

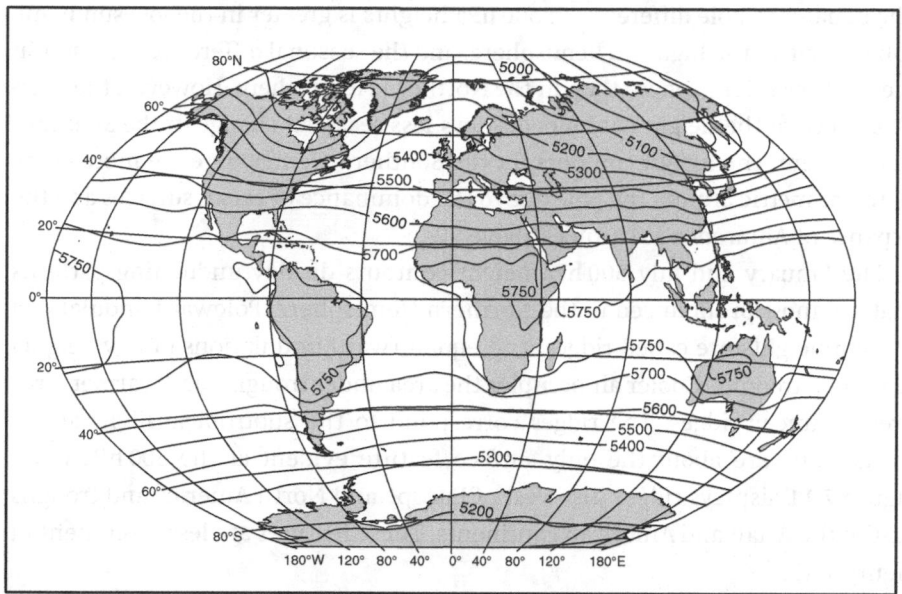

Fig. 7.11. Global January mean 500 hPa geopotential heights. Units in m. (NCEP Reanalysis data courtesy of NOAA/OAR/ESRL PSD, Boulder, Colorado, USA, from their website at http://www.cdc.noaa.gov/.)

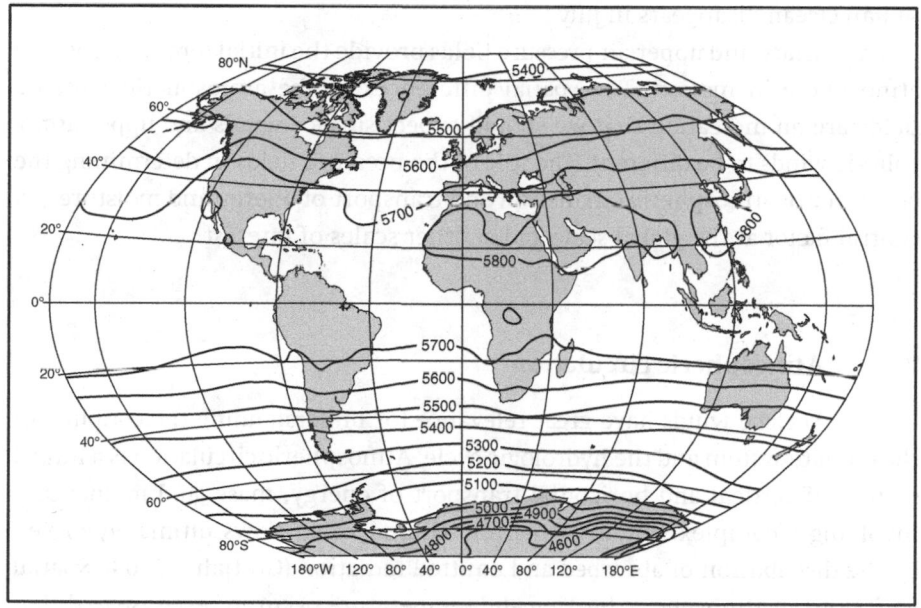

Fig. 7.12. Global July mean 500 hPa geopotential heights. Units in m. (NCEP Reanalysis data courtesy of NOAA/OAR/ESRL PSD, Boulder, Colorado, USA, from their website at http://www.cdc.noaa.gov/.)

The equator-to-pole difference in 500 hPa heights is greater in the low-sun hemisphere than in the high-sun hemisphere, and the seasonal difference is greater in the Southern Hemisphere than in the Northern Hemisphere. However, the overall pattern of the height contours displays less seasonal change in the Southern Hemisphere. The height contours for the Southern Hemisphere are more circularly symmetric around the pole due to the dominance of ocean surfaces and the expanse of Antarctica at latitudes above 70°.

The January and July 500 hPa height contours display undulating patterns that are most pronounced in the Northern Hemisphere. Poleward undulations of high heights are called ridges, and equatorward undulations of low heights are called troughs. Cooler air occupies the area under troughs, and warmer air is present under ridges. The ridges correspond to the subtropical highs at the surface and are about the only surface feature evident at the 500 hPa level. Figure 7.11 displays ridges just west of Europe and North America and troughs east of the Asian and American continents. These features are less prominent in Figure 7.12.

Wave patterns are evident in the Southern Hemisphere, but the strong zonal influence reduces the wave amplitude. Ridges along the west coasts of southern South America, southern Africa, and Australia are evident in January but less apparent in July. Troughs in January over the eastern Pacific Ocean and eastern Atlantic Ocean are weaker in July, and the January trough over the eastern Indian Ocean disappears in July.

The surface and upper-air pressure fields provide the initial force that sets the atmosphere in motion. The spatial differences in the atmospheric pressure fields are an indication that we should expect surface winds and upper atmospheric winds to be different. The role of the pressure fields in determining the character of atmospheric circulation and transport of energy and moisture is a central factor at the global scale and at other scales of interest.

7.7   Atmospheric circulation

Surface winds have great relevance for understanding the coupling of the climate system and the hydrologic cycle. Atmospheric circulation is a linked system of vertical and horizontal transport of energy, mass, and momentum involving a complex system of meridional and zonal flows ultimately driven by the distribution of absorbed and emitted radiation (Grotjahn, 2003). Spatial variations in atmospheric heating and temperature occur in response to differences in radiation, sensible heat exchange with the surface, latent heat associated with condensation, and the horizontal flux of energy in the atmosphere (Hartmann, 1994). The differential heating between the low and high latitudes is the primary driving force for atmospheric circulation on all time scales (Trenberth and Solomon, 1994). Atmospheric circulation responds to temperature and humidity gradients while helping to determine these gradients by transporting energy and moisture at various spatial scales.

The atmosphere is able to fulfill the transport role because of the ease with which it can be heated and set in motion. Without horizontal transfer, the temperature at a given latitude would be dictated by radiation alone. Summers would be warmer for much of the globe, and winter temperatures in the tropics would be warmer. Temperatures would decrease very rapidly at higher latitudes, and winter latitudinal temperature gradients would be severe.

Combining the latitudinal imbalance of radiation with the hydrologic cycle imbalances establishes the significant role of atmospheric motion in heat transport. The atmosphere accounts for 80% of the global heat transport between the equator and the poles and in the process provides the mechanism for redistributing moisture. Poleward ocean heat transport is dominant only between 0° and 17° N. The peak total poleward transport in each hemisphere is at 35° latitude. Atmospheric transport at this latitude accounts for 78% of the heat total in the Northern Hemisphere and 92% in the Southern Hemisphere (Trenberth and Caron, 2001). The conventional method for viewing how this transfer is accomplished is by depiction of global atmospheric circulation consisting of observed wind systems and their annual and seasonal variations. However, it is important to recognize that the large-scale circulations are produced by imbalances in global radiation that create temperature gradients the atmosphere strives to eliminate. The resulting atmospheric circulation is limited by radiation, energy, mass, and angular momentum balances (Grotjahn, 2003). Perhaps the most evident link is the relationship between the general circulation and the distribution of atmospheric pressure that is an expression of the atmospheric mass.

Changes in atmospheric pressure horizontally and vertically characterize the continuous pressure field created by the envelope of atmospheric gases. Differences in the mass of the gas molecules in the atmosphere above a site give rise to changes in atmospheric pressure. Variations in atmospheric pressure over linear distance establish a pressure gradient that exists horizontally across the Earth's surface and vertically from the surface upward into the atmosphere. Pressure gradient (PG) is expressed as

$$PG = \frac{\Delta p}{d} \qquad\qquad (7.2)$$

where Δp is the change in pressure and d is the linear distance. For horizontal pressure variations a force is created acting from high pressure to low pressure. The horizontal pressure gradient force (F_{PG}) per unit mass is given by

$$F_{PG} = -\frac{1}{\rho}\frac{\Delta p}{\Delta n} \qquad (7.3)$$

where ρ is air density ($1.293\,\mathrm{kg\,m^{-3}}$ for air at $0°\,C$ and $101\,325\,\mathrm{Pa}$), Δp is defined above, Δn is the distance between isobars, and the negative sign indicates the force operates from high to low pressure. The pressure gradient force acts perpendicular to the isobars and the magnitude of the force is inversely proportional to the distance between isobars. The vertical component of the pressure gradient force is calculated in a similar fashion by substituting Δz, the vertical distance separating the pressure surfaces, for Δn in Equation 7.3.

Reference points within the pressure continuum provide a basis for establishing changes in pressure over some specified distance that defines the pressure gradient. The pressure gradient plays a dominant role in atmospheric motion because it is the factor that initiates air motion. Understanding the influence of the pressure gradient is aided by recognizing that the pressure gradient force is always directed from high pressure to low pressure, it is directed perpendicular to the isobars, and it is inversely proportional to the spacing of the isobars.

Other factors act on the atmosphere once it is set in motion, and these factors make an important contribution to the observed characteristics of atmospheric circulation. The general atmospheric circulation is constrained by the need to maintain the global water balance, the balance of the atmospheric mass, and the balance of angular momentum. The global water balance and the conservation of atmospheric mass have been addressed in earlier sections, and only angular momentum is treated here.

Angular momentum is a measure of the rotational motion intensity that is a basic physical quantity for any rotating system. For the Earth, angular momentum is a conserved property and any change in the angular momentum of one component of the climate system is balanced by a corresponding change in the angular momentum of the other climate system components. Much of the historic development of modern meteorology is traced to study of the transport of angular momentum in the atmosphere and its exchange with the oceans and the solid earth (Peixoto and Oort, 1992). The angular momentum (M) of an air parcel is

$$M = mvr \qquad (7.4)$$

where m is the mass of the parcel, r is the perpendicular distance from the axis of rotation of Earth, and v is the Earth's rotation velocity. The more common form is

$$M = \Omega a \cos \Phi \qquad (7.5)$$

where Ω is the Earth's rotation rate (angular velocity of the Earth, $7.292 \times 10^{-5} \, \text{rad s}^{-1}$), a is the mean radius of Earth ($6.371 \times 10^6 \, \text{m}$), and Φ is the latitude. Since the depth of the atmosphere is small relative to the radius of the Earth, latitude has a strong influence on angular momentum because increasing latitude decreases the distance to the axis of rotation. Distance to the axis of rotation is the radius of Earth times the cosine of latitude.

Conservation of angular momentum is particularly significant in explaining the behavior of air moving poleward or equatorward. To maintain a constant total angular momentum, a parcel of air moving poleward in the atmosphere must increase its relative eastward zonal velocity. Theoretical calculations indicate that poleward angular momentum transport easily accounts for the existence of high winds associated with the subtropical jet stream at about 30° N or 30° S. In addition, large-scale eddies at these latitudes transport momentum into the middle latitudes and downward to the surface (Hartmann, 1994).

Atmospheric motion required to satisfy the angular momentum constraint includes tropical easterly flow that is slower than the Earth's surface and transfers angular momentum to the atmosphere. The convergence of easterly winds near the equator supports an upward and poleward flow of the atmosphere and angular momentum contributes to the formation of a vertical circulation known as the Hadley cell. Atmospheric eddies transport angular momentum poleward and downward into the mid-latitude westerly surface winds. Westerly surface winds rotate faster than the Earth's surface and angular momentum is returned to Earth at these latitudes. In this way, surface zonal winds cannot be of the same sign everywhere because eastward angular momentum flows into the atmosphere where surface winds are easterly and returns to the surface where winds are westerly (Grotjahn, 2003).

Annual surface atmospheric circulation features mask important details responsible for explaining much of the hydroclimatic character of different locations. Consequently, average patterns of pressure and winds for January (see Fig. 7.9) and July (see Fig. 7.10) are helpful in defining the coupling of pressure and winds and the driving force of net radiation. The seasonal changes in the mean sea-level atmospheric winds reflect the influence of two major contributors. Inequalities in radiation distribution over the Earth's surface drive the global circulation. The Earth's west to east rotation determines the shape of the global circulation. These thermal and dynamical factors are both evident in the global pattern of atmospheric pressure and winds. Thus, mean sea-level atmospheric pressure serves as an important indicator of the vertical and horizontal characteristics of the general circulation.

The relative permanence and transience of different pressure features were discussed earlier. Global wind features are illustrated in a similar manner in

Figures 7.9 and 7.10. A prominent element in both illustrations is the oceanic cells of high pressure at about 30° N and 30° S and the related surface flows of wind on the equatorward and poleward margins of these systems. These sub-tropical high-pressure cells are positioned beneath the subsiding air on the poleward limb of the Hadley cell that plays a major role in transporting angular momentum into the mid-latitudes (Grotjahn, 2003). The diverging air associated with the core of the subtropical high-pressure cells spirals outward in a clock-wise pattern in the Northern Hemisphere and counterclockwise in the Southern Hemisphere. The resulting surface winds are easterly flow converging on the equator in both hemispheres and predominantly westerly flow on the poleward side of the subtropical high-pressure cells. Clockwise flow around these systems in the Northern Hemisphere feeds the trade winds (easterlies) on the equator-ward limb and the westerlies on the poleward limb. Flow around these systems is counterclockwise in the Southern Hemisphere. Trade winds from both hemispheres converge on the equatorial low-pressure trough and its related circulation feature the intertropical convergence zone (ITCZ). Seasonal changes in the position of the ITCZ are evident in response to the greater warming in the summer hemisphere, but the ITCZ mean position is in the Northern Hemisphere. Subtropical highs display a seasonal latitudinal shift also. Latitudinal pressure gradients and winds are stronger in the winter hemisphere.

The influence of the subtropical high seasonal shifting is evident in surface air flow conditions in North America. North America is dominated by air streams from three distinct sources. A Pacific air stream dominates West Coast locations, a tropical air stream from the Atlantic Ocean and the Gulf of Mexico is the most common influence for the eastern United States and southeastern Canada, and an Arctic air stream is active in the North American interior as far south as 40° N. The tropical and Arctic air streams have a dominant meridio-nal character while the Pacific air stream is basically zonal flow. In addition, the water vapor content of the Atlantic and Gulf air is approximately twice as great as the water vapor in the cooler Pacific air as seen in Figure 7.6.

The Pacific air stream can be subdivided into two components. A northern branch enters the Pacific from Asia predominantly in the winter, and a southern element emerges from the oceanic anticyclone over the Pacific and is most dominant during the summer. The northern Pacific westerlies arrive on the West Coast cool, with a near moist-adiabatic lapse rate and with high humidity to a considerable depth due to a relatively long traverse over the ocean. The influence of these air streams is evident in precipitation quantity and season-ality along the entire North American west coast. Precipitation is greatest at the highest latitudes and a relatively dry summer season begins at about 40° N. The precipitation quantity decreases and the dry season is longest in Baja California due to the persistent influence of the southern portion of the westerlies, which are fed by subsiding air from the subtropical high that is stable and has a shallow moist layer. During the winter, the Pacific southern air stream may penetrate west–east trending mountains in southern California and northern Mexico and contributes to warm, dry conditions in the region from the Pacific Coast to Texas (Lydolph, 1985).

In the atmosphere above the boundary layer, the large-scale atmospheric motions are basically horizontal. In the vertical dimension, the pressure gradient almost balances gravity so that vertical accelerations are negligible and the vertical component of velocity is small everywhere. In the horizontal dimension, the principal forces in the free atmosphere are the pressure gradient and the Coriolis forces leading to a quasi-geostrophic equilibrium (Grotjahn, 2003). The motions are practically parallel to the pressure contours and the wind speeds are inversely proportional to the spacing between the isobars. The isobars are thus approximate streamlines for the flow outside the equatorial region.

7.8   Global terrestrial hydroclimate

The vertical and horizontal flux variables comprising climate of the second kind and the terrestrial branch of the hydrologic cycle display global patterns indicating the influences of state variables modified by surface features. The resulting distributions of hydroclimatic variables contain great complexity that is moderated by the broad view provided by the global scale.

7.9   Precipitation

The spatial distribution of mean annual precipitation illustrates the summation of the various mechanisms producing moisture at the Earth's surface. While the presence of atmospheric moisture and the cooling of ascending air parcels by frontal, convectional, and orographic lifting mechanisms are requisites for producing precipitation, the spatial variation of annual precipitation is so complex that it appears to defy explanation. In reality, the spatial variation of annual precipitation is the combined expression of the global hydrologic cycle, global atmospheric circulation, and the influence of continental landforms.

The subtropical oceans are major source regions for atmospheric moisture that is supplied by uniformly high evaporation in the regions between 15° and

40° latitude (Hartmann, 1994). Global atmospheric circulation is responsible for exporting water vapor out of these regions to support precipitation maxima at other latitudes. Regions of meager rainfall are related to atmospheric subsidence, topographic features, and distance from major atmospheric moisture sources. The global mean precipitation rate is $2.6\,\mathrm{mm\,d^{-1}}$ with $2.8\,\mathrm{mm\,d^{-1}}$ occurring over oceans and $2.1\,\mathrm{mm\,d^{-1}}$ over land (Adler et al., 2003).

Global precipitation based on NCEP Reanalysis data (Kanamitsu et al., 2002) is shown in Figure 7.13. Fekete et al. (2004) compared NCEP and five other precipitation datasets and concluded the overall pattern was fairly similar in each dataset. Figure 7.13 shows the greatest mean annual precipitation occurs in near-equatorial areas where atmospheric water vapor is abundant and the ITCZ provides a nearly continuous mechanism for promoting ascending air. Precipitation in this region exceeds 2000 mm and may reach 3000–5000 mm (Lydolph, 1985). Even larger values of mean annual precipitation occur in some tropical coastal settings where prevailing winds consistently converge and are forced to ascend mountain slopes. Also, heavy precipitation is common in mid-latitude coastal locations where synoptic weather systems embedded in the prevailing westerlies encounter mountain barriers and forced ascent. In south

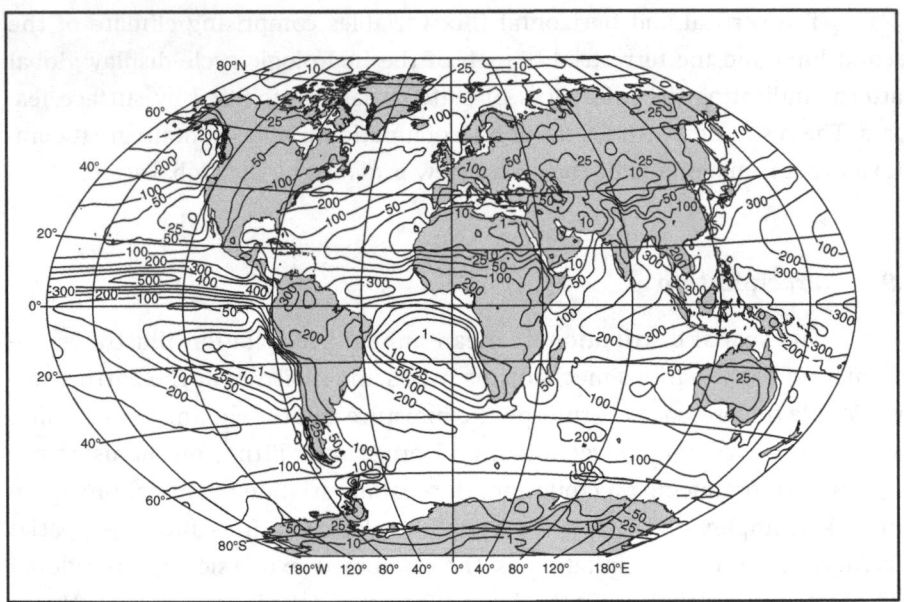

Fig. 7.13. Global annual mean precipitation. Units in cm. (NCEP Reanalysis data courtesy of NOAA/OAR/ESRL PSD, Boulder, Colorado, USA, from their website at http://www.cdc.noaa.gov/.)

Asia and India, the summer monsoon delivers abundant precipitation that is locally enhanced by mountain barriers.

The ITCZ influence as a precipitation delivery mechanism on the continents declines at latitudes approaching the Tropic of Capricorn and the Tropic of Cancer. This declining pattern of rainfall is particularly evident in North Africa. Meager precipitation is the rule in the subtropics where atmospheric subsidence in the general circulation is dominant. Land areas in these latitudes are particularly influenced by the absence of available water vapor.

Precipitation increases to modest levels at higher latitudes because of the increased frequency of synoptic-scale storm systems. Zonal precipitation differences result from the presence of mountain ranges that trigger heavy precipitation on windward slopes and suppress precipitation on leeward slopes. In addition, the expanse of continents in these latitudes promotes atmospheric drying as distance from marine moisture sources increases.

In polar regions, mean annual precipitation is small because the entire hydrologic cycle is slowed by the limited availability of energy. Evaporation is slowed and the water-carrying capacity of the atmosphere is low (Hartmann, 1994). Precipitation, commonly in the form of snow, is conserved at these latitudes and even small annual precipitation amounts may appear to be greater because of the accumulated snow (see Fig. 4.3).

The seasonal occurrence of precipitation is a second major consideration influencing its hydroclimatic significance. At most locations, precipitation is unevenly distributed throughout the year. In general, seasonal changes in precipitation are traced to shifts in atmospheric circulation driven by latitudinal variations in solar radiation. Latitudes near the center of a circulation feature remain under the control of a single system throughout the year and display relatively little precipitation variability. This influence is evident in the spatial characteristics of January (Fig. 7.14) and July (Fig. 7.15) precipitation. Locations near the equator dominated by the ITCZ receive abundant precipitation in both months. Areas dominated by subtropical highs (e.g. North Africa, central Australia) receive little rainfall in either month. Latitudes near the margin of major circulation features display the greatest seasonal precipitation changes as they are influenced by contrasting wind systems during the course of the year (Shelton, 1988). The west coast of North America poleward of 40°N and northern Australia equatorward of 20°S display these seasonal contrasts.

The convergence of the trade winds in the equatorial region provides a relatively consistent supply of warm, moist air and convective forces persist all year. Precipitation is relatively abundant in all months at locations in the latitudinal zone traversed by the ITCZ, but seasonal differences are evident. The

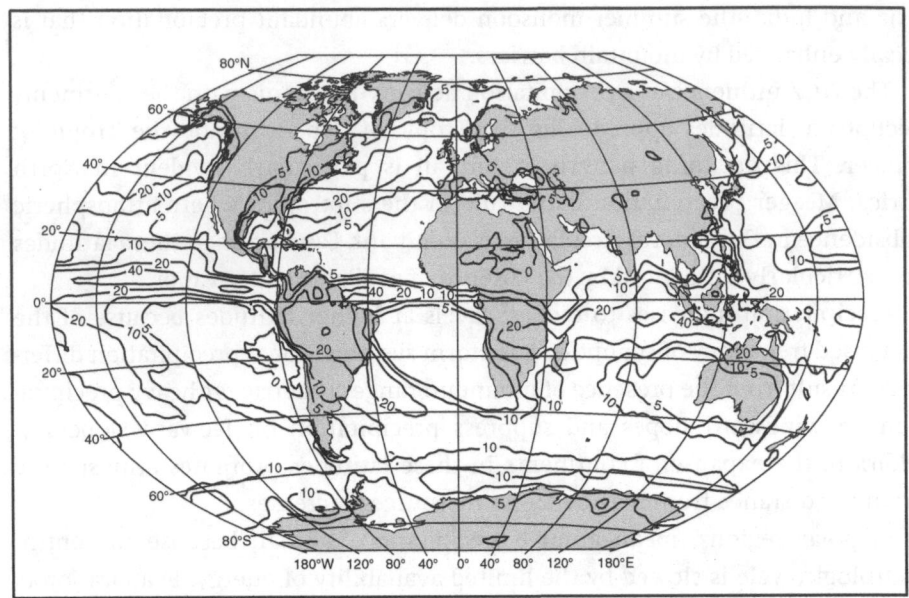

Fig. 7.14. Global January mean precipitation. Units in cm. (NCEP Reanalysis data courtesy of NOAA/OAR/ESRL PSD, Boulder, Colorado, USA, from their website at http://www.cdc.noaa.gov/.)

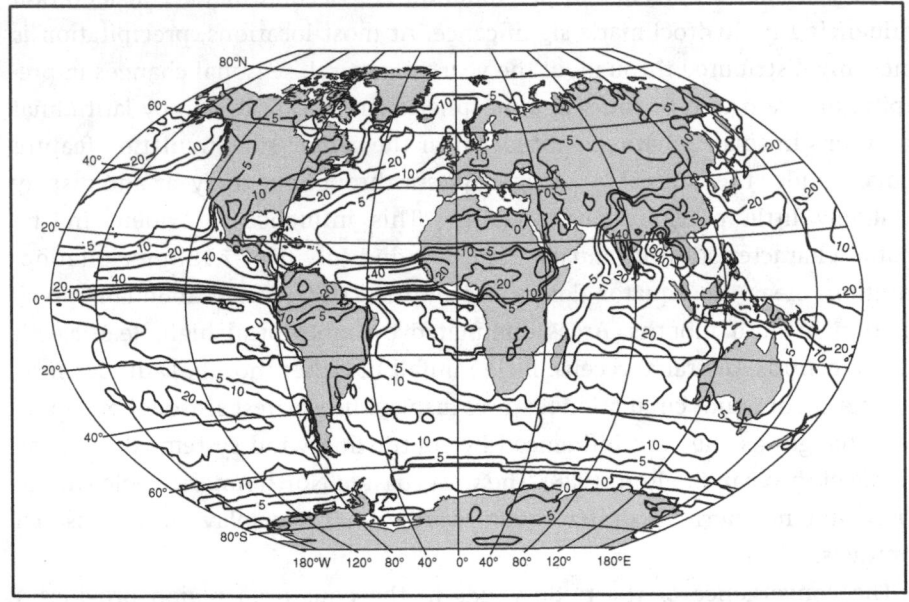

Fig. 7.15. Global July mean precipitation. Units in cm. (NCEP Reanalysis data courtesy of NOAA/OAR/ESRL PSD, Boulder, Colorado, USA, from their website at http://www.cdc.noaa.gov/.)

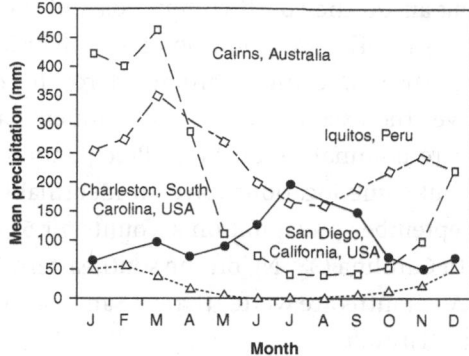

Fig. 7.16. Monthly mean precipitation for four stations showing the influence of latitude on rainfall seasonality. (Data courtesy of NOAA and the National Climate Data Center from their website at http://www.ncdc.noaa.gov/oa/climate/ghcn-monthly/index.php.)

precipitation regime at Iquitos, Peru (3° S), in Figure 7.16 displays a double maximum that is stronger in March than in November and a dry period in July and August. The ITCZ movers further south of the equator in January over South America than it does north of the equator in July in response to changes in the general circulation driven by temperature gradients. The southward movement of the ITCZ is accompanied by warm, unstable northeasterly winds off the Atlantic Ocean. In July, the position of the ITCZ just north of the equator promotes zonal easterly winds that traverse the Amazon Basin before reaching Iquitos.

Seasonal precipitation contrasts become more pronounced at latitudes approaching 20° as the influence of the annual movement of the ITCZ produces a single rainy season. In these realms, precipitation is most abundant during the high-sun season and least during the low-sun season. This effect is evident in the monthly precipitation regime at Cairns, Australia (17° S) (see Fig. 7.16). The difference in mean monthly precipitation between March and July is 419 mm. The effect of the seasonal shift in the ITCZ is magnified at Cairns by its location relative to the confluent zone between the warm, moist air from the northeast and the southeast.

For latitudes between 20° and 35°, precipitation amount and seasonality are dependent upon continental location. Precipitation is meager and seasonality is of little consequence for the western portion of continents at these latitudes. Subsiding air associated with subtropical highs is dominant and most rain-producing storm systems are prevented from entering the region except in the winter. San Diego, California (33° N), has a precipitation regime that illustrates the effect of persistent subsiding air on the southwestern coast of the United States (see Fig. 7.16). Eight months receive precipitation of 25 mm or less, and 75% of annual precipitation occurs from December through March. In contrast, subtropical highs are weaker over the eastern portion of continents at these latitudes and prevailing winds are dominantly onshore. Precipitation is relatively abundant in all months, but some locations have an identifiable warm season concentration. June to September precipitation accounts for 51% of the annual total at Charleston, South Carolina (33° N), on the southeastern coast of the United States (see Fig. 7.16). All months are wetter than at San Diego which is on the west coast and at the same latitude.

Poleward of 35° latitude, precipitation results from cyclonic activity related to storms imbedded in the westerlies. The core regions of the oceanic subtropical high-pressure cells display seasonal shifting of 5° to 7° of latitude as illustrated by the conditions in the Northern Hemisphere. This characteristic has important significance for precipitation at mid-latitude locations on the west coast of continents because it is related to seasonal variations in frontal tracks associated with the most vigorous and frequent mid-latitude cyclones (Hartmann, 1994). In January, the oceanic subtropical highs in the Northern Hemisphere retreat equatorward as the zone of maximum solar radiation is near the Tropic of Capricorn in the Southern Hemisphere and the global circulation responds by shifting southward. The equator to pole temperature and pressure gradients in the Northern Hemisphere are greatest at this time, the polar front jet stream is stronger and makes more equatorward excursions, and the westerlies are more intense. By July, the oceanic subtropical highs assume a more poleward position as the zone of maximum insolation moves toward the Tropic of Cancer and the Hadley cell circulation strengthens in the Northern Hemisphere. In their higher latitude location, the subtropical highs support weaker westerly winds because the hemispheric temperature and pressure gradients are weakened. Cyclonic storms imbedded in the westerlies are forced poleward by the more northerly location of the subtropical high pressure and a distinct seasonal variation in storm frequencies is observed. The result is predominantly winter rainfall along the west coast of continents poleward to about 50° latitude. The length of the wet season decreases at lower latitudes. The eastern margin of the subtropical highs dominates the east coast of continents poleward of 30° latitude with the result that the eastern portions of continents receive rainfall in all months.

Lisbon, Portugal (39° N), and Washington, D.C. (39° N), in Figure 7.17 show representative precipitation regimes for opposite coast locations. Lisbon receives 30% less annual precipitation than Washington, D.C., and October to March precipitation at Lisbon accounts for 78% of the annual total. April to

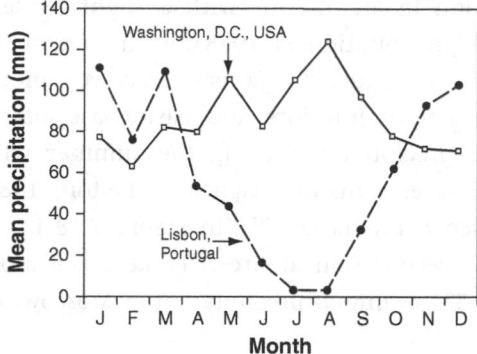

Fig. 7.17. Monthly mean precipitation for two stations illustrating the influence of subtropical high pressure on rainfall seasonality. (Data courtesy of NOAA and the National Climate Data Center from their website at http://www.ncdc.noaa.gov/oa/climate/ghcn-monthly/index.php.)

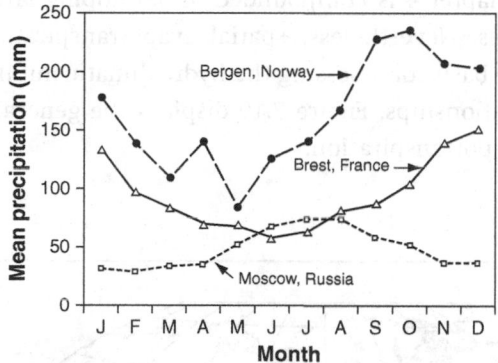

Fig. 7.18. Monthly mean precipitation for three stations displaying the influence of coastal and inland locations on precipitation seasonality at high latitudes. (Data courtesy of NOAA and the National Climate Data Center from their website at http://www.ncdc.noaa.gov/oa/climate/ghcn-monthly/index.php.)

September accounts for 57% of annual precipitation at Washington, D.C. August is the wettest month at Washington, D.C., averaging 124 mm while the August average in Lisbon is 4 mm.

At 50° latitude and higher, precipitation seasonality is largely controlled by location influences. Continental west coasts between 40° and 50° latitude have a slightly dry summer like the regime for Brest, France (48° N), in Figure 7.18. However, the summer dryness is much less severe than at Lisbon (see Fig. 7.17) or San Diego (see Fig. 7.16). Increased cyclonic activity at latitudes above 50° delivers precipitation in all months with a slight tendency for winter to be wetter. Annual precipitation is 1958 mm at Bergen, Norway (60° N), and each month from August to January receives more than 150 mm (see Fig. 7.18). Continental interiors and east coast locations at these latitudes receive more precipitation during the summer when the warmer atmosphere has a greater moisture capacity (Shelton, 1988). The precipitation regime for Moscow, Russia (56° N), in Figure 7.18 is a nearly inverted pattern of monthly precipitation at Brest, France. The months of May to October account for 65% of annual precipitation at Moscow and 40% at Brest.

7.10   Evapotranspiration

The difficulty in measuring and/or estimating evapotranspiration at a specific site discussed in Chapter 4 is compounded in developing large-scale evapotranspiration estimates. Nevertheless, spatial evapotranspiration estimates provide a comparable basis for assessing the hydroclimatic implications of energy and moisture relationships. Figure 7.19 displays the general global pattern of land potential evapotranspiration.

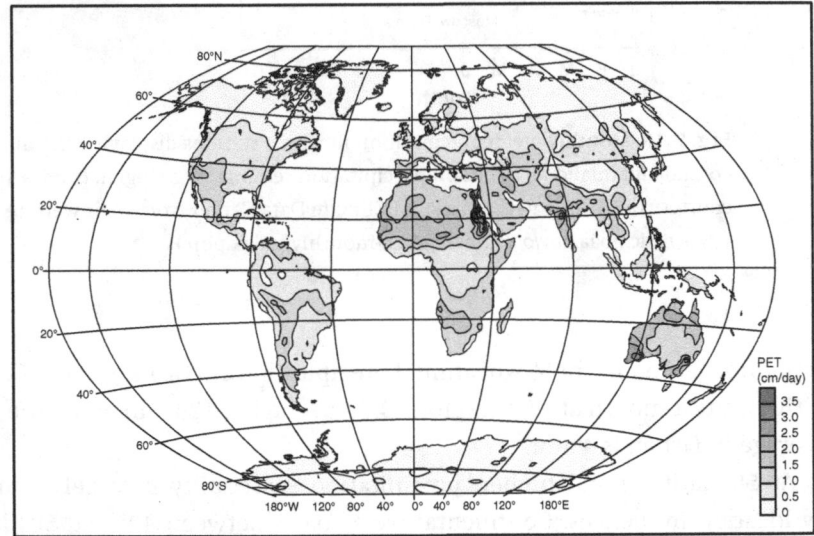

Fig. 7.19. Global annual mean potential evapotranspiration. Contour units in cm. (NCEP Reanalysis data courtesy of NOAA/OAR/ESRL PSD, Boulder, Colorado, USA, from their website at http://www.cdc.noaa.gov/.)

Potential evapotranspiration varies relatively smoothly with a broad maximum in the tropics and decreasing with increasing latitude. Northern Hemisphere ETp is about 59% greater than the Southern Hemisphere total due to the greater land area in the Northern Hemisphere. The major exception to the general hemispheric pattern is that ETp for the latitudinal band from 0° to 20° S exceeds ETp for the band from 0° to 20° N by 7%. ETp is greater in the Northern Hemisphere for all other latitude bands.

The pattern within latitudinal bands displays complexity as factors other than solar radiation exert an influence on the maximum evapotranspiration quantity. The region in Africa between 10° N and 20° N is a good illustration of how the systematic influence of solar radiation is altered by seasonal variations in atmospheric circulation and cloud cover that influence available energy and humidity. Available energy is especially abundant in this zone due to the directness and duration of sunlight relative to latitudes further north or south.

ETp indicates the maximum moisture loss from the land surface, but the moisture flux that incorporates climate, soils, and vegetation influences is ETa. In most instances, ETa is less than ETp as discussed in Section 4.8.3, and it represents the actual moisture flux to the atmosphere from land areas. However, the ET process is limited by soil moisture and plant conditions for most land areas, causing the total moisture flux from the continents to be less than from the oceans (see Table 2.2). Oceans contribute 6-times the moisture input to the atmosphere by evaporation compared to the volume land areas provide through ETa. The oceans benefit from their surfaces providing an unlimited moisture supply and from the abundant energy over the topical oceans to drive the evaporation process.

7.11   Soil moisture

Soil moisture is one of the important physical state variables of the hydroclimatic system because it affects the partitioning of energy and water balances at the Earth's surface. It influences evapotranspiration, albedo, and thermal fluxes, and recent studies have shown a positive correlation between soil moisture levels and the probability of precipitation occurrence (Salvucci et al., 2002). The Global Soil Moisture Data Bank contains the largest collection of in situ observational data (Robock et al., 2000), and satellite retrievals of surface soil moisture by both active and passive sensors are increasingly available. However, Reichle et al. (2004) found a large number of land surface models produce widely different global-scale soil moisture output using identical meteorological forcing inputs, and they conclude that errors in present global-scale soil moisture observations and modeling datasets are so large that a universally agreed climatology cannot be identified with confidence.

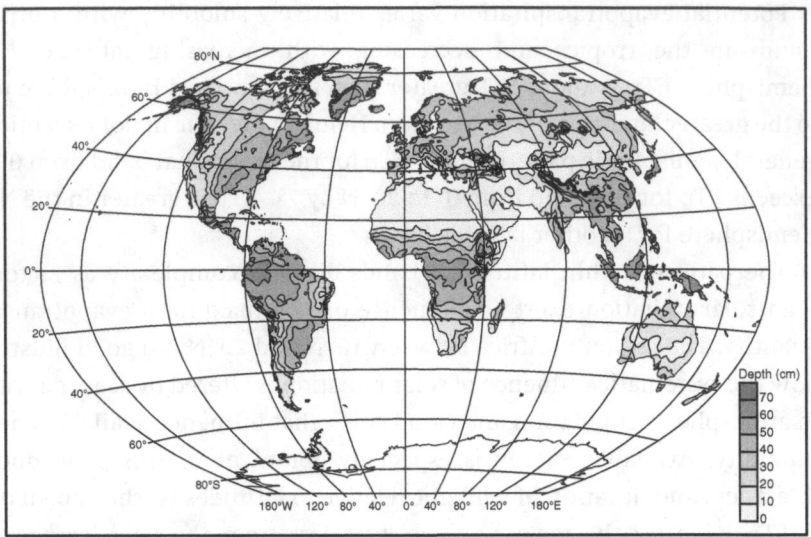

Fig. 7.20. Global annual mean soil moisture. Contour units in cm. (NCEP Reanalysis data courtesy of NOAA/OAR/ESRL PSD, Boulder, Colorado, USA, from their website at http://www.cdc.noaa.gov/.)

An estimate of mean annual global soil moisture variation derived from the NCEP-NCAR Reanalysis data (Kalnay *et al.*, 1996) is shown in Figure 7.20. At the global scale, soil moisture variability is related to fields of precipitation and evapotranspiration. Soil moisture is highest in the equatorial latitudes where precipitation is abundant. It decreases poleward toward the Tropics as precipitation decreases and evapotranspiration increases. This pattern is most evident in Africa. In North America, Europe, and Asia, the soil moisture pattern displays greater variability and is more complex as the precipitation and evapotranspiration fields are complicated due to topographic factors and seasonally varying atmospheric influences. Nevertheless, the general pattern is that the latitudes between 0° and 20°S have the greatest soil moisture values, and the region between 40°N and 60°N has the greatest quantity of moisture stored in the soil due to the vast expanse of land area at these latitudes.

7.12   Runoff

The unequal runoff distribution among the continents (see Table 2.2) conceals an even greater runoff complexity resulting from the coupled spatial

Table 7.1. *Large rivers of the world listed according to annual discharge*

River	Continent	Discharge into	Length (km)	Drainage area (10^3 km^2)	Average annual discharge (km^3)
Amazon	South America	Atlantic Ocean	6308	6915	6923
Ganges-Brahmaputra	Asia	Bay of Bengal	2897	1621	1386
Congo	Africa	Atlantic Ocean	4370	3457	1320
Orinoco	South America	Atlantic Ocean	2740	948	1007
Yangtze	Asia	East China Sea	6300	1959	1006
La Plata	South America	Atlantic Ocean	4700	3100	811
Yenisei	Asia	Kara Sea	5540	2580	618
Lena	Asia	Laptev Sea	4345	2490	539
Mississippi	North America	Gulf of Mexico	5971	3221	510
Mekong	Asia	South China Sea	4500	810	505

patterns of precipitation and evapotranspiration for individual continents. Global continental runoff is an estimated 35% of terrestrial precipitation, and the world's 50 largest rivers account for 57% of the global runoff. However, large differences exist among the discharges into individual ocean basins. The Atlantic Ocean receives over twice the runoff delivered to the Pacific Ocean and more than four times the river discharge into the Indian Ocean (Dai and Trenberth, 2002). The data in Table 7.1 illustrate the disparity of river discharge into ocean basins.

Figure 7.21 reveals latitudinal differences in runoff evident by river discharge into the oceans. The discharge spike for the zone from the equator to 5° S is due to the Amazon and Congo rivers discharging into the Atlantic Ocean at these latitudes. These rivers are the world's two largest by volume (Dai and Trenberth, 2002). The domination of the Northern Hemisphere zones at latitudes higher than 5° results from the relatively greater land areas at middle and high latitudes in the Northern Hemisphere. Another notable feature is the runoff minimum at 27.5° N which corresponds to the expanse of arid lands at these latitudes. However, seasonal runoff variations related to snowmelt at higher latitudes and high altitudes (Lee *et al.*, 2004), the onset of evapotranspiration during the growing season in middle latitudes (Czikowsky and Fitzjarrald, 2004), and runoff changes at the close of the twentieth century from Arctic and sub-Arctic watersheds (McClelland *et al.*, 2006) are not evident in Figure 7.21. These influences are evident in the comprehensive analysis of streamflow seasonality and variability for over 1300 sites globally reported by Dettinger and Diaz (2000).

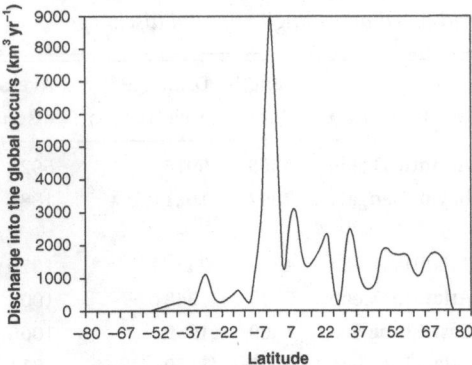

Fig. 7.21. Annual mean runoff into the global oceans by latitude zone using a 5° latitude running mean. (After Dai and Trenberth, 2002, Fig. 8. Used with the permission of the American Meteorological Society.)

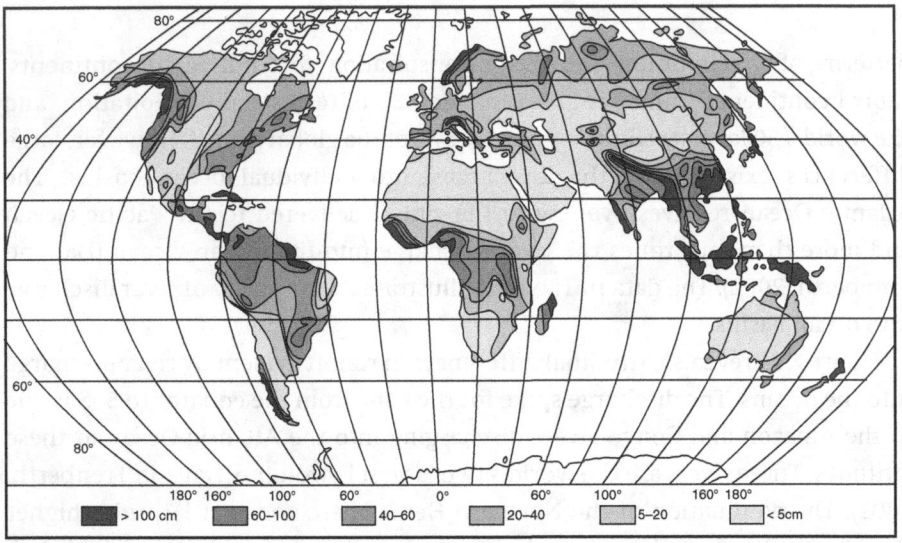

Fig. 7.22. Global annual mean runoff. (From Barry, 1969, Fig. 1.1.8. Methuen and Co. Ltd. Reproduced by permission of Taylor and Francis Books UK.)

Figure 7.22 provides a high-resolution view of the variable nature of annual runoff for the continents. This map is based on observational data, which are considered more reliable than model-derived runoff (Kalnay *et al.*, 1996; Coe, 2000). The vast expanse in Africa, Asia, Australia, and North America with runoff less than 5 cm is striking. The low runoff area of South America is modest and overshadowed by the abundant runoff characteristic of the Amazon Basin and the northern coastal zone. In North America, the contrast

Table 7.2. *Large lakes of the world listed according to the lake surface area*

Lake	Continent	Surface area (km^2)	Lake volume (km^3)
Caspian Sea[a]	Asia	374 000	78 200
Superior	North America	82 367	12 232
Victoria	Africa	68 800	2 750
Aral Sea[a,b]	Asia	62 000 (28 000)	1 066 (181)
Huron	North America	59 570	3 540
Michigan	North America	58 016	4 870
Tanganyika	Africa	32 000	17 800
Baikal	Asia	31 500	23 000
Great Bear	North America	31 153	2 240
Nyasa (Malawi)	Africa	30 044	8 400

[a] Considered a saline lake because it is landlocked.
[b] Irrigation diversions from tributary streams beginning in the 1960s reduced inflows. Area and volume are now variable. Current representative values are shown in parentheses.

between the heavy runoff along the Pacific Coast contrasts markedly with the low to modest runoff extending from the central United States into northern Canada. Europe displays moderate runoff increasing at higher latitudes. Peel *et al.* (2001) provide a further assessment of continental differences in annual runoff variability.

7.12.1   Lakes and reservoirs

Lehner and Döll (2004) estimate there are nearly 250 000 lakes and reservoirs worldwide larger than 0.1 km^2 covering a total area of 2.7 million km^2. The volume of water in freshwater lakes is 90 000 km^3, but large lakes account for a disproportionately high percentage of this volume (Table 7.2). The volume of water stored in over 10 000 reservoirs worldwide, not including regulated natural lakes, is estimated at 4286 km^3. The largest reservoirs represent a minority of the total number of reservoirs, but large reservoirs are the common focus of public attention.

Lakes and reservoirs are unequally distributed globally. The majority of lakes occur in the Northern Hemisphere, the maximum number occurs at 35° N, and over 60% of the world's lakes are in Canada. The maximum number of large reservoirs occurs between 30° N and 55° N. However, a significant number of large reservoirs exist between 30° N and 30° S, reflecting major dam construction in South America, Africa, and Southeast Asia (Lehner and Döll, 2004). There are over 40 000 dams worldwide higher than 15 m, and more

Table 7.3. *Large reservoirs of the world listed according to reservoir water storage volume[a]*

Reservoir	River	Country	Continent	Volume (km^3)	Surface area (km^2)
Bratskoye	Angora	Russia	Asia	169.3	5478
Nasser	Nile	Egypt, Sudan	Africa	162	6000
Kariba	Zambesi	Zambia, Zimbabwe	Africa	160	5400
Volta	Volta	Ghana	Africa	148	8502
Guri	Caroni	Venezuela	South America	129	4250
Krasnoyarskiye	Yenisei	Russia	Asia	73.3	2000
Williston	Peace	Canada	North America	70.3	1779
Zeiskoye	Zeya	Russia	Asia	68.4	2119
Kujbyshevskoye	Volga	Russia	Europe	58	5900
Cabora Bassa	Zambezi	Mozambique	Africa	55.8	2739

[a] Natural lakes regulated by a dam are not included.

than half of these large dams are in China (ICOLD, 2003). Table 7.3 shows selected characteristics of several large reservoirs emphasizing the water volume impounded by large dams. It is estimated that 77% of water discharged from rivers in the Northern Hemisphere is from watersheds regulated by dams or other human influences.

7.12.2   Wetlands

Permanent and intermittent wetlands cover 8–10 million km^2 in a variety of hydroclimatic settings. Wetlands have a maximum development around 60° N with a secondary peak near the equator. They are more widespread in the Northern Hemisphere than in the Southern Hemisphere (Lehner and Döll, 2004).

7.13   Regional hydroclimate

The Earth's surface can be divided into regions that have similar hydroclimates due to the global climate system, but regional hydroclimate variations elucidate the role of a second tier of factors responsible for the spatial variation in hydroclimate when viewed at a higher resolution than global patterns. Regional hydroclimates emerge as identifiable entities based on latitude, altitude, and orientation of the surface in relation to water

bodies, mountains, and prevailing winds (Hartmann, 1994). Regional hydroclimates resulting from these second-tier factors are masked by the level of
generalization used to characterize global hydroclimates. However, regional-
scale hydroclimates are components of the global hydroclimatic system, and
they form the mosaic pattern that gives detail to the expression of hydroclimate most apparent for human observers at the Earth's surface. Large river
basins are often employed as a basis for quantifying hydroclimatic variables
of interest. The precise dimensions of a hydroclimatic region are variable, but
they commonly involve areas of several hundred to thousands of square
kilometers, and the time scales used for analysis are weeks, months, and
years (Linacre, 1992).

7.13.1   Atmospheric hydroclimatic variables

Temperature serves to illustrate regional characteristics of atmospheric
hydroclimatic variables due to the convenient availability of appropriate data
for this variable. The temperature examples depict variations commonly
observed at this scale for this class of hydroclimatic variables. Regional temperature characteristics are revealed by graphic portrayal of individual station
data and by maps of monthly temperature.

A station's annual temperature regime is the sequence of mean monthly
temperatures resulting from the multiple factors influencing temperature at a
specific location. The three stations shown in Figure 7.23 span 5° of latitude along
the Texas/Mexico border. The seasonal temperature variation of 23 °C at El Paso,
Texas, is characteristic of the temperature resulting from the rapid heating and

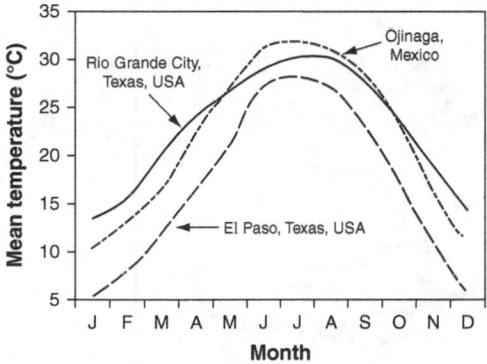

Fig. 7.23. Monthly mean temperature for three stations along the Texas/Mexico
border showing regional temperature variations. (Data courtesy of NOAA and the
National Climate Data Center from their website at http://www.ncdc.noaa.gov/oa/
climate/ghcn-monthly/index.php.)

cooling at inland locations. El Paso's elevation (1194 m) accounts for modest cooling of summer temperatures compared to those observed at Ojinaga, Mexico (elevation 841 m), 330 km southeast of El Paso. The January mean temperature is 5 °C warmer at Ojinaga which is 750 km from the Gulf of Mexico. Mean monthly temperatures at Rio Grande City, Texas, approximately 200 km from the Gulf of Mexico, display a seasonal temperature variation of 17 °C related to the moderating influence of the Gulf. Cool season temperatures are 3 °C warmer at Rio Grande City compared to Ojinaga and summer temperatures are 2 °C cooler at Rio Grande City. An additional indication of the temperature-moderating influence of the Gulf of Mexico is that August is the warmest month at Rio Grande City, but July is the warmest month at Ojinaga.

Maps of monthly temperature provide a spatial summary of regional temperature variations. National, state, or watershed boundaries are commonly employed to define the area examined, but any definable areal construct can be employed. State boundaries, such as California (Fig. 7.24), are a convenient delimiter. California covers 411 000 km² and extends 1280 km north to south. Mountains near the coastline range in maximum elevation from 2300 m in the north to 3600 m in the south, and mountains along the eastern border north of 35° N have

Fig. 7.24. Physical features of California.

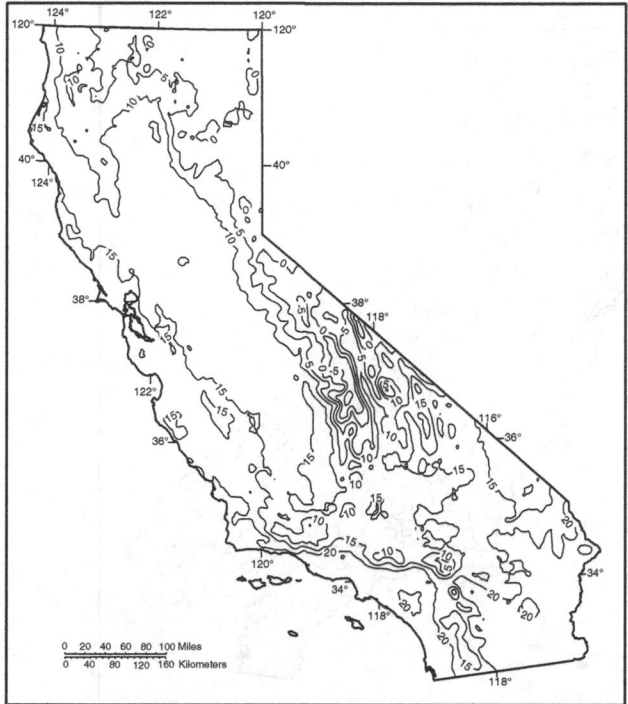

Fig. 7.25. California January mean temperature. Units in °C. (Data courtesy of the
USDA Natural Resources Conservation Service and the Prism Group from their
website at http://www.ncgc.nrcs.usda.gov/.)

peaks above 4400 m. California temperatures in January (Fig. 7.25) and July
(Fig. 7.26) are increasingly warmer from north to south due to solar radiation
being 15% greater at the southern border at 32° N compared to the northern border
at 42° N (Peixoto and Oort, 1992). Temperature differences west to east across
California are related to the presence of the Pacific Ocean that increases January
temperatures and decreases July temperatures along the coastline compared to
more inland areas. Elevation influences due to the north–south trending mountain
ranges are especially evident in greater January cooling than seen elsewhere in the
state. Slope and aspect influences combine with elevation to produce a more
complex pattern of isotherms along the eastern border in July than in January.

7.13.2   Terrestrial hydroclimatic variables

Precipitation illustrates regional features of terrestrial hydroclimatic variables. Precipitation has a dominant role in the runoff process, and precipitation
data are the most readily available of this class of variables. Precipitation data are
readily graphed and mapped to depict regional-scale variations.

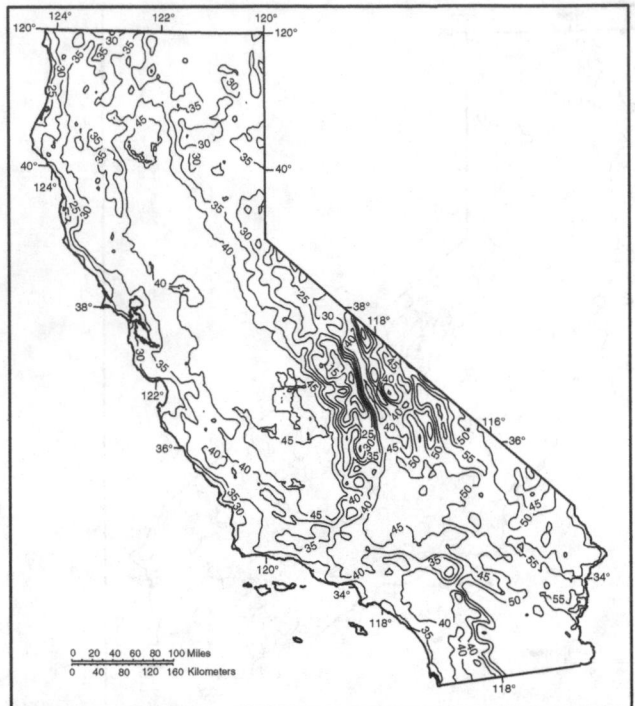

Fig. 7.26. California July mean temperature. Units in °C. (Data courtesy of the USDA Natural Resources Conservation Service and the Prism Group from their website at http://www.ncgc.nrcs.usda.gov/.)

Graphic portrayal of the monthly precipitation sequence reveals valuable insights regarding the moisture supply for hydroclimatic processes. Mean monthly precipitation for four stations in southern Arizona and northern Mexico (Fig. 7.27) displays regional differences in the quantity and monthly occurrence of precipitation. These stations form a trapezoid approximately 350 km between the western apex at Ajo, Arizona, and the eastern apex at Douglas, Arizona. A summer-dominant precipitation pattern is evident at all four stations, but the annual precipitation varies from 377 mm at Douglas to 227 mm at Ajo. The July to September precipitation is attributed to the atmospheric circulation feature known regionally as the North American monsoon (Lorenz and Hartmann, 2006). This phenomenon is an expression of complex atmospheric circulation and water vapor transport that produces considerable temporal and spatial variability across a wide area of the southwestern United States and northwestern Mexico.

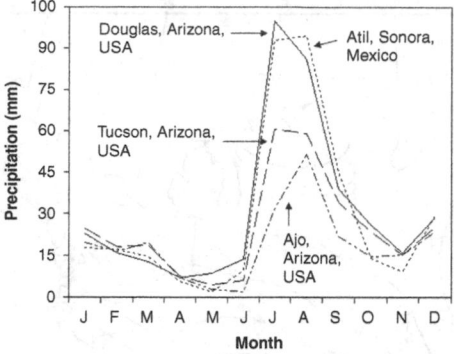

Fig. 7.27. Monthly mean precipitation for four stations along the Arizona/Mexico border showing regional precipitation variations. (Data courtesy of NOAA and the National Climate Data Center from their website at http://www.ncdc.noaa.gov/oa/climate/ghcn-monthly/index.php, and the Oak Ridge National Laboratory, Carbon Dioxide Information Analysis Center from their website at http://cdiac.ornl.gov/epubs/ndp/ushcn/usa_monthly.html.)

The average January (Fig. 7.28) and July (Fig. 7.29) precipitable water vapor over the United States portrays the large-scale atmospheric moisture field spatial characteristics. The general patterns evident in both months are decreasing values from south to north, decreasing values inland from the west coast and the south coast, and the effect of the Rocky Mountains in producing a relatively dry area centered on an axis at 110° W. These patterns result from temperature variations that influence the atmosphere's capacity to retain moisture, distance from oceanic moisture sources, and prevailing atmospheric motion that transports the moisture from oceanic sources.

The January (see Fig. 7.28) and July (see Fig. 7.29) contrast in mean precipitable water across the Arizona/Mexico region reveals a significant seasonal disparity in atmospheric moisture. The magnitude of the seasonal water vapor variability is ultimately responsible for regional precipitation variability. July precipitable water is three times greater than January values across much of southern Arizona and northern Mexico. July precipitable water over the Arizona portion of the Mexico border region is second only to the Texas portion of the border, which benefits from its proximity to moisture inflow from the Gulf of Mexico resulting in elevated precipitable water values. The contrasting orientation of the precipitable water contours reveals another facet contributing to seasonal precipitation variability. The January contours across the Arizona/Mexico region display a meridional pattern that magnifies east–west moisture variability across the region. The July contours have a zonal pattern that emphasizes north–south moisture differences across the region. Differences in

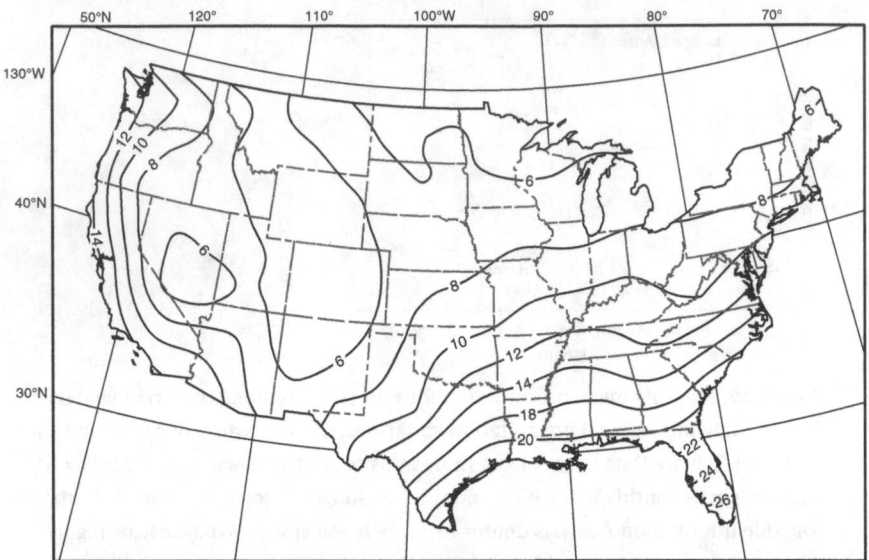

Fig. 7.28. United States January mean precipitable water. Units in mm. (NCEP Reanalysis data courtesy of NOAA/OAR/ESRL PSD, Boulder, Colorado, USA, from their website at http://www.cdc.noaa.gov/.)

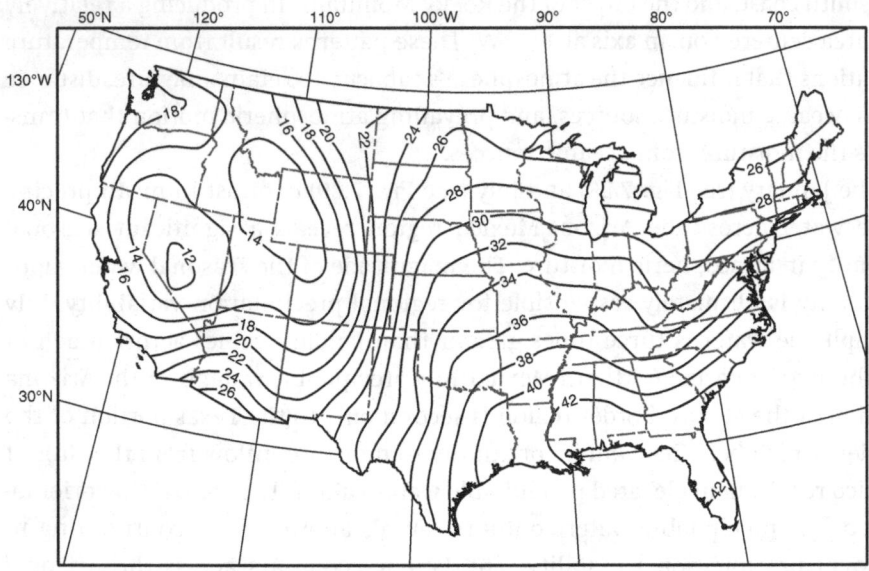

Fig. 7.29. United States July mean precipitable water. Units in mm. (NCEP Reanalysis data courtesy of NOAA/OAR/ESRL PSD, Boulder, Colorado, USA, from their website at http://www.cdc.noaa.gov/.)

precipitable water and surface influences contributing to instability help account for seasonal and spatial precipitation differences at individual sites.

The precipitation regime at Atil, Sonora (see Fig. 7.27), displays the strongest expression of the monsoon influence in that 53% of the annual precipitation of 351 mm arrives in July and August. Further east at Douglas, Arizona, July accounts for nearly 13% more precipitation than August, and the two months combine for 49% of the annual total. July and August precipitation at Tucson, Arizona, is about 60% of precipitation for these months at Atil, and they provide 40% of the 297 mm of annual precipitation at Tucson. On the west side of the region at Ajo, the monsoon influence accounts for a July and August contribution of 37% of annual precipitation. A notable feature of the monthly pattern at Ajo is that August precipitation is 61% greater than July precipitation. At the other three stations, July and August precipitation are relatively similar in magnitude.

Annual precipitation for California (Fig 7.30) displays regional variations related to its proximity to the Pacific Ocean, topography, and atmospheric circulation. Precipitation exceeding 200 cm near the northwest coast results

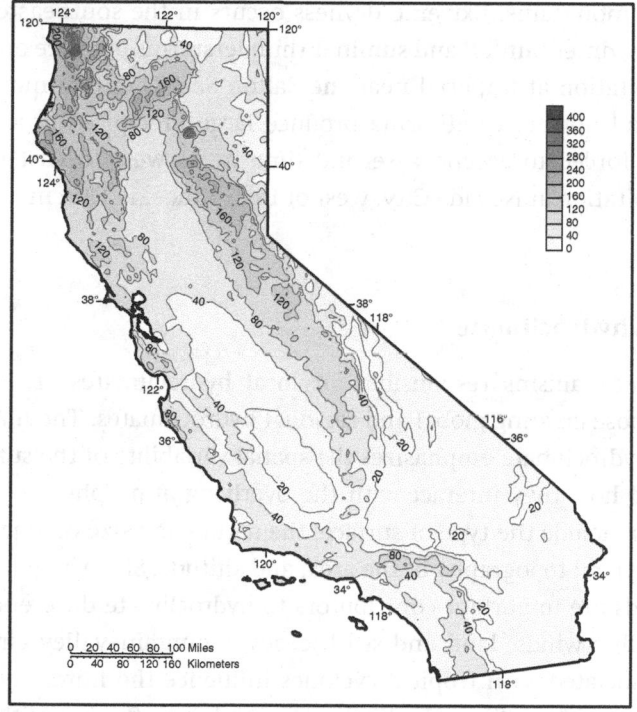

Fig. 7.30. California annual mean precipitation. Units in cm. (Data courtesy of the USDA Natural Resources Conservation Service and the Prism Group from their website at http://www.ncgc.nrcs.usda.gov/.)

from the passage of cold fronts and cyclones imbedded in prevailing westerly winds encountering mountains perpendicular to the atmospheric flow. Much of the moisture in the air from the Pacific Ocean moving onshore is removed by orographic influences. Decreasing precipitation from north to south along the coastline results from seasonal shifts in storm tracks and a decreasing occurrence of rain-producing storms related to latitudinal changes in the position and strength of the Pacific subtropical high-pressure system (Castello and Shelton, 2004). Storm tracks move southward in the fall and winter as the westerlies increase in strength in response to the stronger Northern Hemisphere pressure gradient. Frontal tracks move poleward in the spring and summer as the Pacific subtropical high-pressure cell strengthens, moves poleward, and asserts domination of the general circulation in these latitudes. Annual precipitation on the southern California coast at San Diego is 24 cm.

Inland locations reflect similar north to south precipitation differences, but complexity is added by elevation, slope orientation, distance from the ocean, and the nature of intervening topography. Precipitation at lower elevations in central and southeastern California is 20 cm or less as these areas are downwind of the coastal mountains. Extreme dryness occurs in the southeastern desert where meager winter rainfall and summer thunderstorms produce only 6 cm of annual precipitation at Imperial near the Salton Sea. The high mountains in east-central and northern California produce large precipitation amounts as marine air is forced to ascend a second time in its west to east trajectory. Annual precipitation at Nevada City, west of Lake Tahoe, is 139 cm.

7.14   Local hydroclimate

The mechanisms responsible for local hydroclimates are essentially the same as those creating global and regional hydroclimates. The major difference is local hydroclimate emphasizes the spatial variability of the surface characteristics and how they interact with the overlying atmosphere. The surface characteristics include the type of surface, the nature and size of objects on the surface, the general topography of the area, and altitude. Slope angle and aspect and local winds are important contributors to hydroclimate differences at this scale. Orographic winds, land and sea breezes, mountain–valley circulations, and winds associated with tropical cyclones influence the humidity and temperature characteristics of local hydroclimates, and the influence of these differences can be significant over relatively short distances (Linacre, 1992). The main physical consequence of these characteristics is that they produce variations in the radiation and energy balances and spatial temperature variations. In addition, surface characteristics produce local cloud and precipitation regimes that influence the water balance that is a fundamental expression of hydroclimatology.

Local hydroclimate produces the greatest diversity because it is a representation for a specific site. Another site only a few meters away may be very different due to variations in the site-specific characteristics controlling energy and moisture loadings. The areal range of local hydroclimates is typically less than a few square kilometers, and the time step for expressing these characteristics is minutes or hours (Linacre, 1992). Because measurements of hydroclimate variables are site specific, a complementary relationship exists between local hydroclimate and regional and global hydroclimates. The perception of global and regional hydroclimates is founded on local hydroclimate measurements, and local hydroclimate results from the same mechanisms as those creating global and regional hydroclimates.

7.14.1   Local solar radiation

Solar radiation reaches the Earth's surface by both direct and diffuse pathways. The optical air mass and the amount of water vapor, cloud, and aerosols in the atmosphere all contribute to reducing the solar radiation arriving at the surface to drive hydroclimatic processes. Variations in atmospheric conditions produce differences in hourly solar radiation over short distances that are not apparent when values are averaged for longer time intervals (Linacre, 1992). Local variations in solar radiation establish the foundation for complex differences in hydroclimatic processes not evident at larger time and space scales.

Hourly solar radiation data for three sites in California's Sacramento Valley (Fig. 7.31) illustrate solar radiation differences over relatively short distances.

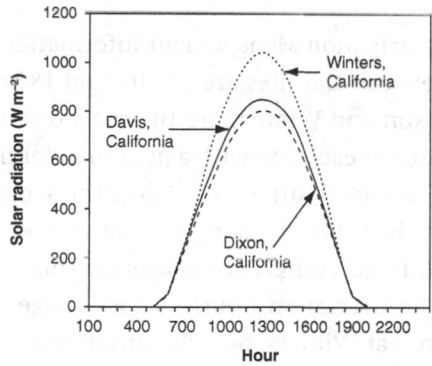

Fig. 7.31. Solar radiation on 7 May 2006 at three California Irrigation Management Information System (CIMIS) sites less than 20 km apart in California's Sacramento Valley. (Data courtesy of the California Department of Water Resources from their website at http://wwwcimis.water.ca.gov/.)

Table 7.4. *Daily averages for selected climatic variables on 7 May 2006 for three California Irrigation Management Information System (CIMIS) sites in California's Sacramento Valley*

Variable	Davis	Dixon	Winters
Solar radiation (W m^{-2})	297	279	353
Vapor pressure (kPa)	1.3	1.5	1.4
Air temperature (°C)	19.1	18.1	20.1
Dew point temperature (°C)	11.0	12.5	12.2
Wind speed (m s^{-1})	1.6	2.6	1.5
Wind direction (0–360)	221	236	145

Fig. 7.32. The California Irrigation Management Information System (CIMIS) instruments at Davis, California. (Photo by author.)

The sites are part of the California Irrigation Management Information System (CIMIS) Network. The distances between the sites are 13, 16, and 18 km. Davis and Dixon are the closest, and Dixon and Winters are the most distant. Total incoming solar radiation is measured at each site with a pyranometer mounted 2 m above a grass-covered and unobstructed surface (Fig. 7.32). The average solar radiation for 7 May 2006 is shown in Table 7.4 along with averages for other selected variables for this date. The largest difference in average solar radiation is 27% for Dixon and Winters. The maximum hourly values range between 802 W m^{-2} at Dixon and 1036 W m^{-2} at Winters, or a 29% difference.

Earth–Sun geometry indicates solar radiation should be similar at the three sites. Observed differences are the product of the interplay of environmental influences acting to attenuate solar radiation. The air at Dixon is cooler and more humid, winds are stronger, and the dominant flow is from the west, southwest. This trajectory promotes the inland flow of marine air from the Pacific Ocean. These conditions are especially strong during the mid-day hours when the solar radiation differences between Dixon and the other sites are greatest. The elevated moisture content of air at Dixon provides greater opportunities for reducing incoming solar radiation below the levels observed at the other sites.

At Winters, winds prior to 1500 are from the north, northeast, the air is drier than at Dixon, and solar radiation is greater each hour as less is attenuated. Davis is the furthest north and east of the three sites, and the air at Davis is the driest. Light winds are variable from the south to north, northwest throughout the day. Between 1200 and 1400, winds are stronger at Davis than at the other sites and are from the north, northwest. Winds from this trajectory often have increased aerosols after flowing across agricultural lands, and these aerosols reduce solar radiation. This influence would account for the reduced mid-day solar radiation values at Davis, but Winters does not experience a similar influence because winds during these hours are weak and from the north and do not provide similar aerosol loadings.

7.14.2   Local precipitation

Precipitation, evapotranspiration, and soil moisture readily illustrate the character of local hydroclimatic variations when portrayed as point data. Point values avoid the issue of interpolation arising from the uncertain spatial characteristics of variables. Precipitation is measured with high space and time frequency and is recommended as an observed value for comparison at the local scale.

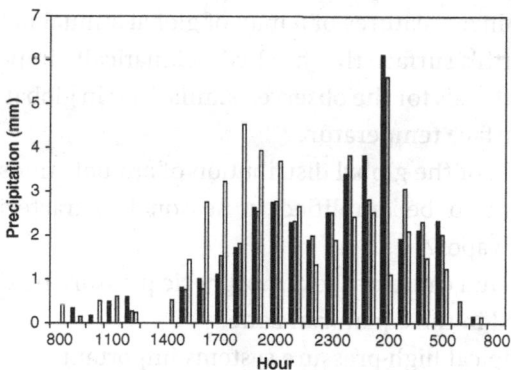

Fig. 7.33. Precipitation on 5–6 March 2006 at three California Irrigation Management Information System (CIMIS) sites less than 20 km apart in California's Sacramento Valley. Solid column is Davis, California, shaded column is Dixon, California, and clear column is Winters, California. (Data courtesy of the California Department of Water Resources from their website at http://wwwcimis.water.ca.gov/.)

Hourly precipitation for 5–6 March 2006 at the three sites described in Section 7.14.1 is shown in Figure 7.33. The Dixon total is 37 mm and the total for both Davis and Winters is 38 mm. All three sites record the major precipitation occurrence during the 15 hours from 1500 on 5 March to 0500 on 6 March. The temporal rainfall occurrence is similar at Davis and Dixon with an initial pulse peaking at 2000, a maximum at 0200, and a comparatively rapid decline after the 0200 maximum. The hourly rainfall occurrence at Winters is distinctly earlier in the evening and contrasts markedly with the pattern at the other two sites. Precipitation reaches a maximum at Winters at 1800, declines through 2200, increases to a secondary peak at 0100, and generally declines thereafter. While Davis and Dixon record 24 mm and 25 mm, respectively, of their total between 2200 and 0600, Winters accumulates 23 mm of its total from 1400 to 2100. Winters has four hours that receive 3 mm or more of rainfall, while Davis and Dixon have three hours each. Winters maximum rainfall is 4.5 mm at 1800, and the maximum at Davis and Dixon of 6.1 mm and 5.6 mm, respectively, is at 0200. In summary, although the rainfall event totals are similar at the three sites, the temporal characteristics illustrate that important differences with hydroclimatic significance occur over small areas.

Review questions

7.1 Why is the global pattern of outgoing radiant energy at the top of the atmosphere more complex than the global pattern of incoming radiant energy at the top of the atmosphere?

7.2 What are the prominent features of a map of global annual net radiation at the Earth's surface that are hydroclimatically important?

7.3 What is the physical basis for the observed similarities in global maps of January and July surface temperature?

7.4 What characteristics of the global distribution of annual atmospheric water vapor appear to be amplified in seasonal characteristics of atmospheric water vapor?

7.5 What is the nature and cause of the atmospheric pressure field commonly observed in the equatorial zone?

7.6 How are the subtropical high-pressure systems important hydroclimatic features?

7.7 How do global maps of 500 hPa heights contribute to understanding the surface patterns of atmospheric pressure?

7.8 How can westerly winds dominate upper air flow poleward of about 20° latitude?

7.9   What is the linkage between the trade winds, the intertropical convergence zone (ITCZ), the Hadley cells, and the subtropical anticyclones?

7.10   What is the role of atmospheric circulation in explaining seasonal changes in atmospheric water vapor over North America?

7.11   What accounts for spatial variability of annual rainfall?

7.12   What is the general equator to pole pattern of precipitation seasonality?

7.13   What is the hydroclimatic significance of the seasonal variability of evapotranspiration and precipitation?

7.14   What makes portrayal of local hydroclimates different from portraying regional hydroclimates?

8

Hydroclimate temporal variations

8.1   Temporal scale

The history of the Earth's hydroclimate is very complex and efforts to understand hydroclimatic processes through time are hindered by the relatively short existence of humans on the planet. The ability to quantify climate variability is fundamental to understanding past and future hydroclimates and to developing a comprehensive perspective of the range and temporal variability of hydroclimatic processes. In the absence of direct observations of climate variables, alternative approaches have been devised to extract climatic information from natural recording systems to develop a climatic record of sufficient duration to expand our understanding of the climate system, climate variability, and hydroclimate variability.

Temporal variability is an inherent characteristic of climate and hydroclimate, and empirical evidence shows that most hydroclimatic processes deviate from stationarity in the long term. However, most historical records are too short to accurately reveal the trends, oscillatory behavior, persistence, and sudden shifts from one stationary state to another that characterize long records (Lockwood, 2001; Sveinsson *et al.*, 2003; Rybski *et al.*, 2006). Examination of climate variability is complicated further by the absence of a universally accepted distinction between the terms "variability" and "change". Both terms refer to fluctuations in climate from some expected or previously defined mean climate state. Variability refers to the oscillations, or the pattern of fluctuations, about some specified mean value while change refers to a secular trend producing a displacement of the average. Distinctions can only be made relative to the time scales of concern. On longer time scales, change is viewed as an instance of climate variability. Aspects of temporal climate variability include those associated with day-to-day changes, interseasonal and interannual variations, and the spatial variability associated with horizontal gradients. Stochastic mechanisms create what is thought of as "white noise" and account for a large proportion of climate variation. Therefore, a substantial component of climate variation cannot be predicted.

A pervasive problem for assessing temporal climate variability is the establishment of a reference level about which to calculate the variability. The selection of the baseline for comparison is often a critical decision. Results vary depending on the baseline selection. Conceptualizing climate as the summation of energy and moisture conditions at a site implies the presence of a temporal record that reflects the full range of conditions characteristic of the location.

Variations in the operation of the climate system over time define the temporal variations of interest to climatologists. Divisions of the time continuum are products of natural characteristics of physical processes, or they are the product of human perception or human convenience. The 24-hour day has an obvious physical reality for its basis, but the summation of precipitation for a particular month is defined by a human imposed boundary on the data. Nevertheless, an approximate linear relationship between the size of atmospheric features and their time scales is widely recognized. Small-scale turbulence has a time scale of seconds or minutes, but turbulence that ultimately produces a hurricane lasts for days or weeks.

Most regions experience hydroclimatic variation from month to month and season to season. The frequencies of extreme events are more sensitive to changes in the variance than to changes in the mean. Studies indicate there have been significant changes in both interannual and day-to-day climate variability in historical times, but simple or distinct relationships between changes in mean climate conditions and changes in variability have not been established.

What has been learned is that the causes of climate variability are largely time-scale dependent and can be divided into the two major categories of external and internal forcings. External forcings are related to natural climate variability resulting from changes in forcing traceable to solar irradiance. For nearly a century, scientists have recognized a significant statistical correlation between the periodic variation of many climate variables and either the 11-year sunspot cycle or the 22-year double-sunspot cycle. These forcings tend to be periodic and can be anticipated with some degree of success.

The second set of forcings arises from internal dynamics of the climate system, and they produce stochastic or random fluctuations and possible chaotic behavior. Natural random variability is related to forcings such as volcanic activity or the feedback between land and water surfaces. Variability may be due to "chaos", which is the random and erratic behavior of a deterministic or nearly deterministic system that can arise without a forcing influence.

The combined effect of these influences is that climate varies on all time scales in response to both random and periodic forcing factors. There is reason to believe that different forcings may result in different patterns of climate variability. Daily variability of a non-periodic nature largely results from variations in synoptic-scale weather processes. The daily climatic variability of a given location is a function of the frequency of occurrence of different air masses. The air mass frequency is partially determined by a site's position with respect to the air mass source regions, orography, and the mean locations of circulation features such as cyclones and anticyclones. Variations of climate on a year-to-year basis (interannual variability) can arise from external forcings or from slowly varying internal processes including interactions between the atmosphere and oceans, soils, and variations in sea ice extent. On decadal time scales, precipitation is more sensitive than temperature to atmospheric circulation variations.

8.2   Earth's climate in perspective

Climate variability is meaningful only when there is a standard of comparison for assessing the perceived deviation. Establishing the climate standard is complicated by the relatively short history of human existence on Earth. The instrument record is still shorter. An appreciation of the climate record is needed to define the context for hydroclimate variability in the past, present, and future. A comprehensive analysis is beyond the scope of this book, but an overview of selected concepts is presented to illustrate the nature of the time scale differences.

The Earth's age is about 4.5 billion years, and the vast majority of this history lies outside of the human perspective (Saltzman, 2002). Little or no information is known about the earliest parts of the Earth's history because the climate system has been a major agent in sweeping away evidence. The Earth's surface changed through time as its plates moved into different alignments and erosion and tectonic activity altered continental configurations. Changes in atmospheric composition and the radiative properties of the atmosphere accompanied the progressive changes in terrestrial and oceanic environments. The changes in the world's continents, oceans, and atmosphere produced a global climate that varied over geologic time.

A general understanding of ancient climatic conditions, known as paleoclimates, has been assembled for specific parts of the geologic past. Since paleoclimates existed before humans began collecting weather data, they are based on reconstructed climates inferred from natural environmental records rather than the observational data that form the framework for expressing present hydroclimates. Because paleoclimate data are highly specific for a given site, they may represent local anomalies rather than large-scale variations. Nevertheless, these reconstructed climates provide a wide range of useful knowledge about the climate system. By extrapolation, we can use this knowledge to expand our understanding of hydroclimate. Paleoclimate insights that are especially transferable to hydroclimate are knowledge of how the climate system works, the range over which climate can vary and has actually varied, the relative stability and/or instability of the climate system, and how fast and to what degree climate can vary.

Paleoclimate evidence indicates that global temperatures varied widely during the last 3.8 billion years, and the global climate was warmer and wetter than today for most of Earth's history. Although periods with widespread glaciation were common early in the Earth's history, the Earth's basic climate pattern has existed for the past 1 billion years even though the position of continents has changed. An exception is the Mesozoic era, between 245 million years ago and 65 million years ago, when global climate was warmer and drier and continents were clustered around the South Pole. The Cretaceous period during the last 80 million years of the Mesozoic era had a globally averaged surface temperature 6–12 °C warmer than the present temperature. Modern tropical to subtropical conditions extended to about 45° N, and higher latitudes were warmer than today. Tropical surface water temperatures were similar to or slightly warmer than temperatures today, but oceanic deep water was warmer. Warmer polar regions supported abundant high-latitude floras and faunas, and there is no direct evidence of polar ice caps (Burroughs, 2001; Saltzman, 2002).

Warm and wet conditions continued into the Tertiary period, but a cold and arid climate resembling today's was dominant by the end of the Oligocene epoch about 25 million years ago. A slight climatic amelioration occurred during the Miocene epoch, but temperature and precipitation continued downward during the Pliocene epoch. By the end of the Pliocene about 2 million years ago, the climate was colder and drier than today. This set the stage for the most recent and best known of the glacial epochs spanning the Pleistocene. Temperature changes during ice ages were greatest at the high latitudes and small near the equator. Tropical land areas were drier on average during glacial maxima than during warm periods. The global mean temperature was about 5 °C colder than now 20 000 years ago during a full-glacial episode. Temperature increases during the present Holocene interglacial have returned global temperatures to about the level of those occurring 125 000 years ago, and altered precipitation, evaporation, and runoff during this time have affected the expansion and desiccation of lakes worldwide (Melack, 1992). Since the current climate is very near the warmest that has been observed in the last million years, we have little information about what constitutes a really warm climate and how the environment will respond.

What emerges from this paleoclimate background is that our understanding of past climates degrades with time. We have a very good understanding of climate over the past several thousand years, but a poor understanding of climate beyond several hundred million years. Nevertheless, it is evident that during the past 10 000 years, as civilizations developed, our climate has been remarkably stable. In the context of the paleoclimate record, this stability appears to be highly unusual.

Paleoclimates are reconstructed using geologic data and proxy data that serve as a substitute for actual climatic records. Proxy data are physical, biological, and chemical information present in natural features of the environment that are indicators of a particular climate or climate variable. These natural recording systems provide indirect evidence of climatic trends over geologic time. Proxies include tree rings, pollen, ice cores, lake sediments and shorelines, relic soils, marine shorelines, corals, and deep-sea sediments. These natural climate archives substitute for thermometers, rain gauges, and other modern instruments used to record climate, but they have a wide range of differences in terms of the resolution of the climate element they portray. The utility of proxy climate data is achieved by calibrating the proxy with the available instrument record using established dating methods as described by Bradley (1999). The key to meaningful paleoclimatic reconstruction is the interaction between proxy phenomena and present climate.

For hydroclimatic purposes, the most extensive and highest resolution natural record is provided by annual tree rings from temperate and boreal forests (Saltzman, 2002). However, this record is relatively short compared to the time line available using other proxy data. Combining information from tree rings, ice cores, lake sediments, corals, and deep-sea sediments provides improved understanding of climate and the climate system for the past several thousand years. Reconstructing precise atmospheric conditions is hindered by the limited spatial data on which to base definition of a pattern that can be related to reasonable atmospheric circulation features, but some success has been achieved. The stable hydrogen isotope composition of bristlecone pine tree rings has yielded information on temperature and air mass trajectories for North America during the past 8000 years (Feng and Epstein, 1994). Ocean sediment cores have revealed evidence linking shifts of the ITCZ to climate change in Central and South America during the last 14 000 years (Haug et al. 2001). Ice cores from Greenland and West Antarctica have contributed to developing temperature patterns for the past 90 000 years through analysis of gases trapped in the annual ice layers at different depths and inference of a relationship between temperature and the changing composition of the atmosphere through time (Blunier and Brook, 2001). Similarly, pollen in lake sediments has been employed to estimate regional temperature and precipitation variations for the last 130 000 years (Adam and West, 1983). For periods covering the past several hundred million years, a high-resolution record is being sought from deep-sea sediments. However, this task is difficult because most of the sea floor older than 200 million years has been subducted as part of tectonic recycling, and the hydroclimatic information contained in these sediments is lost (Saltzman, 2002).

8.2.1 Historical hydroclimate

The relatively brief record of instrument data severely hinders efforts to develop a comprehensive understanding of hydroclimate over the past 1000 years. For the immediate pre-instrument period, seasonal and monthly temperature and precipitation can be gleaned from a variety of written and illustrated sources ranging from crop harvest dates to artistic works. These historic data are in documentary form rather than quantitative and are commonly grouped into four categories. First is the observations of specific weather phenomena, such as the first day of frost or the occurrence of snow, which are frequent entries in personal journals and diaries. A second category is the dates of floods, drought, and other weather-dependent natural phenomena, sometimes called parameteorological phenomena, which appear in municipal and governmental reports. The third group is phenological records that deal with the timing of recurrent weather-dependent biological phenomena, such as crop harvest dates or crop yield data (Bradley and Jones, 1995). A fourth group is records of forcing factors, such as sunspot activity, that influence climatic conditions. Utilizing this historical information requires recognition that the information is subjective, sketchy and limited to a few geographical areas. In addition, extreme events that have a high probability of appearing in historical accounts may not be representative of the overall climate (Bradley, 1999).

8.2.2 Documentary data

Reconstruction of past weather and climate using anthropogenic climatic data in the form of written and illustrated records found in archives, libraries, and museums constitutes hydroclimatic documentary data. In one sense, this is another form of proxy data, but documentary data are derived through a human filter rather than being produced by a natural biological or chemical response to climate. The accurate and appropriate use of documentary sources involves a rigorous methodology for isolating reliable materials from those less faithful in conveying climatic information. Content analysis has been useful as a technique for assessing this information in quantitative terms, allowing the documentary data to be calibrated with instrumental measurements to estimate specific meteorological variables (Bradley and Jones, 1995).

Archives of central and local governments have been the source of some of the most useful documentary data. An illustrative case is the dates of tithe auctions in Western and Central Europe. Tithe auction dates are highly correlated with wine harvest dates and with Central England temperature time-series. The tithe dates are earlier than the mean during warm periods and are later than the mean during cold periods (Pfister, 1980). Similarly strong relationships are found between wine harvest dates and wine yields in France, Switzerland, and other parts of Europe. Grapevines are good climatic indicators because the plant remains the same for 25 to 50 years and does not require annual planting. Also, the entire length of the growing season from March or April to October is needed to bring grapes to maturity. This means that harvest date, yield per acre, and wine quality can be used as climatic proxy evidence for three different periods of the growing season, even though the date of harvest is the most popular proxy and the least subject to interpretation controversy (Ladurie and Baulant, 1980).

Although the overwhelming majority of documentary data come from written records, pictorial documents have proven useful. Paintings of Alpine glaciers in Europe record changes in glaciers since the sixteenth century, and cloud cover depicted by Dutch and British landscape painters since 1550 shows summer atmospheric characteristics. However, the use of such materials involves major interpretative difficulties related to subjective biases and the fallibility of the untrained observer. This presents major problems in quantifying the record so we really know what the event is telling us.

The utility of documentary data for developing time-series of hydroclimatic data is constrained because the period of observation is variable and often incomplete and/or discontinuous. Spatial coverage further constrains documentary data because the availability of these data is linked to settlement histories (Bradley, 1999). Documentary data suitable for climatic interpretation are available for about 700 years for Europe and Asia, but the earliest data for Australia begins about the mid to late 1700s. For western North America, a catalog of historical agricultural disasters in Mexico provides a basis for identifying historical droughts from 1450 to 1899 (Mendoza et al., 2006). The spread of Spanish presidios and missions into California between 1776 and 1834 provides an important source of documentary data. Annual reports of the crops planted and harvested at the presidios and missions have been transformed into estimates of rainfall variability in southern California for this period (Rowntree, 1985).

Derived climate histories using all available sources of documentary data can provide very useful numerical expressions of climate and valuable information

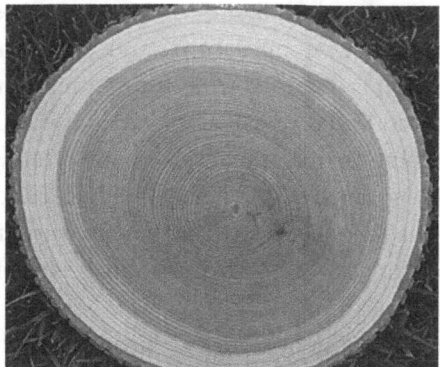

Fig 8.1. Cross-section of a hickory stem showing annual growth rings in the heartwood (darker wood in the center) and the sapwood (lighter wood near the bark). The small variation in tree-ring widths indicates the tree experienced relatively little environmental stress each year. (Photo by author.)

on climate trends. Lamb (1982) used documentary data to estimate temperature and precipitation for England and Wales for 50-year blocks beginning in AD 1100. What stands out in these data is the certainty of a warmer period that lasted several centuries in the Middle Ages (1100–1300) that is commonly identified as the Medieval Warm Epoch or Little Optimum. An equally long colder period follows from 1500–1700 corresponding to the Little Ice Age, characterized by wetter summers and more frequent severe winters in Europe. This record emphasizes that the Little Ice Age (c. 1500–1800) occurred at the transition from documentary data to instrument data. Instrument data are reliable for Central England beginning in about 1700.

Documentary evidence is a promising source of data regarding the details of short-term shifts and fluctuations for the millennium or so immediately preceding the era of modern instrumentation. Documentary data may be a particularly important source of information where other proxy data, such as tree rings, are not available. Bradley (1999) discusses the nature of documentary data and emphasizes its use in regional climate studies.

8.3   Tree-ring reconstructions

The use of tree rings to reconstruct hydroclimate is called dendroclimatology and dendrohydrology. Most temperate forest trees display concentric annual deposits of tree trunk material forming alternating lighter and darker bands of seasonal growth increments around the tree's circumference. The annual couplets of earlywood and latewood comprise an annual growth increment known as a tree ring (Fig. 8.1). The mean ring width in any tree is a function of tree species, tree age, nutrients available in the tree and soil, and climatic factors including precipitation, temperature, and solar radiation (Saltzman, 2002). Climatic information is obtained from interannual variations in ring widths and from variation in wood density, which is used like ring widths to identify annual growth increments. The earlywood and latewood of the tree rings vary markedly in average density, and density variations contain a strong climatic signal (Bradley, 1999). Tree-ring width and density variations are translated via a statistical transfer function into hydroclimatic variables that reveal temporal patterns of the variables.

Tree rings are the basis for pre-instrument climatic reconstructions for a variety of locations worldwide, including some limited tropical sites. Only a few tropical tree species form distinct annual rings, and tree growth in the tropics is less susceptible to interannual climate variability than at other latitudes (Pumijumnong, 1999). However, subtropical montane forests experience moderate temperature seasonality and large precipitation seasonality that induce dormancy and production of annual rings similar to temperate regions (Villalba et al., 1998). The basic strategy of dendroclimatology is to identify regions where trees are most sensitive to climatic stress so climatic differences are evident in the character of the tree rings. Such locations are commonly at the limit of the natural ranges of temperature and precipitation for a specific tree species. The longest continuous tree-ring records extending over several thousand years are achieved by overlapping the records for individual trees. Long records based on bristlecone pine in the White Mountains in California and oaks and Norway spruce in western Europe have been particularly successful (Hughes, 1996; Wilson et al., 2005).

The width and structure of the tree ring provide information on the temperature and precipitation when the tree ring was formed. In dry regions, the width or thickness of the annual ring is often controlled by the availability of moisture. In cool, moist regions, ring width or maximum wood density is determined by summer temperatures. Tree-ring widths in tropical tree species are most often related to precipitation during the transition from the dry season to the rainy season (Pumijumnong, 1999). The key is that the tree ring is an indicator of the factor that is restricting growth or is the most variable for that environment. By correlating tree-ring characteristics with temperature and precipitation data, a transfer function is developed to convert tree-ring characteristics into weather information. An extensive literature addresses the techniques used to extract climate information from tree rings and how to apply rigorous methods for paleoclimate reconstruction (e.g. Fritts, 1976, 1991; Hughes et al., 1981; Hughes, 1996; Tessier et al., 1997), and Loáiciga et al. (1993) focus on tree-ring based hydrologic reconstructions.

It is very important to remember that tree rings provide an indication of the energy and moisture condition and not a direct temperature or precipitation recording. The tree ring is the biological response to the energy or moisture condition. Time resolution is a significant issue involving tree rings. Floods related to high precipitation years may be inferred, but a flood produced by a single storm is not evident in the tree-ring record. Tree rings are more effective for identifying drought than for floods because of the longer time resolution of drought. Temperature reconstructions using ring widths of trees growing in cold environments usually show the influence of warm-season temperatures on growth most strongly (Esper *et al.*, 2002).

Hughes and Graumlich (1996) summarize the extensive tree-ring chronologies for the western United States which form a better tree-ring record for the past 1000 years for this region than for anywhere else in the world. These chronologies include an extraordinary wealth of climatically sensitive tree-ring records that can be used to place the instrumental period in a wider perspective. The most frequently used tree-ring variable for drought and other hydroclimatic studies has been the ring-width index, which measures the departures from normal of annual diameter growth of the tree (Hughes *et al.*, 1981). Both the growth increment of a tree and the annual or seasonal flow of a river are closely related to the water balance of the soil integrated over days, weeks, or months. Statistical studies have repeatedly shown that hydrologic variables and annual growth indices from properly selected trees are highly correlated (Woodside, 2001).

Streamflow reconstructions using tree rings are an extension of the research developed for temperature and precipitation reconstructions. Tree rings are correlated with streamflow records, and streamflow is estimated based on the tree rings. Tree-ring analysis has been employed to examine various facets of the time and space characteristics of streamflow in a variety of global settings (e.g. Meko *et al.*, 2001; Pederson *et al.*, 2001; Gedalof *et al.*, 2004; Davi *et al.*, 2006).

The Colorado River Basin in the western United States is an excellent example of the contribution provided by reconstructed streamflow. More water is diverted out of the Colorado River Basin than any other river basin in the United States, and an international treaty controls delivery of water from the Colorado River to Mexico. The Colorado River Compact was signed in 1922 dividing the flow of the river between upper basin and lower basin states and Mexico based on the average flow at Lees Ferry, Arizona. Streamflow data available in 1922 and beginning around 1900 were used in the Compact for dividing the water among the various states and Mexico. Runoff reconstruction for the Colorado River to 1490 is reported by Woodhouse *et al.* (2006). The streamflow reconstruction (Fig. 8.2) reveals that the Compact was based on a period of persistently high

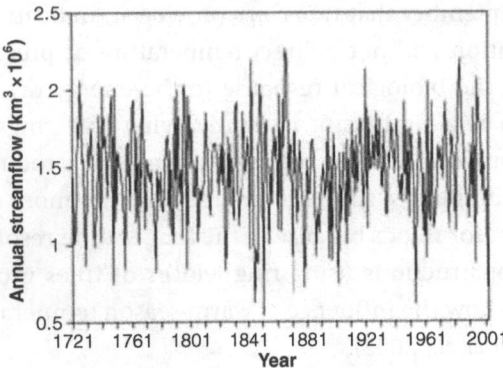

Fig. 8.2. Reconstructed annual flow of the Colorado River at Kremmling, Colorado, for 1721–2002 derived from tree-ring analysis and a statistical model. (Data courtesy of NOAA's National Climate Data Center from their website at http://www.ncdc.noaa. gov/paleo/streamflow/kremmling.html.)

flows and subsequent actual water allocations have been faced with the problem of dividing a small and more variable water supply than all of the participants expected. Perhaps the most important point revealed by the streamflow reconstructions is that the modern gauged streamflow record may be an unrepresentative sample. The high flows on the Colorado River in the twentieth century are among the highest in the record, and the twentieth century's most severe dry period is exceeded by several longer duration events in the reconstructed streamflow.

8.4   Ocean–atmosphere influences on hydroclimate

Hydroclimatic analysis benefits from the availability of instrument data for the most recent era only. However, climatic and hydrologic instrument data are almost always collected for a purpose other than hydroclimatic analysis. Daily meteorological observations at many sites are collected to support weather forecasts, and streamflow data are collected to provide information on navigation or flood control rather than analysis of past, present, or future hydroclimates. The disparity between the primary purpose for data collection and eventual data application is a particularly acute issue for developing the time-series most useful for hydroclimatic analysis. Meteorological data are collected at a variety of space and time scales dominantly at low elevation sites, and hydrologic data are collected at relatively fewer sites often at higher elevations so that space and time coordination are a challenge. Time-series of hydroclimatic data provide a rich source of information on climatic trends and variations, but they also contain the influence of forcing factors. Interpretation of hydroclimatic data requires awareness of the forcings embedded in the data.

The general circulation of the atmosphere is critically important for defining the hydroclimatic characteristics of a specified location. Planetary atmospheric circulation transports both heat and moisture zonally and meridionally. The mobility of the atmosphere permits it to provide rapid communication of energy and mass between places. Large-scale circulation features are imbedded in the global atmospheric circulation scheme, and the large-scale features play an important role in defining year-to-year and interdecadal hydroclimatic variations. These large-scale circulation features arise from internal atmospheric dynamics, or they can be forced by atmosphere–ocean–land interactions that are frequently linked to sea surface temperature (SST) variations. Bigg et al. (2003) discuss tropical and extratropical atmosphere–ocean couplings and the interaction between ocean basins. The variability of these circulation features can have especially strong impacts on regional temperature and precipitation differences and on creating definable cycles in these variables.

The interconnectedness of the atmospheric and surface systems and their hydroclimatic variables is expressed as teleconnections or the correlation between variables in one region with those in another region. Teleconnection patterns are recurring and persistent, large-scale pressure and circulation associations that span entire ocean basins and continents (Barnston and Livezey, 1987). These atmospheric features reflect large-scale changes in atmospheric wave patterns, the jet stream, and storm tracks, and a shift in storm tracks associated with teleconnections can change temperature and rainfall patterns in the mid-latitudes (Trenberth, 1998). Teleconnections are an important mechanism for transporting energy and moisture manifested as cycles at a given location. A distinctive characteristic of teleconnections and the related circulation features is that they occur with repeating patterns of time-space variability known as modes. In climatology, a mode is a spatial structure with at least two strongly coupled centers of action, and its polarity and amplitude are represented by the mode index (Wang and Schimel, 2003). Several identifiable atmospheric modes are recognized as factors in producing cyclic influences on hydroclimatic variables.

8.5   Madden–Julian Oscillation

The Madden–Julian Oscillation (MJO) is the most significant form of tropical atmospheric variability at time scales in excess of one week but less than one season (Slingo et al., 1999). The MJO is characterized by eastward-moving, planetary-scale disturbances in zonal winds and convective rainfall with a domi-nant mode of circulation variability at time scales of 40 to 50 days. However, it displays both pronounced seasonality and substantial interannual variability in intensity. The oscillation is the result of large-scale circulation cells oriented in the equatorial plane that move eastward from the Indian Ocean to the central Pacific Ocean (Madden and Julian, 1994). The convection cells are accompanied by intense precipitation. Interactions between MJO convection and large-scale circulation are strongest over the Indian Ocean and western Pacific Ocean where the oscillation exhibits its greatest variability and typically reaches its maximum amplitude (Waliser et al., 1999).

While the MJO occurs throughout the year, it is strongest near the equator during the Northern Hemisphere winter and spring and weakest in the summer. Also, the strongest MJO signal migrates from 5°–10° S in the winter to 5°–10° N in the summer (Zhang and Dong, 2004). The MJO enhances or suppresses convec-tion in the convective centers over the Amazon and Congo basins and Indonesia and subsidence in the strong semipermanent high-pressure centers over the eastern equatorial Pacific. The oscillation affects the onset, breaks, and with-drawal of the monsoon systems of Asia and Australia. The dynamical founda-tions of the MJO remain unclear and key issues are the manner in which cumulus convection and the wind oscillations interact and the basis for the slow eastward movement. However, theoretical and modeling studies show that the MJO prefers weak zonal winds at low levels and at the surface (Zhang and Dong, 2004), and its origins and modulation are related to interaction among large-scale circulation, latent heating, and surface evaporation (Kemball-Cook and Weare, 2001).

There is evidence that the MJO and related variations in the tropical circula-tion are coupled to extratropical circulation through zonal wind anomalies and Rossby waves propagating from certain tropical regions (Bond and Vecchi, 2003). Consequently, the MJO may influence the evolution of extratropical weather and precipitation on time scales of months to seasons, and it may be linked to El Niño and the Southern Oscillation (ENSO) and the onset, breaks, and withdrawal of the Asian monsoon. Teleconnection patterns show a weak rela-tionship between the MJO index and ENSO events, suggesting that the MJO is more active during ENSO cold phases (Gualdi et al., 1999; Slingo et al., 1999). The MJO has been related to winter extreme precipitation in California (Jones, 2000), cool-season precipitation variability in Oregon and Washington, and flooding enhancement in western Washington (Bond and Vecchi, 2003). A time-lagged MJO has been suggested as the trigger for a northward moisture surge in the Gulf of California that produces above-normal precipitation related to the North American monsoon (Lorenz and Hartmann, 2006).

8.6   El Niño, La Niña, and the Southern Oscillation

ENSO is probably the most widely recognized low-frequency cyclic event associated with atmospheric circulation. This phenomenon is a coupling of the atmosphere and ocean with a non-periodic recurrence of 2–5 or 3–7 years that has been suggested as the strongest source of natural variability in the Earth's climate system (Lockwood, 2001). It is particularly relevant for hydroclimate because it highlights the link between the climate system and the hydrologic cycle and the recurrence of floods and drought and other extreme conditions (Rajagopalan *et al.*, 2000). In simple terms, El Niño is an oceanic circulation component consisting of a huge pool of anomalously warm water along the equator in the eastern Pacific Ocean. La Niña is another oceanic component identifiable as anomalously cool SSTs in the equatorial Pacific Ocean. The Southern Oscillation is an alteration of the trade wind circulation resulting in a reversal of flow over the tropical ocean. Typical ENSO events tend to develop during summer to early fall, mature during winter, and terminate the following spring (Deser and Wallace, 1990). In the following discussion, ENSO is used to refer to the general system that comprises both the El Niño warm phase and the La Niña cold episodes that together constitute SST extremes in the Central Pacific.

El Niño originated in reference to the local occurrence of a warm water current off the coast of Peru and Equador around Christmas and lasting for one to two months. Eventually, scientists began using the term El Niño in a broader context for describing the appearance of abnormally warm water across the entire equatorial Pacific lasting for a year or longer. El Niño events are so important in the year-to-year weather variability that signs of a developing event are routinely incorporated into long-range seasonal weather forecasts as well as in regional agricultural strategies.

8.6.1   *ENSO development*

The normal distribution of SSTs across the tropical Pacific Ocean includes a warm region in the western Pacific, mostly to the west of the International Date Line (IDL). Temperatures over most of this area exceed 28 °C. Further east, the ocean becomes colder and colder until the temperature along the coast of South America is less than 23 °C. The cold coastal water is due to upwelling that is especially active along the west coast of South America. Ocean temperatures decrease at higher latitudes along the coast. This climate state of the ocean is determined primarily by the stress placed on the ocean surface by winds that blow east to west (Fig. 8.3). Cold water is advected westward along the equator. The easterly winds weaken considerably as they approach the IDL and the waters of the western Pacific remain warm year-round.

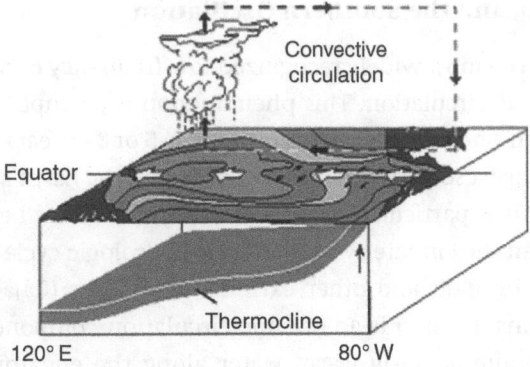

Fig. 8.3. Schematic illustration of normal conditions of sea surface temperatures and atmospheric circulation in the tropical Pacific Ocean. (Drawing courtesy of NOAA and the Pacific Marine Environmental Laboratory from their website at http://www.pmel.noaa.gov/.)

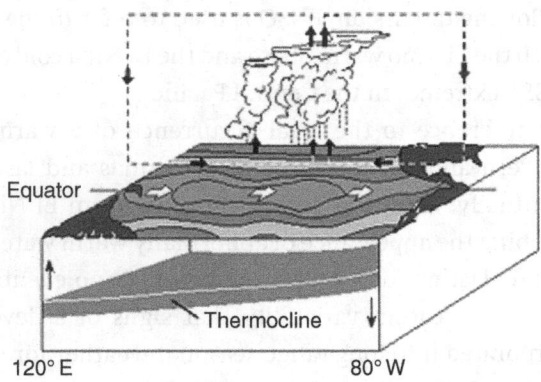

Fig. 8.4. Schematic illustration of sea surface temperatures and atmospheric circulation in the tropical Pacific Ocean during an El Niño event. (Drawing courtesy of NOAA and the Pacific Marine Environmental Laboratory from their website at http://www.pmel.noaa.gov/.)

The expansion of the area of warm water in the equatorial Pacific is now the indicator for an El Niño event. The anomalously warm water for 1982 was the largest warming in the last 100 years. The expansion of warm water is related to a weakening of the easterly flow of the atmosphere and a strengthening of westerly surface winds in response to changes in atmospheric circulation, ocean currents, and SST changes (Fig. 8.4). Warm water normally driven by the trade winds toward the western tropical Pacific drifts slowly eastward and may go as far poleward as British Columbia and central Chile. The ocean water drift is slow and may take several months to reach the west coasts of North and South

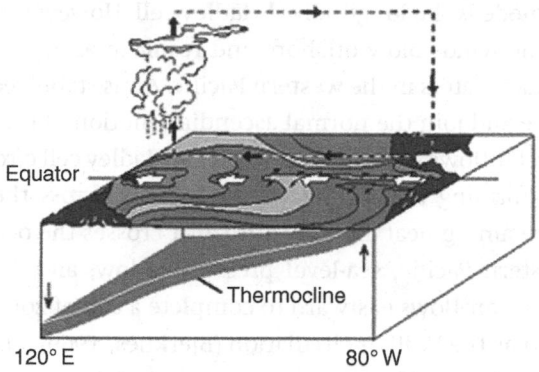

Equator

Thermocline

120° E   80° W

Fig. 8.5. Schematic illustration of sea surface temperatures and atmospheric circulation in the tropical Pacific Ocean during a La Niña event. (Drawing courtesy of NOAA and the Pacific Marine Environmental Laboratory from their website at http://www.pmel.noaa.gov/.)

America. The tropical Pacific thermocline rises in the west and sinks in the east effectively cutting off upwelling along the coast of South America.

At times, an El Niño is followed by abnormally cold water along the coast of Peru and the equatorial Pacific. This cold water and the accompanying SST anomalies are basically opposite those observed during El Niño and they signal a La Niña event. La Niña begins with a period of unusually strong trade winds and vigorous upwelling over the eastern tropical Pacific (Fig. 8.5). Extremes of weather are often the opposite of those observed during El Niño. The atmosphere over the central equatorial Pacific tends to be less cloudy with less rainfall when the SSTs are colder than normal.

The Southern Oscillation is the surface pressure and atmospheric circulation component related to the warming of the equatorial Pacific Ocean that occurs with an El Niño event. The association between the warm water and an altered atmospheric circulation was first recognized in the 1950s, but variations in air pressure between the eastern and western tropical Pacific were recognized as early as the 1890s (Walker, 1923). The common depiction of the Southern Oscillation as an interannual fluctuation of atmospheric conditions over the tropical Pacific and Indian oceans that results in a reversal of winds over the equatorial Pacific Ocean obscures a complex interaction of processes. A significant facet of this circulation is a meridional component that supports a latitudinal transfer of energy and mass accounting for the ability of the tropical Southern Oscillation to have worldwide ramifications.

It remains unclear whether the ocean or atmosphere initiates the conditions that become an ENSO event. In the south equatorial Pacific, the normal atmospheric circulation mode is the longitudinal Hadley cell. However, near the coast of South America the winds blow offshore and result in an upwelling of water that is 5 °C colder than waters in the western Pacific. Air is stabilized by the cold water and cannot rise and join the normal ascending motion of the Hadley cell circulation. The stable air flows westward between the Hadley cell circulation of the two hemispheres forming the southeast trade winds across the South Pacific (see Fig. 8.3). After gaining heat and moisture as it crosses the ocean, the air ascends over the western Pacific, sea-level pressure is low, and rainfall is plentiful. Some of the rising air flows eastward to complete a cell of zonal atmospheric circulation known as the Walker circulation (Bjerknes, 1969). The atmosphere over the eastern tropical Pacific at this time is cold and dry.

At intervals of 1 to 5 years, the Walker circulation weakens and reverses direction and the Hadley circulation intensifies. The trade winds are relaxed, the zone of warm surface water and heavy precipitation shifts eastward, and sealevel pressure rises in the west while it falls in the east (see Fig. 8.4). The eastward shift of the region of heavy precipitation brings plentiful rainfall to the eastern tropical Pacific while inflicting drought on northern Australia and parts of southeastern Asia and Indonesia. The related seesaw variation in atmospheric mass and air pressure between the eastern and western tropical Pacific is known as the Southern Oscillation. This variation in atmospheric conditions is quantified by the Southern Oscillation Index (SOI), which is based on the difference in surface air pressure over Tahiti Island (17° S, 149° W) and Darwin, Australia (12° S, 131° E), compared to normal conditions. When the difference in pressure at these two stations is high the SOI is persistently negative, an El Niño event is in progress, and the low-pressure area near Darwin has moved eastward toward Tahiti. La Niña events are associated with an opposite array of conditions (see Fig. 8.5). The SOI is persistently positive, and atmospheric pressure is low in the western Pacific and high in the central Pacific near Tahiti (McCabe and Dettinger, 1999).

The Multivariate ENSO Index (MEI) evaluates both atmospheric and oceanic conditions to quantify ENSO events. A comprehensive dataset incorporates air temperature, SSTs, sea-level pressure, zonal and meridional surface winds, and cloudiness to define the coupled ocean–atmosphere state. Spatial filtering, principal component analysis, and standardization are used to keep the MEI comparable (Wolter and Timlin, 1998). Positive MEI values represent El Niño events, and MEIs greater than 1 are significant events (Fig. 8.6). Negative MEIs indicate La Niña events and MEIs below −1 correspond to significant events. The MEI was developed for research purposes, but it has broad appeal because it specifically addresses the coupled ocean–atmosphere phenomenon and incorporates more information than other indices. Kiem and Franks (2001)

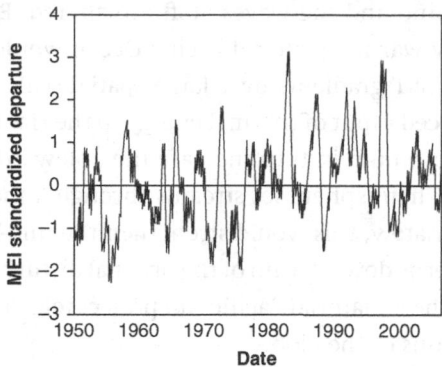

Fig. 8.6. The monthly Multivariate ENSO Index (MEI) for January 1950 to August 2006 plotted as standardized departures from the 1950–93 reference period. (Data courtesy of NOAA and the Earth System Research Laboratory from their website at http://www.cdc.noaa.gov/people/klaus.wolter/MEI/mei.html.)

demonstrate that eastern Australia rainfall and runoff are better estimated using the MEI compared to estimates using other ENSO indices.

8.6.2   Global ENSO impact

ENSO is part of a global-scale variability in climate that occurs periodi-cally, but not regularly enough to be predicted reliably. The global consequences can be traced to shifts in tropical rainfall which affect global wind patterns. The ITCZ in the Pacific is at about 8° N to 15° N. The South Pacific Convergence Zone (SPCZ) is a somewhat broader region of heavy rains extending southeastward from New Guinea towards the Polynesian region. The SPCZ is interrupted on its eastern margin by the relatively cold SSTs of the South Pacific dry zone, which is air influenced by the Humboldt Current flowing equatorward along the west coast of South America. Both convergence zones exist throughout the year, but they are notably stronger during their respective summer seasons. During warm ENSO events, both convergence zones shift equatorward and appear to become merged over their western portions near the IDL. The merged state of the con-vergence zones results in wetter than normal conditions along the equator with anomalously dry conditions in their usual positions. Drier than normal condi-tions further west affect a broad region of the tropics bordering the eastern Indian Ocean, such as Australia, Indonesia, and south Asia (Trenberth et al., 2002). Dense tropical rain clouds distort atmospheric flow aloft much as rocks distort the flow of a river, but on a horizontal scale of thousands of miles. Waves in the atmo-sphere determine the positions of monsoons, jet streams, and storm tracks. Rainfall areas over Indonesia and the far western Pacific move eastward into the central Pacific during an El Niño and the waves aloft are altered. Bjerknes (1969) proposed that an unusually warm equatorial Pacific Ocean would create anomalous zonal and meridional SST gradients over large spatial scales. These gradients would provide an enhanced input of thermal energy to the Hadley cells in that quadrant of the globe. This would in turn increase the poleward flux of angular momentum to the winter hemisphere jet streams through a more efficient meridional circulation. Ultimately, this would strengthen the mid-latitude westerlies and affect weather patterns downstream of the original disturbance. In this way, the El Niño warming in the equatorial Pacific can project teleconnected climatic anomalies to remote regions of the globe.

The year-to-year SST changes in the tropical Pacific are on the order of 0.5 °C to 1.0 °C, but ENSO affects climate and weather conditions in a number of locations in different ways. It is the dominant mode of interannual variability in global and hemispheric land precipitation (New et al., 2001). The El Niño signal is strongest in the eastern tropical Pacific Ocean, but it has far-reaching consequences throughout a large area of the tropics and into the mid-latitudes. ENSO has severe local impact along the west coast of South America, and its influence in middle latitudes shows up clearly during the winter (Kane, 1997; Hoerling and Kumar, 2003). ENSO is the single largest cause of global extreme precipitation events, but the seasonality is region dependent (Dai et al., 1997). Global annual average precipitation for 1979–2004 shows most variations are associated with ENSO and have no trend (Smith et al., 2006).

The strongest ENSO precipitation signal over North America affects the Gulf coast region of the United States and parts of northern Mexico, Texas, and the Caribbean islands where wetter than normal conditions occur during the winter. This signal is one of the most consistent extratropical teleconnections associated with ENSO. These signals are related to a strengthening of the subtropical jet over the Gulf of Mexico and are associated with an active storm track to the north. The southwestern United States is drier than normal during La Niña events. In general, opposite ENSO effects are observed in the northwestern United States, but the ENSO influence on United States West Coast precipitation largely depends on SST patterns in the central Pacific Ocean (Cayan and Webb, 1992; Hidalgo and Dracup, 2003). ENSO-related influences are most evident in months at the beginning and end of the winter season in the Upper Rio Grande basin of Colorado and New Mexico (Lee et al., 2004). The collective result of ENSO on annual precipitation in the United States is to increase precipitation variability by 5%–25% for stations in ENSO-influenced regions (Peel et al., 2002). In southern and eastern Mexico, ENSO is associated with reduced precipitation, but the ENSO signal is weaker (Mendoza et al., 2006). The ENSO influence on twentieth century precipitation in Europe and North Africa is strongest in the spring, and it displays three distinct periods. During the first and third periods, precipitation is enhanced in northern and central Europe and precipitation is reduced over the Iberian Peninsula and northwestern Africa. Precipitation during the second period is reversed with central and western Europe experiencing low rainfall and little rainfall influence evident over the Iberian Peninsula and northwestern Africa (Knippertz et al., 2003).

The North American runoff response to ENSO is a complex pattern with a negatively correlated northern region and four distinct positively correlated regions in the west, south, central, and east (Chen and Kumar, 2002). For the western United States, McCabe and Dettinger (2002) found that Niño-3 SSTs explained only a small percentage of the 1 April snowpack variability. In the Pacific Northwest region, advances in the timing of spring peak streamflow appear related to enhanced ENSO activity, which results in more precipitation occurring as rain instead of snow (Regonda et al., 2005). Such alterations in the moisture input may underlie changes in the probability of floods in a given year (Jain and Lall, 2001). However, spatial scale is a consideration as shown by Twine et al. (2005) who report no ENSO signal in streamflow for the Mississippi River at Vicksburg, Mississippi, but the presence of significant correlations in certain regions within the basin. Coulibaly and Burn (2005) found a strong ENSO signal in spring-summer and winter streamflow in both eastern and western regions of Canada. In southeastern South America, the Negro and Uruguay rivers have enhanced streamflow coincident with El Niño events (Robertson and Mechoso, 1998). The major influence of ENSO on the hydroclimate of the North Atlantic region is to initiate circulation changes that produce identifiable temperature anomalies in the British Isles and Europe (Pozo-Vázquez et al., 2001). Daily extreme winter temperatures in the United States are reduced during El Niño events and increased during La Niña and ENSO-neutral years (Higgins et al., 2002). At the global scale, ENSO mechanisms account for a 0.06 °C temperature increase for 1950–98 and contribute to the spatial character of the temperature lag relative to tropical Pacific Ocean SSTs (Trenberth et al., 2002).

Increasing evidence indicates that the strength, duration, and frequency of ENSO events have varied significantly over the past two centuries, and that no two El Niño events are the same (Bonsal and Lawford, 1999; Rajagopalan et al., 2000; Verdon and Franks, 2006). May 1982 to July 1983 produced the largest area of warm water anomalies and was one of the most intense ENSO events of the twentieth century (see Fig. 8.6). The 1990–3 ENSO was one of the longest ENSO events in history, but it was of modest intensity. El Niño occurrences back to 1525 have been documented by Quinn and Neal (1992) using various proxy data sources.

ENSO provides a glimpse into the complex workings of the hydroclimatic system afforded by few other phenomena. Because it occurs every few years, the variability of the hydroclimate system components can be viewed with relative frequency. From these observations it is evident there is no single weather pattern associated with ENSO, and time variations of ENSO amplitude and period occur on decadal and longer time scales (Mokhov *et al.*, 2004).

8.7   North Atlantic Oscillation

The dominant mode of climate variability in the North Atlantic basin is the North Atlantic Oscillation (NAO). This meridional oscillation of atmospheric pressure and winds is considered less pervasive than ENSO, but the variation of the westerly winds over the North Atlantic Ocean bears some similarity to the ENSO phenomenon in the equatorial Pacific Ocean. NAO is recognized as a factor influencing weather patterns throughout the Northern Hemisphere (Hurrell and van Loon, 1997), and it may be a more influential weather factor than ENSO in regions surrounding the North Atlantic Ocean.

8.7.1   NAO characteristics

NAO is characterized by atmospheric pressure variations related to north–south atmospheric mass oscillations. These oscillations are due to changing Arctic air masses near Iceland and subtropical air masses over the Atlantic Ocean from the Azores to the Iberian Peninsula. The NAO signature is strongly regional and is represented by an index defined as the difference between the normalized mean December to March sea-level pressure anomalies at Lisbon, Portugal, and Stykkisholmur, Iceland (Hurrell, 1995). The atmospheric pressure and wind variations associated with NAO alter heat and moisture transport between the Atlantic Ocean and surrounding continents by influencing the number and path of winter storms (Hurrell *et al.*, 2001; Trigo *et al.*, 2004). A strong Icelandic low and Azores high produce a strong south to north pressure gradient, strong westerly winds, and stronger winter storms following more northerly tracks. This positive phase of the NAO delivers warm, moist air and milder maritime winters over the European continent and above average temperatures in the eastern United States. The negative NAO phase is associated with a weak pressure gradient, weaker westerlies, fewer and weaker winter storms, and colder than normal winter temperatures in Europe and the United States.

The NAO is most pronounced during the winter when it accounts for more than one-third of the total variance of the sea-level pressure field over the North Atlantic Ocean (Hurrell and van Loon, 1997). However, both NAO phases are associated with changes in the location and intensity of the North Atlantic jet stream and storm tracks. The NAO varies at time scales of days to centuries without a clear cyclical pattern, but it exhibits a tendency to remain in one

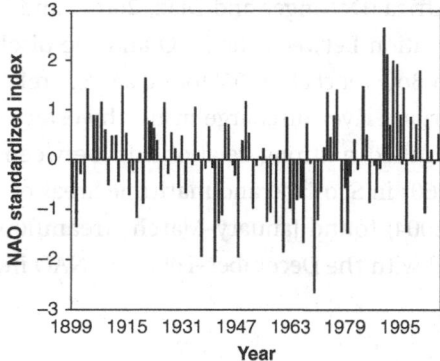

Fig. 8.7. The North Atlantic Oscillation (NAO) index for 1899–2005 expressed as normalized December to March principal component values. (Data courtesy of NOAA and the Climate Prediction Center from their website at http://www.cpc.ncep.noaa.gov/products/precip/CWlink/pna/nao.shtml.)

phase for intervals lasting less than 15 years (Hurrell, 1995; Appenzeller *et al.*, 1998). The NAO was in a generally positive phase from about 1900 to 1930 (Fig. 8.7) resulting in warmer winter temperatures across much of Europe. A negative NAO phase from the early 1940s to the early 1970s was associated with markedly cold winters in Europe. Over the last 30 years of the twentieth century, the NAO abruptly changed to a highly positive phase containing the highest positive values since 1864 (Hurrell and van Loon, 1997; Hurrell *et al.*, 2001). During this recent period, warmer winters returned to Europe accompanied by anomalously dry conditions over southern Europe and the Mediterranean and wetter than normal winters over northern Europe and parts of Scandinavia (Dai *et al.*, 1997; Knippertz *et al.*, 2003). However, correlations between NAO and precipitation are strongest in western and central Europe at locations near the coast and decrease inland (Trigo *et al.*, 2004; Bouwer *et al.*, 2006).

The NAO is most directly associated with changes in the surface westerlies across the Atlantic Ocean into Europe, but NAO plays a role in the Northern Hemisphere planetary wave system structure and is a factor in defining global atmospheric circulation. The planetary waves are important in defining temperature patterns, storm tracks, and time and space characteristics of precipitation. Therefore, precipitation patterns and winter temperatures in North America and North Africa, in addition to Europe, are attributed to NAO phases (Uppenbrink, 1999), and the NAO has been related to monsoon rainfall in India. A significant part of the winter variability in spatially averaged Northern Hemisphere mid-latitude precipitation is accounted for by NAO (New *et al.*, 2001). Seasonal NAO variability is evident in streamflow variability in the eastern United States, Europe, and tropical South America and Africa (Dettinger and Diaz, 2000), and Peterson *et al.* (2002) report a positive correlation between the NAO and the discharge of northern Eurasian rivers. Although Bouwer *et al.* (2006) found a weak relationship between NAO and December–February river discharge in northwestern Europe, Hannaford and Marsh (2006) identified a strong relationship between annual runoff and NAO since the early 1960s in Scotland and maritime areas of western England and Wales. Trigo *et al.* (2004) found January–March streamflow in the Iberian Peninsula highly correlated with the December–February NAO index.

8.7.2   NAO initiation

The process or processes responsible for the low-frequency variations of the NAO and its unprecedented trend over the past 30 years are poorly understood, but a complex ocean–atmosphere link involving ENSO is likely (Hoerling *et al.*, 2001; Hurrell *et al.*, 2001; Pozo-Vázquez *et al.*, 2001). Identification of processes is hindered by the superimposition of decadal NAO variability on interannual variability. However, it is well established that much of the NAO atmospheric variability arises from processes internal to the atmosphere (Hoerling *et al.*, 2001). The difficulty in defining the complex processes related to NAO is emphasized by emerging evidence that it might be a seesaw of atmospheric mass between the polar cap and the middle latitudes in both the Atlantic and Pacific ocean basins. This hemispheric high-latitude circulation with links to the stratosphere is called the Arctic Oscillation (Ambaum *et al.*, 2001; Higgins *et al.*, 2002), and Wang and Schimel (2003) suggest the NAO and the Arctic Oscillation may represent two paradigms of the same phenomenon.

8.8   Pacific–North American teleconnection pattern

The Pacific–North American (PNA) pattern is a large-scale atmospheric teleconnection between the North Pacific Ocean and North America that occurs as a wave train of middle and upper tropospheric geopotential heights and height anomalies (Wallace and Gutzler, 1981; Barnston and Livezey, 1987). The PNA north–south seesaw of pressure is most expansive in the winter, but it is identifiable in all months except June and July. The persistence and strength of this teleconnection makes it the most prominent mode of low-frequency atmospheric variability of two weeks to two months in the Northern Hemisphere extratropics (Wallace and Gutzler, 1981; Ambaum *et al.*, 2001). Temporally, the teleconnection is evident on both meteorological and climatological time scales extending from periods of days, weeks, months, and seasons to decades (Leathers and Palecki, 1992), but it displays substantial interseasonal, interannual, and interdecadal variability.

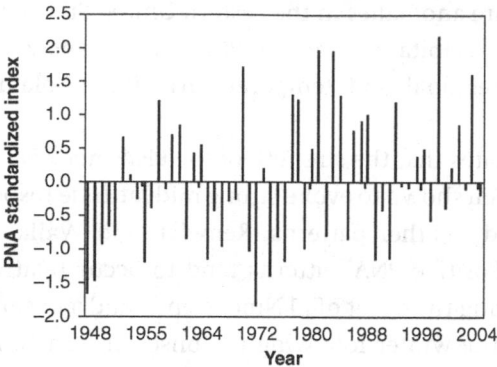

Fig. 8.8. The Pacific–North American (PNA) index for 1948–2004 expressed as standardized November to March values for the period of record. (Data courtesy of NOAA and the University of Washington Joint Institute for the Study of Atmosphere and Ocean from their website at http://jisao.washington.edu/data_sets/pna/.)

The PNA index is defined as a linear combination of the normalized geopotential height anomalies at four pattern centers over the eastern Pacific Ocean, western Canada, and the southeastern United States. The sign of the PNA index is positive during seasons with strong ridging over western Canada and above normal geopotential heights along the West Coast of North America. This ridging is juxtaposed to strong negative geopotential height anomalies in the mid Pacific near 45° N and over the southeastern United States (Wallace and Gutzler, 1981). The PNA sign is negative during seasons with strong ridges over the North Pacific Ocean and the southeastern United States and a low-pressure trough centered over the central United States. The positive PNA phase is associated with enhanced meridional flow across North America while atmospheric flow associated with the negative phase is more zonal (Leathers *et al.*, 1991). The winter PNA index (Fig. 8.8) was generally negative from 1948 to 1957 and from 1965 through 1976. The winter index was dominantly positive from 1958 to 1964 and from 1995 to 2001.

The PNA pattern is strongly linked with the lower atmospheric circulation through its influence on the position of storm tracks and the movement of warm and cold air masses. Consequently, the PNA pattern displays a significant relationship with regional temperature and precipitation variability (Barnston and Livezey, 1987; Leathers *et al.*, 1991; Konrad, 1998) and to streamflow variability (Anctil and Coulibaly, 2004; Coulibaly and Burn, 2005). In general, positive PNA values are related to positive temperature anomalies in the west and negative temperature anomalies in the eastern United States. Negative PNA values display a reversed pattern with negative temperature anomalies in the west and positive temperature anomalies in the eastern United States (Leathers *et al.*, 1991; Konrad, 1998). Precipitation and the PNA index display a weaker relationship with greater regional and temporal variability (Coleman and Rogers, 2003).

Increasing evidence indicates that the strength of the PNA index is linked to the ENSO cycle. ENSO has been shown to evoke strong mid-latitude responses in atmospheric circulation and weather patterns (Renwick and Wallace, 1996; Pozo-Vázquez *et al.*, 2001). Positive PNA patterns tend to occur relatively frequently in the winter following the onset of El Niño events, and a negative PNA index frequently occurs in the winter following the onset of a La Niña event (Renwick and Wallace, 1996; Bonsal and Lawford, 1999). The ENSO-related changes in pressure and circulation are especially evident in alternating patterns of precipitation anomalies. However, modeling studies have shown that the PNA pattern can be generated by internal dynamics alone without the support of SST anomalies (Mo *et al.*, 1998).

8.9   Pacific Decadal Oscillation

The Pacific Decadal Oscillation (PDO) is a North Pacific Ocean climate state that varies on a multi-decadal time scale between two modes with distinct spatial and temporal SST characteristics. The SSTs of central interest correspond to a large area in the northern and western Pacific Ocean and a smaller region in the eastern tropical Pacific (Mantua *et al.*, 1997). Since PDO is identified on the basis of SSTs and is associated with periodicities ranging from 20 to 30 years, it has been described as a long-lived El Niño-like pattern of Pacific climate variability. The PDO is in a "warm or positive phase" when temperatures are anomalously warm in the eastern tropical Pacific and anomalously cool in the central North Pacific. Warm PDO phases dominated from 1925 to 1946 and from 1977 to at least 1998 (Hare and Mantua, 2000). A "cool or negative" PDO phase has anomalously cool temperatures in the eastern tropical Pacific and anomalously warm temperatures in the central North Pacific. The PDO was in a cool phase from 1890 to 1924 and from 1947 to 1976 (Mantua *et al.*, 1997). Although the PDO is much slower to switch from one phase to another than the fast-fluctuating ENSO, SSTs characterizing the PDO undergo fairly abrupt changes from one phase to another termed regime shifts (Miller *et al.*, 1994). Accumulating evidence indicates an important climate regime shift occurred in the North Pacific Ocean in the winter of 1976–7 (Trenberth, 1990; Ebbesmeyer *et al.*, 1991; Zhang *et al.*, 1997). Another regime shift may have occurred in 1998 coincident with the end of the 1997–8 El Niño and the beginning of the subsequent La Niña event (Hare and Mantua, 2000). Verdon and Franks (2006) study of paleo-records

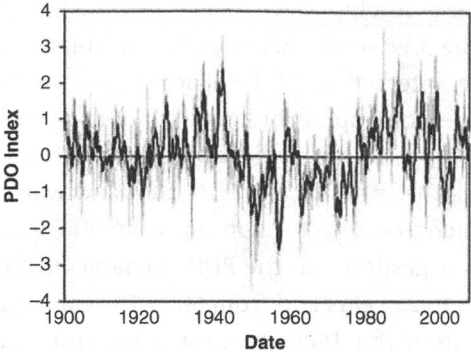

Fig. 8.9. The Pacific Decadal Oscillation (PDO) index for January 1900 to July 2006. The gray line is the monthly value, and the bold line is a 7-month moving average. (Data courtesy NOAA and the University of Washington Joint Institute for the Study of Atmosphere and Ocean from their website at http://jisao.washington.edu/pdo/.)

suggests that a relationship between ENSO frequency and PDO phase changes has existed for the last 16 phase changes, and Biondi *et al.* (2001) and Hidalgo (2004) conclude that PDO and ENSO interactions have occurred throughout the last four centuries.

8.9.1   PDO index

The PDO is portrayed by an index derived from the leading principal component from an un-rotated normalized empirical orthogonal function analysis of gridded mean November through March SSTs poleward of $20°$ N latitude in the Pacific Basin (Mantua *et al.*, 1997). Positive PDO index values identify a warm phase and indicate months of above normal SSTs along the west coast of North and Central America and along the equator (Fig. 8.9). Below normal SSTs are present in the central and western North Pacific Ocean at about the latitude of Japan. Negative PDO values indicate a cool phase and are associated with below normal SSTs along the equator and above normal SSTs east of Japan (Trenberth and Hurrell, 1994). Although the PDO shares some SST features with ENSO, the PDO is set apart from ENSO by two main characteristics. Typical PDO events show greater persistence than ENSO events, and evidence of the PDO is most visible in the North Pacific and North America. The ENSO signature features an opposite pattern of SSTs with stronger evidence in the topics (Mantua *et al.*, 1997; Hare and Mantua, 2000). The possibility that the appearance and strength of ENSO events may depend on how PDO dominates ocean circulation and temperature patterns is receiving increasing attention.

The physical mechanisms responsible for the PDO have not been identified, but several possible mechanisms involving SSTs, mean oceanic currents and

Rossby waves, and atmospheric forcing have been proposed (Miller and Schneider, 2000). Identifying the precise mechanisms responsible for PDO variations is complicated by the interaction of feedback processes involving ocean–atmosphere and tropical–extratropical linkages (Trenberth and Hurrell, 1994). This is illustrated by the reported intensification of North Pacific winter cyclones since 1948 that spans the most recent PDO cool and warm phases (Graham and Diaz, 2001). An additional factor complicating efforts to understand PDO mechanisms is the suggestion that the PDO is characterized by two general periodicities of from 15 to 25 years and from 50 to 75 years (Chao et al., 2000). However, PDO indices from the 1600s reconstructed from tree rings indicate an average regime duration of 23 years (Biondi et al., 2001). Even in the absence of theoretical understanding, empirical PDO climate information is useful for season-to-season and year-to-year climate forecasts for the Pacific, Australia, and the Americas because of its strong tendency for multiseason and multiyear persistence. Also, proxy sources have yielded a clear pattern of symmetric atmospheric circulation changes associated with the PDO that signal an influence of this phenomenon on climate in the tropics and the Southern Hemisphere (Linsley et al., 2000).

8.9.2   PDO influences

Over the past century, the amplitude of the PDO climate pattern has varied irregularly at interannual-to-interdecadal time scales (see Fig. 8.9). The difference in SSTs from positive to negative phases is not more than 1 to 2 °C, but the affected ocean area is so huge that it can impact North American temperature and precipitation patterns through its effect on steering storms across the North Pacific Ocean (Mantua et al., 1997). PDO warm phases produce above average winter and spring temperatures in northwestern North America and below average temperatures in the southeastern United States. The southern United States and northern Mexico receive above average winter and spring rainfall during warm PDO phases, but precipitation is below average in the interior Pacific Northwest and the Great Lakes region. Cool phase PDO temperatures and precipitation are broadly the reverse of these climate anomaly patterns across North America (Mantua et al., 1997; Nigam et al., 1999). Severe and sustained droughts in the western United States over the past 200 years show a relationship to SSTs indexed by the PDO (Gedalof et al., 2004; Hidalgo, 2004), and streamflow variability in the United States displays a significant correlation with PDO warm and cool phases (Tootle and Piechota, 2006). McCabe and Dettinger (2002) found that the 1 April snowpack in the western United States is more highly correlated with PDO than the Niño-3 SST, but Stewart et al. (2005) conclude that evidence is lacking to support warmer temperatures related to PDO as the major contributor to a one-to-four week advance in peak streamflow in snowmelt dominated water-sheds across a broad region of the western United States.

In general terms, North American climate anomalies associated with PDO warm and cool phases are broadly similar to El Niño and La Niña patterns. However, closer inspection reveals that positive PDO phases enhance El Niño conditions and weaken La Niña effects while negative PDO phases weaken El Niño events and enhance La Niña events (Gershunov and Barnett, 1998; McCabe and Dettinger, 1999). Consequently, the coincidence of the PDO phase and ENSO events produces a complex array of wet and dry conditions that complicates generalizations addressing the regional wetness and dryness associated with El Niño and La Niña events. Emerging evidence indicates that the commonly por-trayed El Niño and La Niña temperature and precipitation patterns are only valid during years in which a warm PDO coincides with an El Niño event and a cool PDO coincides with a La Niña event (Gershunov and Barnett, 1998; Nigam *et al.*, 1999).

8.10   Recent temperature trends

The oscillatory form of time-series variations portrayed by hydroclimatic variables depicts values that move gradually and smoothly between successive maximum and minimum values. For convenience of description, the dominant characteristics of the undulating pattern are identified as cycles and/or trends. Wave physics provides the basic terminology to describe cyclic changes in terms of the duration and magnitude of the wave pattern. The sinusoidal pattern characteristic of cycles is the product of variable fluctuations about some speci-fied mean value. If there is an increasing or decreasing change in the mean value during the period of record the resultant time-series is time dependent and is described as a trend. Trends in hydroclimate variables are important because they signal a continuing change over time that may be attributed to either natural or human factors. Our ability to detect trends in most hydroclimatic variables is hindered by the relatively short record available for analysis. Record length is a primary factor in the study of trends because an apparent trend in a short record may be part of an oscillation in a long record for a specific variable. In general, research on hydroclimatic trends has been geographically dispersed, sporadic in time, and stimulated by individual events (Hunt, 2001).

8.10.1   *Global temperature*

A multitude of problems surround efforts to define global temperature, and these problems are fundamental to hydroclimate and the global climate change debate. The annual global temperature anomaly time-series is recom-mended by the IPCC (2001) because it is widely recognized as a representative

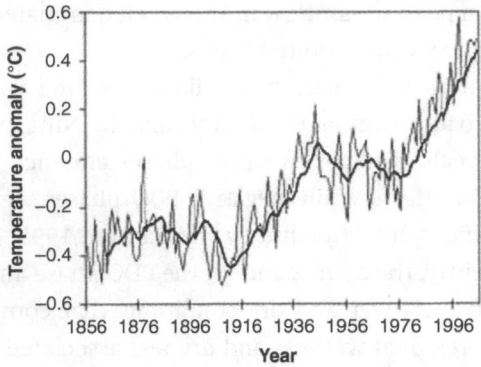

Fig. 8.10. Global temperature anomalies for 1856–2005 relative to the 1961–90 mean. The gray line is the annual value, and the bold line is the 9-year moving average. (Data courtesy of NOAA's National Climate Data Center and the Oak Ridge National Laboratory, Carbon Dioxide Information Analysis Center from their website at http://cdiac.ornl.gov/trends/temp/jonescru/jones.html)

depiction of global surface conditions (Fig. 8.10). These records have been adjusted to take into account urban heating effects, instrument changes, instrument location changes, and other factors that influence the reliability of the instrument record. This time-series for 1856 to 2005 indicates an increasing global temperature trend of 0.6 °C ± 0.2 °C. The overall trend is composed of short-term increasing and decreasing temperature trend segments that form the complete record. Probably the most commonly recognized segments are the warming trend from 1856 to 1945, the cooling trend from 1946 to 1975, and the warming trend from 1976 to the present. Several intermediate points in the time-series support a perception of trend that is different from the trend evident in the period from 1856 to 2005. Temperature reconstructions indicate that twentieth century global temperatures are approaching the warmest temperature that occurred around AD 990 (Esper *et al.*, 2002), and at least a part of the recent warming cannot be solely related to natural factors (Rybski *et al.*, 2006).

At scales smaller than the global scale, annual temperatures display a complex pattern of changes that elude generalization. The Northern Hemisphere pattern is more similar than the Southern Hemisphere pattern to the global temperature time-series. The annual temperature pattern for land areas displays greater variability than the global time-series, while the SST time-series displays closer similarity with the global temperature. Seasonal temperature anomalies and anomalies calculated for different latitudinal zones reveal temperature trends that are increasing, decreasing, and unchanged (IPCC, 2001).

The year-to-year variability of temperature in the high latitudes is several times larger than that observed in other areas. However, the statistical significance of

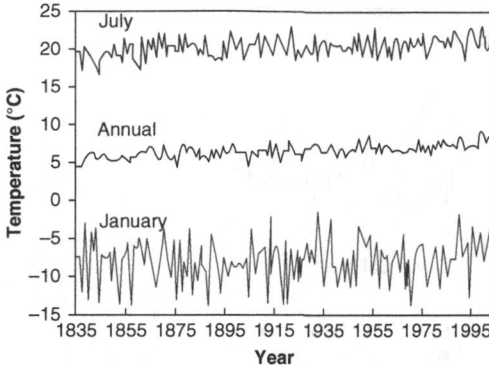

Fig. 8.11. January, July, and annual temperatures for 1835–2002 at Hanover, New Hampshire (44° N). (Data courtesy of NOAA's National Climate Data Center and the Oak Ridge National Laboratory, Carbon Dioxide Information Analysis Center from their website at http://cdiac.ornl.gov/epubs/ndp/ushcn/usa_monthly.html.)

the warming is actually greater in the lower latitudes. Variability as measured by the standard deviation is about 0.04 °C greater in the Northern Hemisphere for all latitudes. There appears to be little relation between interannual variability and the relative warmth or coldness of decadal averages except in winter at high latitudes. Karl *et al.* (1993) and Easterling *et al.* (1997) suggest that strong evidence exists for a widespread decrease in the mean monthly diurnal temperature range (DTR) over the past several decades. DTR is derived from an average of daily maximum and minimum temperatures. The rise in minimum temperatures (0.84 °C) occurred at a rate three times that of the maximum temperature (0.28 °C) during 1951–93. The DTR decrease is approximately equal to the increase in global mean temperature, and urbanization alone cannot account for the widespread DTR decrease. Change is detectable in all seasons and most regions, but the trend of change is inconsistent. Some regions, such as the British Isles, the Iberian Peninsula, India, central Canada, and certain coastal areas of North America have experienced DTR increases (Durre and Wallace, 2001).

8.10.2 Individual station temperatures

The behavior of individual stations reveals a broad array of annual temperature trends. The diversity found in the station records is a reminder that spatial and temporal averaging may mask important information.

Hanover, New Hampshire (44° N), is in the Upper Connecticut River Valley in the northeastern United States at an elevation of 183 m. It is 153 km inland from the Atlantic Ocean, and it is climatically influenced by continental factors more than marine conditions due to the dominance of westerly atmospheric flow at this latitude. The annual temperature at Hanover (Fig. 8.11) displays an

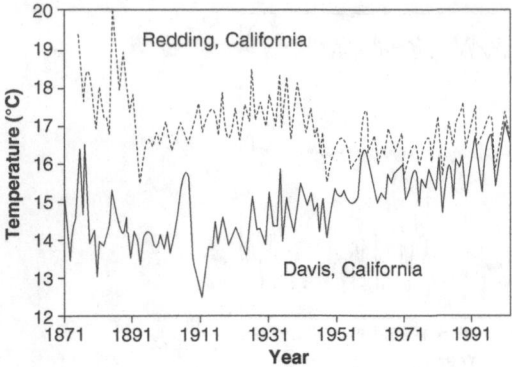

Fig. 8.12. Annual temperatures for Davis, California, and Redding, California, for 1871–2002. (Data courtesy of NOAA's National Climate Data Center and the Oak Ridge National Laboratory, Carbon Dioxide Information Analysis Center from their website at http://cdiac.ornl.gov/epubs/ndp/ushcn/usa_monthly.html.)

increasing trend of $0.01\,°C\,yr^{-1}$ for the period 1835 to 2002 and ranges from a low of $4.3\,°C$ in 1875 to a high of $8.8\,°C$ in 1999. The mean January temperature at Hanover displays no trend from 1835 to 2002 even though the mean January temperature ranges between $-14\,°C$ in 1857 and 1888 and $-1.4\,°C$ in 1932. The mean July temperature at Hanover for 1835 to 2002 is characterized by an increasing trend of $0.01\,°C\,yr^{-1}$, and it ranges from a low of $16.3\,°C$ in 1844 to a high of $23.2\,°C$ in 1999. These data reveal that the annual warming trend at Hanover is not occurring equally in all months nor is the response of each year similar. In addition, the Hanover temperature time-series show little similarity with the general pattern of the global temperature anomaly time-series in Figure 8.10. In marked contrast to the Hanover data is the temperature time-series for Hohenheim University in Stuttgart, Germany, beginning in 1878, which closely resembles the trends in the global data (Wulfmeyer and Henning-Müller, 2006).

Another complication regarding temperature is illustrated by the temperature trends at two California stations (Fig. 8.12). Redding (41° N) and Davis (39° N) are both inland stations 125 to 150 km from the Pacific Ocean. Redding is 225 km north of Davis and 1736 m higher in elevation. Annual temperatures at Davis from 1871 to 2002 increased $0.01\,°C\,yr^{-1}$, but Redding annual temperatures decreased by $0.01\,°C\,yr^{-1}$ from 1876 to 2002. Both stations are included in the USHCN dataset and the records are considered to be reliable. The temperature time-series for these two stations are representative of regional characteristics that extend to approximately halfway between the two sites. Also noteworthy in the Davis and Redding data is the absence of the alternating increasing and decreasing temperature trend segments evident in the global temperature anomaly time-series.

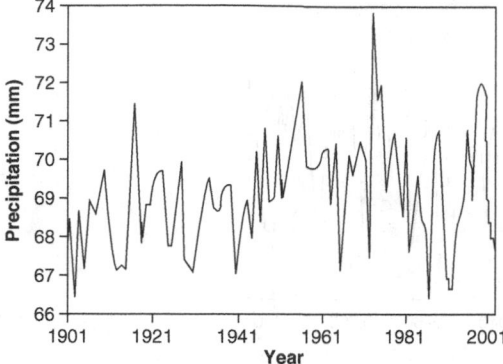

Fig. 8.13. Global land surface precipitation for 1901–2002. (Data courtesy of NOAA's National Climate Data Center from their website at http://www.ncdc.noaa.gov/gcag/ gcag.html.)

8.11   Recent precipitation trends

Rain and snow are pivotal variables in the hydrologic cycle, but reliable estimates of global precipitation are difficult to achieve. An obvious contributor to the difficulty in achieving accurate quantitative documentation of global precipitation is that precipitation is discontinuous in time and space. The variation in the spatial character of precipitation is exacerbated by the fact that most oceanic and unpopulated land areas are inadequately represented in existing data (Xie and Arkin, 1997). In addition, errors and inhomogeneities in precipitation measurements and time-limited observational programs contribute to uncertainty in the record. Consequently, global precipitation time-series commonly represent land areas only (IPCC, 2001). Efforts to combine various data sources that include both land and ocean areas have produced promising results, but these datasets are limited to the last two decades of the twentieth century (Huffman *et al.*, 1995; Xie and Arkin, 1997).

8.11.1   Global precipitation

The time-series of annual precipitation for the twentieth century displays a slight increasing trend of 0.89 mm per decade for land precipitation (New *et al.*, 2001) and 2.4 mm per decade for global precipitation (Dai *et al.*, 1997; IPCC, 2001). This increasing trend is accompanied by changes in precipitation characteristics (Trenberth *et al.*, 2003) and a tendency toward increases in the frequency of high precipitation amounts (Katz *et al.*, 2002; Fowler and Kilsby, 2003). These conditions are evident in Australia, the UK, the United States, and in Germany in the fall and spring (Kundzewicz, 2002). The first and last 20 years of the land surface record (Fig. 8.13) are dominated by relatively dry years, and

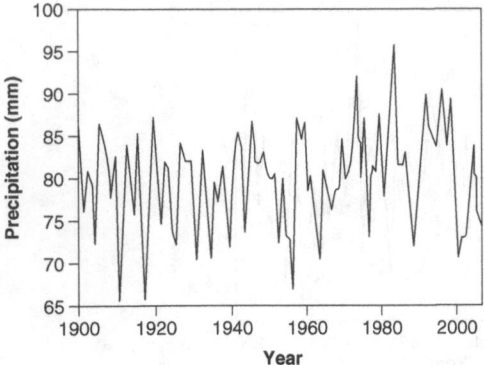

Fig. 8.14. Annual precipitation for the contiguous United States for 1900–2005. (Data courtesy of NOAA's National Satellite and Information Service and the National Climate Data Center from their website at http://www.ncdc.noaa.gov/oal/climate/ research/cag3/na.html.)

sharp increasing precipitation trends characterize the mid 1940s to mid 1950s and the mid 1960s to mid 1970s. Global precipitation for 1979–2002 shows increases in some regions and decreases in others, but the global average change is near zero (Smith *et al.*, 2006). Time-series for selected latitude zones display a variety of trends. Substantial increasing precipitation trends characterize the Northern Hemisphere middle and high latitudes, and strong trends in moisture recycling are especially evident over northern Europe and North America (Dirmeyer and Brubaker, 2006). Precipitation trends for the Southern Hemisphere and the tropics in both hemispheres are relatively flat, but they display decreasing trend tendencies in recent years (Dai *et al.*, 1997). One exception is a precipitation increase in the lowland tropics of South America (Villalba *et al.*, 1998). In a long-term view, a precipitation reconstruction for central Chile indicates precipitation was greater in the nineteenth century than in the twentieth century (LeQuesne *et al.*, 2006).

Important year-to-year differences in annual precipitation for the contiguous United States (Fig. 8.14) are related to shifts in rainfall patterns, increases in annual precipitation, and changes in the seasonal distribution of precipitation related to ocean–atmospheric influences. On decadal time scales, precipitation is more sensitive than temperature to atmospheric circulation variations. The average precipitation for 1900 to 2005 is 80 cm, but a trend toward greater precipitation and greater precipitation variability is indicated by the data. Precipitation increases by 3 mm per decade and the coefficient of variation is 2-times larger for the first half of the record compared to the second half. The wettest year is 1983 (95.7 cm), and 12 of the 16 years with precipitation greater

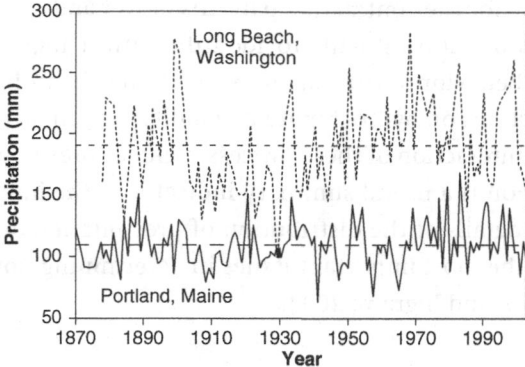

Fig. 8.15. Annual precipitation for Long Beach, Washington, and Portland, Maine, for 1870–2002. The broken horizontal lines are the 1971–2000 average for each station. (Data courtesy of NOAA's National Climate Data Center and the Oak Ridge National Laboratory, Carbon Dioxide Information Analysis Center from their website at http://cdiac.ornl.gov/epubs/ndp/ushcn/usa_monthly.html.)

than 86 cm occur after 1952. The driest year is 1917 (66 cm), and 8 of the 13 years receiving less than 73 cm occur before 1952.

8.11.2   *Individual station precipitation*

Station data reveal that the Northern Hemisphere mid-latitude increasing precipitation trend reported by Dai *et al.* (1997) is not evident at all locations. Portland, Maine (44° N), and Long Beach, Washington (46° N), on the east and west coasts, respectively, of North America have data available as early as 1870 (Fig. 8.15). The average annual precipitation at Portland is 109 cm and the average annual precipitation at Long Beach is 190 cm. Portland has an increasing trend of 16 mm per decade, but Long Beach has an increase of less than 1 mm per decade. Furthermore, the data for the two stations display little similarity in the occurrence of wet and dry years. The wettest year at Portland is 1983 which recorded 169 cm of precipitation and the driest year is 64 cm received in 1941. The wettest year at Long Beach is the 283 cm received in 1968, and the driest year is 1929 when only 109 cm were recorded. Both stations display groups of years predominantly wet or dry, but most of these groupings do not occur at the same time in the station's records. The one evident exception is the period of predominantly dry years from 1903 to 1930. However, predominantly dry years continue to 1952 at Long Beach while wet years predominate from 1931 to 1945 at Portland. These two stations highlight that spatial and temporal precipitation trend differences occur at specific locations even in a latitudinal zone characterized by a strong increasing precipitation trend.

Mechanisms responsible for the contrasting patterns of wet and dry years at east and west coast locations are difficult to identify with a high level of confidence. The dynamic behavior of the climate system and the influence of ocean–atmosphere interactions on hydroclimate variability are primary considerations, but the complex interaction of these processes eludes precise explanation. Long-term precipitation is a useful summary indicator of the intensity of the hydrologic cycle, and details of the distribution of precipitation over time and space are likely to be the most important issues in determining impacts of precipitation changes (Allen and Ingram, 2002).

8.12   Recent streamflow trends

Rivers integrate the hydroclimatic variables within the watershed they drain. Streamflow results from the interaction of the hydroclimatic variables in both time and space, and watershed physiography exerts temporal influences on transformation of the residual precipitation into streamflow in a specific drainage basin. Precipitation and temperature are major climatic factors determining runoff, and both are influenced by ocean–atmosphere processes to produce streamflow variability on interannual and decadal time scales (Robertson and Mechoso, 1998). However, the water balance (Equation 2.13) indicates that annual precipitation variability drives annual streamflow variability and changes in precipitation are amplified in runoff changes. For this reason it is surprising that the observed increasing frequency of extreme high precipitation is not being detected in annual peak flow data and the duration of low-flow events (Katz *et al.*, 2002). In North America, the evidence of increase in extreme precipitation events may be masked in streamflow records by concurrent advances in the timing of peak spring-season flows by as much as one to four weeks due to the earlier onset of spring snowmelt (Regonda *et al.*, 2005; Stewart *et al.*, 2005; Hodgkins and Dudley, 2006). Low-flow indicators in the United Kingdom show little evidence of sustained change since the 1960s and are characterized by relative stability (Hannaford and Marsh, 2006).

Annual mean streamflow is the mean flow for a given year, and observed annual streamflow serves as a pertinent indicator of interannual variability (Anctil and Coulibaly, 2004). A time-series of annual mean streamflow displays the annual streamflow variation related to hydroclimatic variability resulting from both natural and human influences within the watershed. The extent of human impact on the land and vegetation in many large watersheds has had important hydrologic implications that mask natural processes. Therefore, identifying natural hydroclimatic variability in streamflow records is most successful for watersheds where human impacts are minimized. Changing

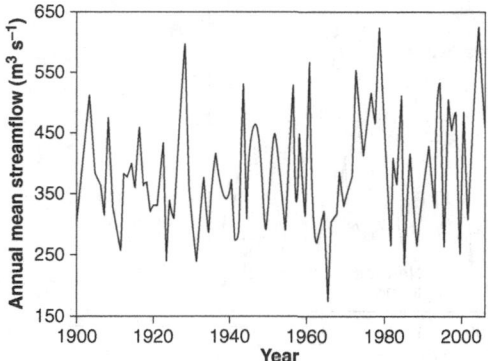

Fig. 8.16. Annual mean streamflow for the Susquehanna River at Wilkes-Barre, Pennsylvania, for 1900–2005. (Data courtesy of the U.S. Geological Survey from their website at http://waterdata.usgs.gov/nwis/.)

conditions within the watershed during the period of streamflow observations is especially likely in regions where dam construction and water diversions for irrigation are common. Pre-instrument data periods are assessed using reconstructed streamflow discussed in Section 8.3.

8.12.1   The Susquehanna River

The Susquehanna River (Fig. 8.16) rises as the outlet of Otsego Lake in central New York at an elevation of 360 m. It flows southeast across Pennsylvania and through Maryland to enter Chesapeake Bay. It is the longest river entirely within the United States that drains into the Atlantic Ocean. The watershed at Wilkes-Barre, Pennsylvania (41° N), covers 25 900 km^2 of the Allegheny Plateau, and the elevation of the gauge at Wilkes-Barre is 155 m. The Susquehanna River is unregulated above Wilkes-Barre as it flows through dairy farm land and a former anthracite coal industrial area in the ridges of northeastern Pennsylvania. The mean annual streamflow at Wilkes-Barre is 385 m^3 s^{-1}, the highest annual mean streamflow is 622 m^3 s^{-1} in 1978, and the lowest is 175 m^3 s^{-1} in 1965.

The Susquehanna River annual mean streamflow displays a generally similar pattern of wet and dry years as the precipitation record for Williamsport, Pennsylvania (Fig. 8.17), 90 km west of Wilkes-Barre. However, the details of the two time-series are different in that the years of maximum and minimum precipitation at Williamsport are not coincident with the maximum and minimum Susquehanna River streamflow at Wilkes-Barre. Both high and low flows lag precipitation maxima and minima by a year or more, and the nature of the streamflow response is related to conditions during intervening years.

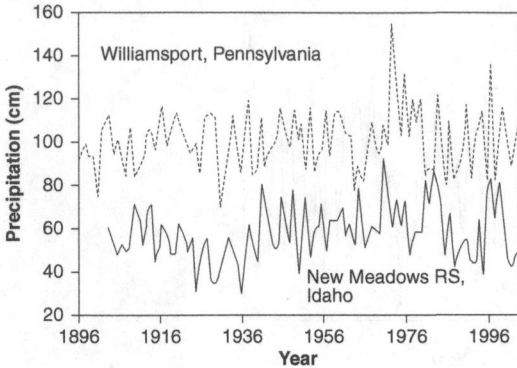

Fig. 8.17. Annual precipitation for New Meadows Ranger Station, Idaho, and Williamsport, Pennsylvania, for 1896–2002. (Data courtesy of NOAA's National Climate Data Center and the Oak Ridge National Laboratory, Carbon Dioxide Information Analysis Center from their website at http://cdiac.ornl.gov/epubs/ndp/ushcn/usa_monthly.html.)

Annual mean streamflow for the Susquehanna River displays drier conditions for the first half of the record compared to the second half. Thirty-five years before 1953 have annual mean streamflow less than the period of record average. The second half of the record contains both the highest and lowest annual mean streamflow, there is greater variability among annual values, and the consecutive years of above and below average streamflow are longer in the more recent record. Land-use changes involving vegetation removal may account for some of the increased variability, but the Williamsport precipitation record displays a generally similar pattern of wet and dry years and increasing precipitation variability in the recent record.

8.12.2   The Salmon River

The Salmon River, in the Pacific Northwest of the United States, drains a region of high mountains and deep canyons in central Idaho. The watershed area at White Bird, Idaho (46° N), is 35 230 km^2 and elevation differences of over 2120 m occur between the river's headwaters in the Sawtooth Mountains and its confluence with the Snake River a few kilometers below White Bird. Snowmelt from the high mountains in the headwaters area provides a significant portion of the Salmon River streamflow, but winter precipitation at lower elevations contributes to streamflow as well. The Salmon River is the longest free-flowing river in the contiguous United States, and considerable effort has been devoted to preserving the natural state of the Salmon River watershed. The Middle Fork of the Salmon River is one of the premier recreational rafting and kayaking rivers in the world.

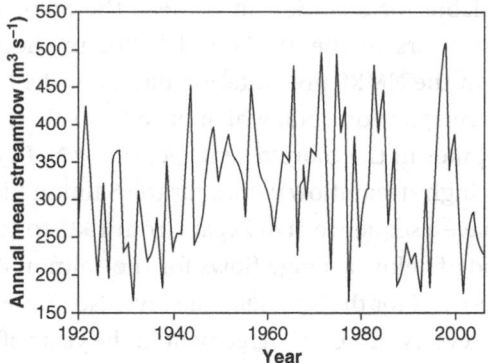

Fig. 8.18. Annual mean streamflow for the Salmon River at White Bird, Idaho, for 1920–2005. (Data courtesy of the U.S. Geological Survey from their website at http://waterdata.usgs.gov/nwis/.)

Cattle, forest products, gold mining, tourism, and hay are the primary economic activities in the Salmon River watershed. Diversions to irrigate 66 000 ha occur upstream from White Bird. The Salmon River annual mean streamflow at White Bird (Fig. 8.18) is 313 m^3 s^{-1}, the highest annual mean streamflow is 506 m^3 s^{-1} in 1997, and the lowest is 165 m^3 s^{-1} in 1931 and 1977. The drier setting of the Salmon River watershed compared to the Susquehanna River is evident in that the mean annual streamflow for the Salmon River is less than that for the Susquehanna River even though the Salmon River watershed area is one-third larger.

The first 47 years of the Salmon River streamflow record are drier than the more recent years largely due to the dominance of below average years from 1923 to 1946. Only six years during the 24 years have annual mean streamflow greater than the period of record average. Since 1958, year-to-year annual mean streamflow displays greater variability with abrupt changes from high values to low values. However, the 24 years from 1963 to 1986 are dominated by annual mean streamflow greater than the average, and this period includes the second to fifth highest values and one of the lowest values in the record. The longest consecutive departure from the average is the eight dry years from 1987 to 1994. Another six consecutive years of below average flow begins in 2000. The relative dryness in the first half of the record and the absence of multiyear droughts from 1950 to 1987 are characteristics shared with the entire Columbia River Basin (Cedalof et al., 2004).

The New Meadows Ranger Station (NMRS), Idaho (45° N), is 100 km south of White Bird and just outside the western boundary of the Salmon River watershed. However, the record for NMRS (see Fig. 8.17) provides the longest precipitation time-series available for a station in or near the Salmon River basin. The predominantly dry years in the 1920s and 1930s in the Salmon River streamflow are evident in the NMRS precipitation data, but the low flow in 1931 is not present in the precipitation data, which have low values in 1924 and 1925. The precipitation spikes in the more recent record for NMRS do not occur in the same years as the high streamflow values for the Salmon River, but the NMRS precipitation data have a sequence of dry years from 1985 to 1994 that coincides with a similar period of below average flows for the Salmon River.

Comparing streamflow time-series for the Susquehanna River (see Fig. 8.16) and the Salmon River (see Fig. 8.18) reveals almost no agreement in the years of highest and lowest flows. The single occurrence of similar conditions is that both watersheds have notable low flows in 1931. The dominant period of below average flows in the first half of the century begins 18 years earlier for the Susquehanna River and lasts 2 years longer for the Salmon River. The Susquehanna River has below average flows in the 1960s while below average flows in the Salmon River are more prevalent in the late 1980s and early 1990s and the early 2000s. Flows are consistently above average in the Susquehanna River during the 1970s, but the longest run of above average flows for the Salmon River occurs in the 1990s.

8.13   Recent lake level trends

Lake levels provide a hydroclimate indicator that is an integrating factor of precipitation and streamflow anomalies for the land area contributing runoff into the lake. Changes in inflow and evaporation from the lake will affect lake volume and water level in ways related to lake geomorphology.

The lake response to changes in atmospheric, land surface, and subsurface water fluxes is expressed by Equation 6.14. A lake's role as a natural hydroclimatic indicator on time scales of months to years is most valuable when the lake and watershed are relatively undisturbed (Street-Perrott, 1995), but many lakes and/or their watersheds have experienced significant human influences. Since few continuous instrumental records of lake levels are available before 1875, selecting examples where disturbances are minimized is important so that natural responses are not masked.

8.13.1   Lake Champlain, Vermont/New York

Lake Champlain (42° N) is an exorheic lake in the Champlain Valley between the Green Mountains of Vermont and the Adirondack Mountains of New York in the northeastern United States. The lake area is 1130 km², and it extends across the United States–Canada border into Quebec. Lake Champlain is drained northward by the Richelieu River which joins the St. Lawrence River

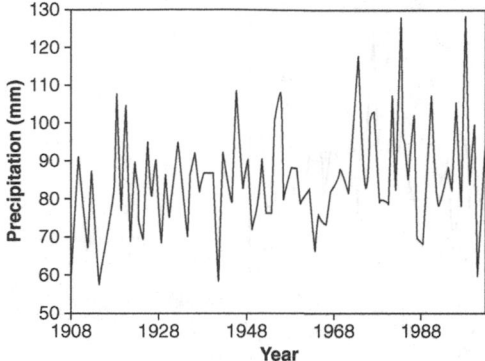

Fig. 8.19. Annual precipitation for Burlington, Vermont, for 1908–2002. (Data courtesy of NOAA's National Climate Data Center and the Oak Ridge National Laboratory, Carbon Dioxide Information Analysis Center from their website at http://cdiac.ornl.gov/epubs/ndp/ushcn/usa_monthly.html.)

east of Montreal. Several cities are located around the lake, but small craft, ferries, and lake-cruise ships constitute the contemporary lake traffic.

The Lake Champlain Basin covers 21 300 km^2, and the largest portion of the watershed is in Vermont. The watershed's hydroclimatology is influenced by the lake's location on the eastern edge of the continent, by the frequency of Arctic, Pacific, and tropical air passing over the region, and by the presence of mountains on both the east and west sides of the watershed. Precipitation amounts vary from 76 cm near the lake to over 127 cm in the surrounding mountains, which rise to elevations exceeding 1300 m. Annual precipitation on the lake's east shore at Burlington, Vermont (44° N), averages 85 cm (Fig. 8.19). The annual values range from 128 cm in 1998 to 58 cm in 1914.

The average depth of Lake Champlain is 19.5 m, but the greatest depth is 122 m. The mean annual lake level is 29.1 m. The highest annual mean lake level is 29.6 m in 1976, and the lowest is 28.6 m in 1915 and 1941 (Fig. 8.20). The distinctive feature of the 97-year time-series is the low levels from 1930 to 1968. Only 6 of these 39 years have an annual mean lake level higher than the 97-year mean annual value. The mean annual lake level for 1930 to 1968 is 28.9 m compared to the mean of 29.2 m for the 36 years from 1969 to 2005.

The lake level data indicate hydroclimatic fluctuations in the Lake Champlain Basin have been modest for the past century. The change in the average lake level from the dry years of 1930 to 1968 to the wet years of 1969 to 2005 is 0.3 m. Furthermore, the variability of the annual mean level indicated by the standard deviation is the same for the two periods (0.17). These characteristics portray a dampened response compared to the variation evident in the Burlington precipitation data.

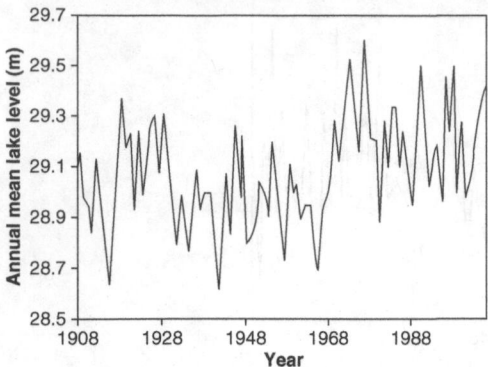

Fig. 8.20. Annual mean lake level above sea level for Lake Champlain at Burlington, Vermont, for 1908–2005. (Data courtesy of the U.S. Geological Survey from their website at http://waterdata.usgs.gov/nwis/.)

8.13.2   Great Salt Lake, Utah

Great Salt Lake (41° N) in northern Utah is the fourth largest endorheic lake in the world and the largest of this type in the United States. It is a small remnant of prehistoric Lake Bonneville that covered much of present-day western Utah and parts of neighboring states. The arid region west and southwest of Great Salt Lake is part of the former Lake Bonneville lake bed, and the low, flat landscape is known as the Great Salt Lake Desert. The Bonneville Salt Flats occupy 180 km^2 of extremely level salt beds on the western edge of the Great Salt Lake Desert.

The Great Salt Lake historic surface area is 4400 km^2, but the lake's area fluctuates substantially due to its shallowness and changes in its water balance. The watershed area is 34 600 km^2, and three major rivers discharge into the lake. All three rivers receive runoff directly or indirectly from the Uinta and Wasatch mountains that form the eastern watershed area, where peaks rise to elevations exceeding 3600 m. The three rivers provide the majority of the lake's water supply, and they provide water for some irrigation and urban use. The majority of the lake's watershed is in an arid region of small mountain ranges and broad basins that produce little runoff. Watershed precipitation varies from less than 30 cm annually near the lake to 140 cm at the highest elevations in the eastern mountains. Winter moisture accompanies cyclonic systems formed over the Gulf of Alaska, summer thunderstorms result from moisture inflows from the Gulf of California, and spring and fall moisture is carried over the region by westerly winds flowing from the Pacific Ocean.

The historical average elevation of the Great Salt Lake is 1280 m. At this elevation, the lake's average depth is 4 m and the maximum depth is 10.7 m.

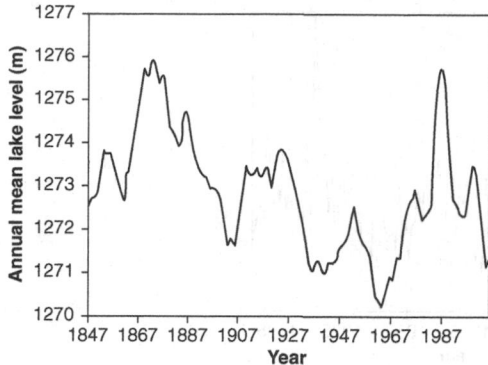

Fig. 8.21. Annual mean lake level above sea level for the Great Salt Lake at Saltair Boat Harbor, Utah, for 1847–2005. (Data courtesy of the U.S. Geological Survey from their website at http://waterdata.usgs.gov/nwis/.)

The lake level fluctuates seasonally about 0.5 m from the highest level in May to July to a low level in October through November. Even modest changes in the lake's level cause relatively large changes in its area and volume due to its shallowness. The salinity of Great Salt Lake varies between 5% and 27% as the lake level changes. The lake's chemical composition is similar to sea water at a salinity of 5%.

The annual mean lake level elevations for historic Great Salt Lake (Fig. 8.21) show that the levels range from highs of nearly 1276 m in 1872 and 1873 and again in 1986 and 1987 to a low of 1270.2 m in 1963. Declining levels predominate from 1873 to 1963 except for the period 1909–29 when lake levels are above the average. The abrupt lake level increase from 1963 to 1986 implies a relatively wet period for these 23 years.

Ogden Pioneer Power House (Ogden PPH) (41° N) at 1326 m in the Wasatch Mountains east of Great Salt Lake is a representative site for precipitation in the runoff-producing area of the Great Salt Lake watershed (Fig. 8.22). Mean annual precipitation of Ogden PPH is 53 cm and ranges from a low of 29 cm in 1928 to a high of 109 cm in 1983. There is a small increasing trend in annual precipitation over the 128 years, and the trend contrasts markedly with the pattern of the Great Salt Lake levels, which show a declining trend to 1963 and recovery after 1963. Another evident contrast is that the Great Salt Lake levels do not show a response to individual years of above or below average precipitation at Ogden PPH. For example, the large precipitation amount received at Ogden PPH in 1983 does not produce an identifiable single-year increase in the Great Salt Lake level in 1983 nor does the meager precipitation at Ogden PPH in 1928 produce a single-year lake level decline in 1928. However, groups of wet years at Ogden

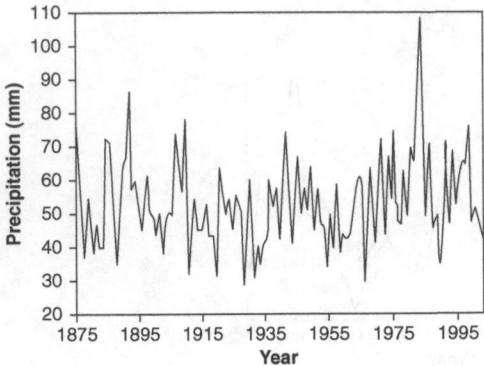

Fig. 8.22. Annual precipitation for Ogden Pioneer Power House, Utah, for 1875–2002. (Data courtesy of NOAA's National Climate Data Center and the Oak Ridge National Laboratory, Carbon Dioxide Information Analysis Center from their website at http://cdiac.ornl.gov/epubs/ndp/ushcn/usa_monthly.html.)

PPH are associated with increasing lake levels and groups of dry years result in declining lake levels. In both cases, a lag of about three years is evident in the lapse between the cumulative precipitation increase or decrease at Ogden PPH and the rising or declining Great Salt Lake level.

Review questions

8.1   What are the hydroclimatic ramifications of short records, a changing climate, and the risk of extreme events?

8.2   What are the relative strengths and weaknesses of proxy data as hydroclimatic indicators?

8.3   How are documentary data and tree rings used to improve knowledge of the hydroclimate of a place or region?

8.4   How are teleconnections a valid aid for understanding hydroclimatic variability?

8.5   What is the relationship between El Niño and the Southern Oscillation?

8.6   How is El Niño warming of the equatorial Pacific Ocean thought to project hydroclimatic anomalies to distant regions of the globe?

8.7   What are the similarities and differences in the atmospheric patterns identified as the NAO and the PNA?

8.8   What is the hydroclimatic relevance of temperature and precipitation changes over time with respect to streamflow?

8.9   How are lakes useful for studying long-term hydroclimatic relationships?

9

Floods: the hydroclimatic extreme of excessive moisture

9.1   Hydroclimatic extreme events

Floods and drought are recognized as extreme hydroclimatic events because they coincide with the tails of a streamflow frequency distribution, but they are recurrent hydroclimate features. These extreme hydroclimatic events are receiving increased attention as the loss of human life associated with these natural hazards grows and the costs related to these events soar exponentially (Easterling *et al.*, 2000). In the last decade of the twentieth century, eight floods in different regions of the world caused more than 1000 fatalities each, and 24 floods resulted in total losses of one billion dollars each (Kundzewicz, 2002). These events have occurred against a backdrop of long-term climate variability and a perceived increase in variability related to climate change. In addition, rapid population growth and changes in land use and economic development in many parts of the world have placed greatly increased numbers of people at risk relative to extreme hydroclimatic events.

Floods and drought account for more damage annually in the United States than any other natural disaster, and there is an increasing trend in both flood damage and drought vulnerability (Lins and Slack, 1999). Concern for extreme events is heightened by an observed increase in atmospheric moisture since 1973 (Trenberth *et al.*, 2003), an apparent increase in the proportion of total annual precipitation occurring as one-day extreme events (New *et al.*, 2001), and a shift of precipitation to more extreme values (Wulfmeyer and Henning-Müller, 2006). While Robson *et al.* (1998) found no detectable trend in annual peak flows in the United Kingdom, increased flood flows on some rivers in Germany and Austria are reported by Kundzewicz (2002). No clear evidence of change in flood flows is evident in Australia, but Milly *et al.* (2002) found a substantial increase in the frequency of floods exceeding 100-year recurrence levels in 29 large basins globally. In contrast, Hisdal *et al.* (2001) conclude that streamflow droughts in Europe since 1911 are not more severe or frequent, and a strong downward trend in tropical storm frequency has occurred in Australia since 1969 (Easterling *et al.*, 2000).

Comprehensive understanding of the complexity of extreme event occurrence requires recognition that extremes can be altered by changes in the mean, the variance, or both the mean and variance of a hydroclimatic variable. Since the frequency of extreme events changes non-linearly with the change in the mean of the distribution, a small change in the mean can result in a large change in the frequency of extremes (Meehl *et al.*, 2000; Katz *et al.*, 2002). Furthermore, identifying changes in the frequency of extremes for hydroclimatic phenomena requires robust statistical tests that address the dependence and long-term persistence characteristic of these complex natural processes (Koutsoyiannis *et al.*, 2007).

Floods and drought have stimulated the search for understanding of hydroclimate extremes since the first civilizations formed along the banks of rivers (Eagleson, 1994; Wohl, 2000). Persian water systems, Egyptian agricultural practices, and Roman aqueducts were all developed to protect societies from extreme hydroclimatic events. The effects of hydroclimate extremes are particularly acute in regions that rely on streamflow for irrigation and urban and industrial water supplies. In addition, sensitivity to hydroclimatic extremes is amplified by growing populations that require larger water supplies, increasing flood protection, and expanded hydroelectric power generation capacities. Understanding the relationship between anomalous meteorological, climatological, and hydrological events is a first step in developing strategies for addressing an array of water resource and management issues related to extreme events.

Discussion of floods in this chapter and drought in Chapter 10 takes the perspective that these events are hydroclimatic extremes. This focus facilitates limiting consideration to weather-related criteria, and it explains the functional linkage of climate of the first kind in delivering moisture and climate of the second kind in processing precipitation at the surface. Examination of floods caused by dam failures, ice jams, tsunamis, and other geophysical events is treated by Hayden (1988) and Matthai (1990), and urban floods related to drainage networks and land-use issues are addressed by Urbonas and Roesner (1993), Akan and Houghtalen (2003) and Mansell (2003). Klemeš (1987) and Walker *et al.* (1991) discuss sociological, economic, and political perspectives employed as human bases for characterizing drought. The emphasis here is to present an overview of the hydroclimatological processes directly and indirectly related to floods and drought.

9.2   Flood hydroclimatology

Most stream channels are incised in relatively flat surfaces of varying widths known as floodplains. These fluvial surfaces are constructed by the alluvial processes of lateral accretion and overbank deposition. In the traditional engineering perspective, a weather-related flood occurs when the drainage basin experiences unusually intense or prolonged rainfall or snowmelt and the resulting streamflow exceeds the channel capacity and overflows onto the floodplain or the land adjacent to the channel that is not usually submerged. The timing and spatial distribution of precipitation are pivotal factors in determining if a flood occurs and the characteristics of the flood, but a mix of causative factors may be responsible for floods at the same site (Merz and Blöschl, 2003). Flash floods result from high to extremely high rainfall rates, and general floods are associated with rainfall events over days or longer (Doswell et al., 1996). Flash floods occur in smaller watershed, and general floods tend to be larger in the largest river basins (Table 9.1). However, the relationship between drainage area and general flood magnitude is not consistent due to other factors. Understanding the origins of weather-related floods requires consideration of atmospheric and hydrologic processes relative to specific drainage basins (Hirschboek et al., 2000) and determination of flood seasonality generated by different atmospheric mechanisms (Cunderlik et al., 2004). In this context, floods are natural events that occur fairly frequently on virtually all unregulated streams every one to three years (Dingman, 1994). Humans have a long history

Table 9.1. *Large rainfall and/or snowmelt floods for selected large river basins*

River	Country	Continent	Station	Drainage area at station (10^3 km^2)	Peak discharge (m^3 s^{-1})
Amazon	Brazil	South America	Obidos	4640	370 000
Lena	Russia	Asia	Kasur	2430	189 000
Yangtze	China	Asia	Yichang	1010	110 000
Orinoco	Venezuela	South America	Puente	836	98 120
Brahmaputra	Bangladesh	Asia	Bahadurabad	636	81 000
Congo	Zaire	Africa	Brazzaville	3475	76 900
Ganges	Bangladesh	Asia	Hardings	950	74 060
Mississippi	United States	North America	Arkansas City	2928	70 000
Mekong	Vietnam	Asia	Kratie	646	66 700

After O'Connor and Costa (2004b), Table 2.

of attempting to control or regulate general floods by structural means, such as dams, levees, and by-pass channels. More recently, various forms of land-use control on the floodplain are employed to prevent and/or limit flood damage (Gruntfest, 2000).

Flood hydroclimatology is closely aligned with hydrometeorology and its emphasis on the probable maximum flood known as the "Noah event" (Kilmartin, 1980). However, flood hydroclimatology seeks to place hydrometeorologic-scale atmospheric activity within the broader time and space perspective of local, regional, and global atmospheric processes and circulation patterns from which the floods develop. In effect, flood hydroclimatology attempts to link isolated and intense convective thunderstorms, squall lines which are lines of rapidly moving thunderstorms including multicellular storms organized along a line (Laing, 2003), extratropical cyclones and major fronts, and rapid snowmelt into an integrated framework of flood-producing mechanisms that can occur as individual events or as parts of organized systems (Hirschboeck, 1988, 1991). This approach recognizes that precipitation systems occurring at one scale are frequently manifestations of larger-scale processes that set the stage for smaller-scale atmospheric activity. By implication, this perspective incorporates the coupling of antecedent conditions, regional relationships, large-scale anomaly patterns, global-scale controls, and long-term trends, along with the seasonal availability of moisture and the role of atmospheric phenomena to release the moisture.

9.2.1   Flood causes

A variety of time and space scales are recognized over which climatological and meteorological activity can generate flooding. However, the key element for producing exceptionally large floods across all scales is the persistence of precipitation related to the slow-moving nature of the precipitation system (Hirschboeck et al., 2000). Small-scale and short-duration downpours related to mesoscale and storm-scale atmospheric processes can produce extreme precipitation amounts that fill culverts and drainage ditches within minutes or hours. Floods produced in this way occupy the local-scale of the continuum. Macroscale and synoptic-scale precipitation processes related to global-scale circulation anomalies that have the ability to steer one major storm after another into an area along the same persistent track operate at larger time and space scales. These atmospheric conditions deliver precipitation to produce general flooding, but the spatial domain of flooding is constrained by the physical characteristics of a given drainage basin.

Watershed characteristics govern how precipitation at the surface is concentrated and transmitted across and through the landscape and the stream channel network. The concentration and transfer of the precipitation input determines whether the runoff can be contained in the stream channel. The time domain of flooding almost always exceeds the duration of the atmospheric input that generates the flood (Hirschboeck, 1988; Fattorelli et al., 1999). This temporal decoupling is related to the physical characteristics of the drainage basin and the flood generation processes identifiable as causative factors for separate flood events (Merz and Blöschl, 2003). The eventual effect of the temporal decoupling is a shift in the flooding domain toward longer durations compared to the atmospheric phenomena at the same spatial scale. Comprehensive reviews of flood hydroclimatology are found in Hirschboeck (1988) and Hirschboeck et al. (2000).

9.2.2   Flood characteristics

Floods are characterized by their magnitude, duration, and frequency, and knowledge of these traits is useful for engineering, design, economic, and floodplain management decisions. Flood magnitude is expressed as a volume per unit time or as a height above a specified reference stage. Duration is the length of time above flood stage, and frequency is the probability of a given flood occurrence during a specific period commonly taken as one year. Flood occurrence is a probability problem because the hydroclimatic factors affecting flood production vary with time to such an extent that the combinations of factors have many characteristics of chance events. The underlying assumption is that floods occurring during a specific period constitute a sample of an indefinitely large population in time.

A widely used means of expressing the magnitude of an annual flood relative to other values is the return period or the probability of exceedance. The return period is the average length of time between occurrences of a specific flood magnitude, and the exceedance probability of that flood magnitude occurring in a given year is the reciprocal of the return period. A flood with a 1% chance of occurring in any year has an annual exceedance probability of 0.01 and a return period of 100 years (Ward and Robinson, 2000).

It is important to recognize that the recurrence interval is not a flood forecast, but a statement of probability. A flood having a recurrence interval of 10 years has a 10% chance of recurring in any year. A 50-year flood in one year can be followed by a 50-year flood in the next year because a new flood-frequency sample distribution is considered. The exceedance probability of any return period flood is unchanged whether or not a flood of that magnitude occurs in a given year. It is desirable for the period of the streamflow record to be twice as long as the greatest recurrence interval for which the flood magnitude is obtained. Several distribution functions are used for computing flood frequency as described by Stedinger *et al.* (1993), but no single statistical distribution fits all data (Jones, 1997; Siccardi *et al.*, 2005). Low-frequency climate change influences on flood frequency are discussed by Jain and Lall (2001).

9.3   Flash floods

A flash flood is sudden local-scale flooding generally produced by a meteorological-scale precipitation event of limited time and space scales that produces rapid runoff. Flash floods are most often related to slow-moving thunderstorms, a sequence of thunderstorms, heavy rains from hurricanes or tropical storms, or convective storms triggered by orographic lifting. In all of these cases, the atmosphere is conditionally unstable and able to produce sustained intense rainfall when atmospheric moisture content is high. When the precipitation-producing atmospheric conditions are organized into muticel-lular convective weather systems with a length scale of 100–1000 km, they are known as a mesoscale convective system (MCS). An MCS with a horizontal scale range of 250–2500 km is identified as a mesoscale convective complex (MCC) (Laing, 2003).

Virtually all flash floods are produced by MCSs, and Doswell *et al.* (1996) suggest that every flash flood shares basic hydrometeorological/hydroclimato-logical ingredients. These ingredients include a suite of atmospheric conditions identified with deep, moist convection combined with antecedent watershed conditions and physiographic features. Deep, moist convection normally occurs during the warm season when high moisture content is possible and buoyant instability promotes strong upward vertical motion. Rainfall rates associated with convection tend to be higher than with other rain-producing weather systems (Doswell *et al.*, 1996).

9.3.1   *The suddenness of flash floods*

Most severe flash floods are produced by quasi-stationary convective systems that produce prolonged heavy rainfall over the same area. Topography, soil, and land-use contribute to producing flash floods when these factors pre-vent water from being absorbed by the soil. The onset of flooding from a few minutes to less than six hours following the rainfall event is often used to define a flash flood (Maddox *et al.*, 1978). This sudden appearance of flooding is an expression of the intensity and duration of the rainfall resulting in a preponder-ance of the rainfall being allocated to rapid surface runoff.

Rainfall that produces a flash flood usually occurs at a rate of 25 mm per hour or greater and lasts for one hour or more (Brooks and Stensrud, 2000). Flash flood forecasting is a serious challenge due to the high annual loss of human life and property damage resulting from these events. The poor spatial resolution of rain gauges and difficulties in estimating rainfall rates with radar exacerbate the problem (Scofield and Kuligowski, 2003). Doswell *et al.* (1996) and Siccardi *et al.* (2005) discuss flash flood forecasting, Scofield and Kuligowski (2003) describe algorithm development for satellite precipitation estimates and flash flood forecasts, and Ferraris *et al.* (2002) address uncertainty issues related to forecasting flash floods.

9.3.2   *Flash flood prone areas*

Flash floods are most prevalent on small streams that generally drain areas ranging in size from a few to several hundred square kilometers. The most dangerous flash floods are usually associated with steep mountain streams, canyons, and desert washes (arroyos) where they are manifested as a huge wall of water traveling downstream. Orographic enhancement of precipitation by mountains often results in flash floods when uplifted moist air produces nearly stationary convective showers and heavy rainfall. Flash floods occur globally except in the high latitudes where the ability to produce intense rainfall is inhibited by atmospheric conditions.

In Europe, the highest rainfall intensities normally occur in the Mediterranean region making the mountainous portions of this area particularly vulnerable to flash floods. However, the flash floods along Russia's Black Sea coast in August 2002 and in Boscastle, England, in August 2004 illustrate that flash floods are not limited to the Mediterranean region. The most severe flash floods in the United States occur most often in small basins in arid and semiarid parts of the West and Southwest. Small watersheds in these regions, like small watersheds elsewhere in the world, are economically impractical and physically impossible to monitor with hydroclimatic gauges, and indirect estimates of peak discharges produced by flash floods are obtained from post-flood surveys and hydraulic equations as described by Costa (1987) and Delrieu *et al.* (2005).

9.4   Mediterranean Europe flash floods

Flash floods in France, Italy, Spain, and Switzerland since 1962 have caused millions of dollars in damage and have resulted in hundreds of fatalities. A flash flood in the Barcelona, Spain, area in September 1962 caused over 500 fatalities, and the most costly flash flood occurred in the Gard region of France in September 2002. Detailed analysis of the Gard area event is provided by Delrieu *et al.* (2005).

The interaction between topography and atmospheric influences in producing flash floods is particularly evident in the Mediterranean region. The Alps, Pyrenees, and Massif Central mountains present pronounced topographic relief in the northern Mediterranean region, and they define the regional physiographic contribution to flash floods. Heavy precipitation in the region is common and is associated with a suite of atmospheric factors. In general, the atmospheric role begins with the Mediterranean Sea serving as a reservoir of energy and moisture for the lower atmospheric layers. The warm, moist air from the Mediterranean Sea is advected toward the coast as a southerly flow along the western limb of an upper-level cold trough and a related cold front at the surface. The advected air is destabilized as it flows inland. The eastward propagation of the cold front is slowed by a persistent large anticyclone over eastern Europe. Finally, the Alps, Pyrenees, and Massif Central mountains channel low-level flow, induce low-level convergence, and trigger convection that contributes to the release of convective instability (Romero *et al.*, 2001; Ferraris *et al.*, 2002; Delrieu *et al.*, 2005).

9.4.1   *August 1996 flash flood in Spain*

Synoptic conditions supporting intense and extensive convection in the Biescas area of Spain in August 1996 displayed the general pattern of atmospheric features associated with heavy rainfall in the Mediterranean region. An upper-level trough was moving over the Iberian Peninsula from the northwest and introducing cold air into the region. Low pressure centered over Spain and a cold front extending from central Europe to Portugal dominated the surface pattern and supported warm air advection east of the front. The warm southeasterly flow was deflected toward the slopes of the Pyrenees, and strong orographic influences produced a diurnally forced mesolow in the Ebro Valley resulting in an intensive convective system. More than 200 mm of rainfall occurred in three hours and caused a flash flood on a basin of 20 km^2 that killed 86 people (Romero *et al.*, 2001). Peak streamflow was estimated at 400–600 m^3 s^{-1} km^{-2}.

9.4.2   *September 2002 flash flood in France*

Southern France along the Mediterranean Sea frequently experiences extreme precipitation and flooding in the fall. Intense rainfall and severe flooding struck the Gard area of southern France on 8–9 September 2002. Maximum cumulative 24-hour rainfall totaled 600–700 mm, and more than 200 mm fell over an area of 5500 km^2. The 24-hour maximum totals are among the highest daily records in the region. Peak discharge was 5–10 m^3 s^{-1} km^{-2} for the area,

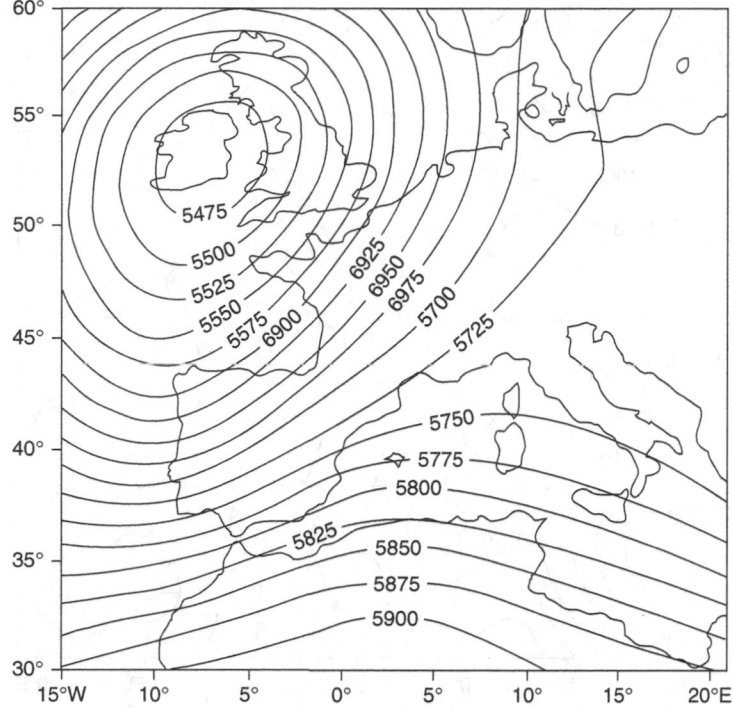

Fig. 9.1. Upper-air conditions indicated by 500 hPa geopotential heights on 8 September 2002 over Western Europe and North Africa when flash flooding occurred in southern France. Heights are in m. (Data courtesy of NOAA and the Cooperative Institute for Research in Environmental Sciences Climate Diagnostics Center from their website at http://www.cdc.noaa.gov/.)

but local values reached $20 \, m^3 \, s^{-1} \, km^{-2}$. The storm killed 24 people and caused damage estimated at $1 billion (Delrieu *et al.*, 2005).

The 8–9 September rainfall that flooded the Gard area resulted from a quasi-stationary MCS that originated as convective cells that formed over the Mediterranean Sea and moved inland. An upper-level cold low was centered over Ireland on 8 September, and a deep trough extended from the low southward to Spain (Fig. 9.1). Southwesterly winds aloft across Spain and southern France were cold and blowing at $22–24 \, m \, s^{-1}$. A radiosonde sounding indicated the air had a high moisture content and was conditionally unstable, and GPS data analyzed by Champollion *et al.* (2004) revealed the maximum water vapor content occurred just south of the maximum rainfall zone. A surface cold front extended from the North Sea through western France and Spain just east of the trough axis (Fig. 9.2). Convection formed east of the cold front as the air moved inland in the warm, moist southeasterly flow of the

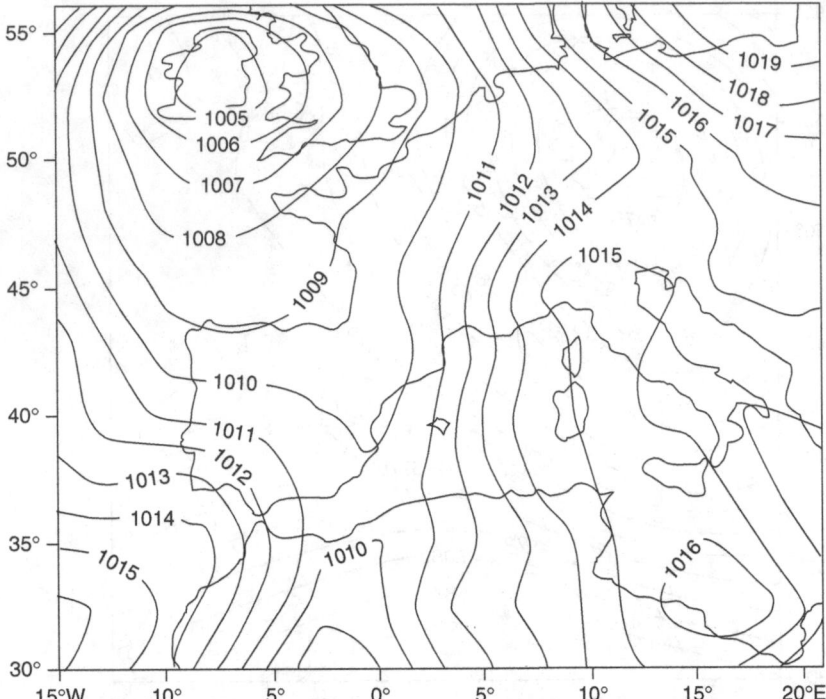

Fig. 9.2. Sea-level pressure over Western Europe and North Africa on 8 September 2002 when flash flooding occurred in southern France. Units are in hPa. (Data courtesy of NOAA and the Cooperative Institute for Research in Environmental Sciences Climate Diagnostics Center from their website at http://www.cdc.noaa.gov/.)

warm sector. Convection produced rather uniform rainfall amounts of 200 mm in less than 12 hours over the area. As the cold front progressed eastward, low-level convergence into southeastern France increased, and high rainfall amounts were induced by the additional uplift occurring over the low-level convergence and promoted by orographic uplift related to funneling of air by the Massif Central and the Alps (Champollion *et al.*, 2004). Sustained rainfall for 6 hours reached 250 mm for some sites during this period. Passage of the cold front over the Gard region produced a final burst of heavy precipitation of about 100 mm, but it also moved convective activity out of the region (Delrieu *et al.*, 2005).

The heavy rainfall produced flash floods in many inland tributary areas and general floods along lower reaches of rivers. However, many stream gauges on larger rivers underestimated observed water levels or were destroyed by the flood. Consequently, post-event surveys of river cross-sections, witnesses, radar data, and hydrologic models were used to estimate stream discharge for much of the region (Delrieu *et al.*, 2005).

9.5   United States flash floods

Flash floods in the United States are particularly evident along the eastern slopes of the Rocky Mountains, in the desert Southwest, and along the foothill slopes of the Appalachian Mountains. These flash flood locations are widely distributed geographically, and Maddox *et al.* (1980) identify differences between eastern and western United States flash floods. However, atmospheric conditions related to some of the most severe recorded regional events display similar features as presented by Maddox *et al.* (1979, 1980) and Doswell *et al.* (1996). Physiography and geology are major factors in maximizing runoff from intense rainfall, and these watershed features produce varying flash flood characteristics in response to similar rainfall intensities (Costa, 1987; Smith *et al.*, 2000).

9.5.1   Northern Great Plains

Flash floods in the northern Great Plains are associated with atmospheric characteristics similar to those resulting in flash floods in the western United States as identified by Maddox *et al.* (1980). A storm over the Black Hills near Rapid City, South Dakota, on 9–10 June 1972 resulted in the most severe flash flood in South Dakota history and devastated Rapid City. Torrential downpours in a narrow north–south band on the eastern slopes of the Black Hills delivered 380 mm of rain in 5 hours at one location and 250 mm or more over an area of 156 km^2. Peak streamflow in Rapid City exceeded 1416 m^3 s^{-1}, which was over 10-times greater than the previous record. Figure 9.3 illustrates the abrupt increase in streamflow driven by the intense precipitation, but it masks the

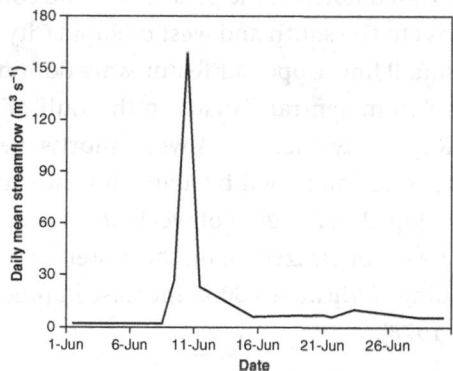

Fig. 9.3. Daily mean streamflow for June 1972 for Rapid Creek at Rapid City, South Dakota. Flash flood occurred on 10 June. (Data courtesy of the U.S. Geological Survey from their website at http://waterdata.usgs.gov/nwis/.)

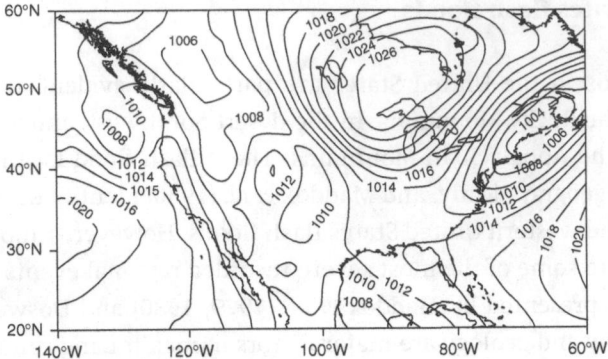

Fig. 9.4. Sea-level pressure over North America on 10 June 1972 when flash flooding occurred in Rapid Creek, South Dakota. Units in hPa. (Data courtesy of NOAA and the Cooperative Institute for Research in Environmental Sciences Climate Diagnostics Center from their website at http://www.cdc.noaa.gov/.)

magnitude of the actual flood peak since daily mean discharge is shown. The 10 June mean discharge is $159\,\mathrm{m}^3\,\mathrm{s}^{-1}$. The high streamflow levels were exacerbated by a reservoir failure that contributed to the flood wave that caused 237 deaths and more than \$164 million in property damage (Nair *et al.*, 1997).

Atmospheric conditions at the surface for 10 June show a deep trough axis extending southward from a high northwest of the Great Lakes (Fig. 9.4). A polar front at the surface coincides with the southern margin of the trough, and the western extension of the cold front lies south of the Black Hills in the extreme southwestern corner of South Dakota. Surface pressure is lower south and west of the Rapid City area, and surface easterly winds in most of South Dakota are blowing at 10–$15\,\mathrm{m}\,\mathrm{s}^{-1}$. A narrow zone of moist air indicated by relatively high dew point temperatures is concentrated north of the polar front, and convective activity in the hotter, drier air mass to the south and west of Rapid City has the potential to develop into a weak squall line. Upper-air features are dominated by a large-amplitude ridge extending from central Canada to the Gulf of Mexico and located 200–$300\,\mathrm{km}$ east of Rapid City (Fig. 9.5). A weak shortwave trough south and southwest of the ridge is accompanied by abundant moisture. The storm causing the flash flood developed in a region of weak positive advection with convection to the south and west organized along the eastern limb of the weak short wave. Upper-air soundings indicated a 300% increase in precipitable water in 12 hours (Maddox *et al.*, 1978).

9.5.2   Southern Great Plains

Texas, in the southern Great Plains, produces some of the largest rainfall accumulations and flood magnitudes in the United States and the world

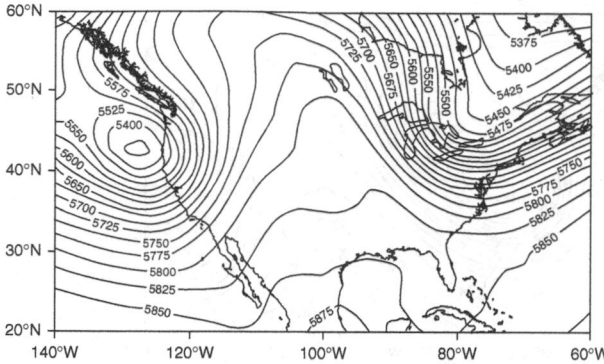

Fig. 9.5. Upper-air conditions indicated by 500 hPa geopotential heights on 10 June 1972 over North America when flash flooding occurred in Rapid Creek, South Dakota. Heights are in m. (Data courtesy of NOAA and the Cooperative Institute for Research in Environmental Sciences Climate Diagnostics Center from their website at http://www.cdc.noaa.gov/.)

(Smith *et al.*, 2000), and the area of the Balcones Escarpment in central Texas has exceptionally high unit discharges (O'Connor and Costa, 2004a). Flash floods in Texas are related to a distinctively different set of atmospheric conditions than those associated with flash floods in the northern Great Plains as defined by Maddox *et al.* (1980). The storm properties that produced record flooding in southeast Texas on 16–17 October 1994 illustrate the atmospheric relationship to flash floods in this region. A complete analysis of this storm is provided by Smith *et al.* (2000).

The October 1994 flooding on the Texas Coastal Plain north of Houston was extreme for three watersheds and was catastrophic in and around the 148 km^2 Kickapoo Creek watershed. Regional flooding caused 22 deaths and more than $1 billion in total storm damage. Peak rainfall accumulations in a 12-hour period of more than 500 mm occurred in the 1085 km^2 Spring Creek watershed and resulted in a peak discharge of 2279 m^3 s^{-1}. Peak rainfall accumulations in the Kickapoo Creek watershed were approximately 60% of record values for the continental United States for time intervals from 15 minutes to 24 hours. The peak discharge for Kickapoo Creek of 2400 m^3 s^{-1} is estimated because the flood destroyed the watershed stream gauge (Smith *et al.*, 2000).

The upper-level atmospheric environment preceding the 1994 Texas storm events was dominated by a long-wave trough over the southern Rocky Mountains and high pressure centered over the Gulf of Mexico (Fig. 9.6). A radiosonde sounding over Corpus Christi on the southwest Texas coast revealed most of the lower troposphere remained nearly saturated following the passage of Hurricane Rosa on 14–15 October (Smith *et al.*, 2000). Surface

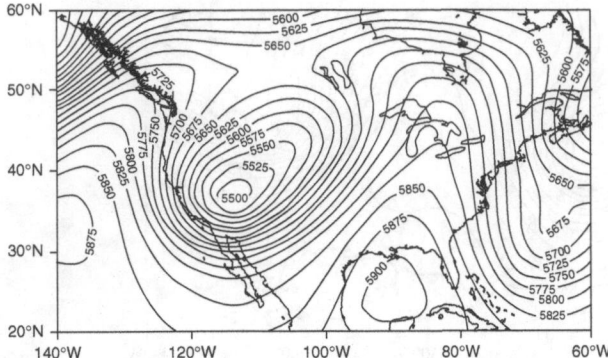

Fig. 9.6. Upper-air conditions indicated by 500 hPa geopotential heights on 16 October 1994 over North America when flash flooding occurred in southeast Texas. Heights are in m. (Data courtesy of NOAA and the Cooperative Institute for Research in Environmental Sciences Climate Diagnostics Center from their website at http://www.cdc.noaa.gov/.)

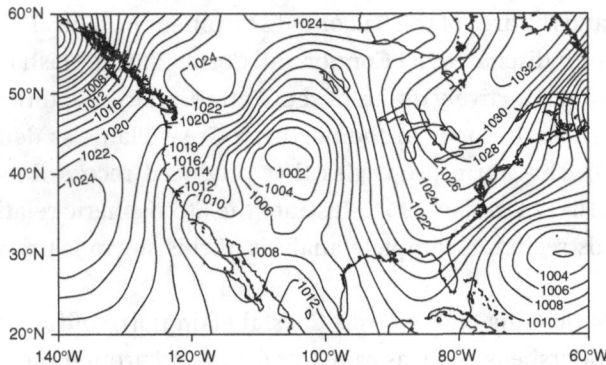

Fig. 9.7. Sea-level pressure over North America on 16 October 1994 when flash flooding occurred in southeast Texas. Units are in hPa. (Data courtesy of NOAA and the Cooperative Institute for Research in Environmental Sciences Climate Diagnostics Center from their website at http://www.cdc.noaa.gov/.)

pressure showed a low centered over Colorado confined by strong higher pressure on three sides (Fig. 9.7). A surface cold front extended from northern Colorado diagonally through New Mexico. Winds across eastern Texas were blowing from the southeast at 7–9 m s^{-1}, and they shifted to the south as the front approached.

The storms that produced flash flooding in Spring Creek and Kickapoo Creek were two elements of a single MCS, but the storm characteristics over the two watersheds were distinctly different. A squall line with very strong storm updrafts, high cloud tops, and a large cold cloud shield produced the heaviest rainfall and the peak intensities in the Spring Creek basin. In addition, near-record storm lightning strikes were recorded in the southwestern corner of the Spring Creek basin. Radar data and surface observations indicated the peak rainfall rates occurred during a relatively short period as the squall line passed over the Spring Creek basin and moved off to the northeast (Smith *et al.*, 2000).

Intense rainfall occurred over the Kickapoo Creek watershed approximately 100 km north of the Spring Creek basin in response to near stationary motion of a non-squall line storm. A plume of low-level cumulus clouds extended from the Gulf of Mexico over the Kickapoo Creek watershed. The flash flood producing storm over the basin had relatively low cloud tops, small cloud shields, and low lightning frequency. These properties are characteristic of warm rain processes with high precipitation efficiencies, and the near stationary storm movement was an important contributor to producing catastrophic total rainfall accumulations of nearly 600 mm and flash flooding (Smith *et al.*, 2000).

9.5.3   The eastern Rocky Mountains

Meteorological features resulting in flash floods along the eastern slopes of the Rocky Mountains are similar to those producing flash floods in the northern Great Plains and across the western United States (Maddox *et al.*, 1980). However, the severe 1976 flash flood in the Big Thompson Canyon, Colorado, and the 1972 Black Hills flash flood both displayed evidence of a strong orographic component that is uncommon to storms of this type in these areas (Caracena *et al.*, 1979).

The Big Thompson Canyon flash flood was less dramatic than the Black Hills flood in terms of fatalities and property loss because it occurred in a dominantly recreational area. The Big Thompson River begins in the Rocky Mountains near Estes Park in north central Colorado and flows eastward about 40 km through a rugged, steep-walled canyon. The river bed drops vertically more than 800 m from its headwaters region to the rolling, forested plain west of Loveland, Colorado. On 31 July 1976 thunderstorms formed along the crest of the Front Range and remained virtually stationary for more than three hours. Although the resulting flood destroyed all stream and rain gauges in the canyon, it is estimated that the thunderstorms delivered more than 305 mm of rainfall over a 182 km^2 area. The most intense rainfall began early on 1 August and fell on slopes in the western end of the canyon.

While the Big Thompson River is normally less than 5 m wide and 2 m deep, the intense and prolonged rainfall produced a torrent of water and debris 6 m deep and driven by a peak discharge of 884 m^3 s^{-1}. The flood wave surged down the canyon as a flood expected once in 10 000 years. Figure 9.8 illustrates the

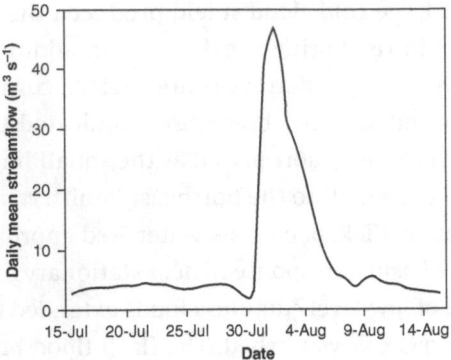

Fig. 9.8. Daily mean streamflow for 15 July 1976 to 15 August 1976 for the Big Thompson River near Drake, Colorado. Flash flood occurred on 1 August. (Data courtesy of the U.S. Geological Survey from their website at http://waterdata.usgs.gov/nwis/.)

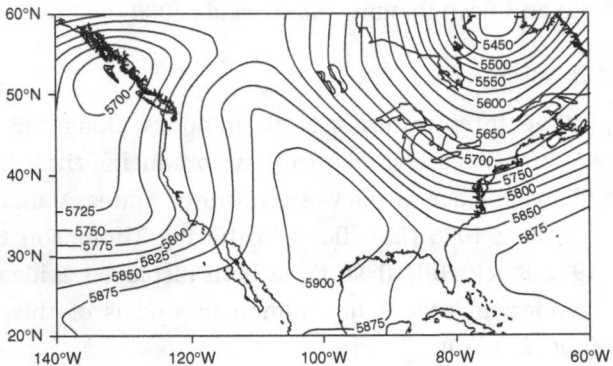

Fig. 9.9. Upper-air conditions indicated by 500 hPa geopotential heights on 1 August 1976 over North America when flash flooding occurred in the Big Thompson River, Colorado. Heights are in m. (Data courtesy of NOAA and the Cooperative Institute for Research in Environmental Sciences Climate Diagnostics Center from their website at http://www.cdc.noaa.gov/.)

dramatic suddenness of the flow increase but the peak is muted by the daily averaging. Since most of the businesses, motels, and campgrounds were located along the scenic river bank, the unexpected flood resulted in 139 known fatalities and 6 additional unaccounted-for people whose bodies were never recovered. Property damage attributed to the flood was $35.5 million even in the sparsely settled canyon (Maddox *et al.*, 1978).

Surface and 500 hPa analysis for the Big Thompson Canyon flash flood storm indicate it developed when surface and upper-air conditions combined to push conditionally unstable and extremely moist air upslope as the air moved westward. Figure 9.9 shows upper-air features dominated by a strong ridge with an axis

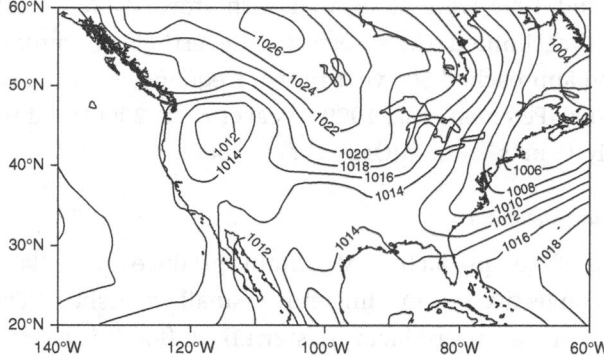

Fig. 9.10. Sea-level pressure over North America on 1 August 1976 when flash flooding occurred in the Big Thompson River, Colorado. Units are in hPa. (Data courtesy of NOAA and the Cooperative Institute for Research in Environmental Sciences Climate Diagnostics Center from their website at http://www.cdc.noaa.gov/.)

oriented northwest to southeast and extending from Canada to Texas. Winds aloft over eastern Colorado are largely from the south and are relatively weak with speeds of 5–$13\,\mathrm{m\,s^{-1}}$. An upper-air sounding 90 km southeast of the Big Thompson Canyon at Denver indicated the precipitable water content from the surface to the 500 hPa level was nearly double the July mean (Maddox *et al.*, 1978).

Surface observations display a southward extension of high pressure across the Midwest associated with slow-moving cold air (Fig. 9.10). A cold front anchored near the Great Lakes coincides with the southern expansion of the cold air and runs diagonally through Wyoming and across eastern Colorado. A small low-pressure area occurred behind the front in western Colorado. Easterly winds were dominant in the region east of the front with speeds of 10–$13\,\mathrm{m\,s^{-1}}$. The moist air revealed by the Denver upper-air sounding was being moved toward the Front Range at a nearly perpendicular intercept angle. The westward moving moist surface air was capped by a frontal inversion, but orographic lifting promoted instability and triggered storm development with minimum downdraft circulation. The high moisture content of the low-level air continued to feed the storm, permitted storm development east of the highest terrain, and promoted convective cell redevelopment to produce a quasi-stationary precipitation system. A high moisture flux coupled with a high precipitation efficiency resulting from warm cloud coalescence processes sustained torrential rains for several hours and resulted in the flash flood in the lower canyon (Caracena *et al.*, 1979).

Less severe flash floods in the Rocky Mountain foothills often have one or more storm features weakly developed compared to the conditions evident in the Big Thompson Canyon storm. For example, flash floods in Fort Collins,

Colorado, in 1997 and 1999 were associated with storm systems with only modest instability that limited their severity. Nevertheless, rainfall totals reached 260 mm and approached 500-year return frequencies during the 1997 Fort Collins flash flood (Petersen et al., 1999; Weaver et al., 2000), and the flood discharge was nearly $15 \, \text{m}^3 \, \text{s}^{-1} \, \text{km}^{-2}$ (Ogden et al., 2000).

9.5.4   The Appalachian Mountains

The foothills of the Appalachian Mountains produce severe flash floods when slow-moving convective systems linger over small watersheds. The atmospheric environment of storms producing eastern flash floods is characterized by different features than the storms associated with flash floods in the western United States, and eastern storms generally produce longer duration but less intense precipitation (Maddox et al., 1979). Nevertheless, flash floods along the Appalachian Mountains are destructive in the foothill watersheds and result from different combinations of atmospheric conditions.

Almost continuous heavy rainfall for 10 hours during the night of 19–20 July 1977 resulted in severe flash flooding in the Johnstown area of western Pennsylvania on the west slope of the Appalachian Mountains. A series of slow-moving thunderstorms associated with part of an MCC produced up to 300 mm of total rainfall on the area with maximum intensities averaging $50 \, \text{mm} \, \text{h}^{-1}$. Record maximum discharges were recorded on 11 stream gauges in the area, and 6 of the record maximums had recurrence intervals of 100 years or greater. Seven earth-fill dams in the area failed contributing to flash floods that caused 77 deaths and property damage estimated at $300 million (Bosart and Sanders, 1981; Paulson et al., 1991).

Streams in southeastern Ohio drain small watersheds, steep slopes, and narrow valleys characteristic of the western foothills of the Appalachian Mountains. A flash flood occurred on several of these streams in the vicinity of Shadyside, Ohio, on 14 June 1990 when moist tropical air from the Gulf of Mexico was pushed eastward toward the Appalachians by an outflow boundary associated with dissipating thunderstorms to the west. This mesoscale forcing promoted rapid thunderstorm development over the foothills, and the convective activity was enhanced by low-level upslope flow. Multiple thunderstorms remained nearly stationary for approximately 90 minutes over three small watersheds that each covered about $30 \, \text{km}^2$. Although no precipitation or streamflow gauges existed in the immediate area, nearby precipitation gauges, radar data, and interviews indicated 127 mm of rain fell in 1 hour, and flooding started about 45 minutes after the heavy rain began. A wave of water 3 to 9 m deep raged down the small streams and caused 26 deaths and an estimated $8 million in property damage (LaPenta et al., 1995; Larson et al., 1995).

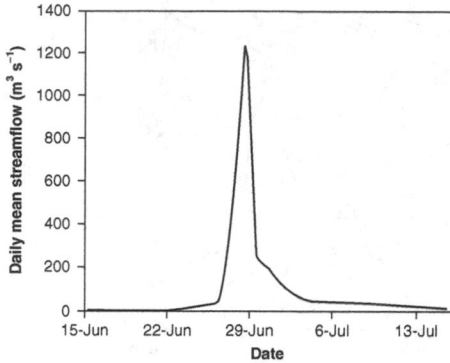

Fig. 9.11. Daily mean streamflow for 15 June 1995 to 15 July 1995 for the Rapidan River near Culpepper, Virginia. Flash flood occurred on 27 June. (Data courtesy of the U.S. Geological Survey from their website at http://waterdata.usgs.gov/nwis/.)

Flash floods on the eastern slopes of the Appalachian Mountains result from rain produced by westward-flowing atmospheric moisture from the Atlantic Ocean and characteristically have high unit discharges (O'Connor and Costa, 2004a). During the morning and early afternoon of 27 June 1995 two storm systems near the Blue Ridge Mountains of Virginia on the eastern flank of the Appalachian Mountains produced record flooding in the Rapidan River Basin. Storm total 6-hour rainfall accumulations exceeded 600 mm. The peak stage of the Ripidan River at Ruckersville, Virginia, was 9.6 m, and the peak discharge was $3000\,\mathrm{m}^3\,\mathrm{s}^{-1}$ or $10.2\,\mathrm{m}^3\,\mathrm{s}^{-1}\,\mathrm{km}^{-2}$. This peak unit discharge is the largest for the United States east of the Mississippi River for basins larger than $100\,\mathrm{km}^2$ (Giannoni *et al.*, 2003). The previous record at Ruckersville was a stage of 6.3 m and a peak discharge of $870\,\mathrm{m}^3\,\mathrm{s}^{-1}$. The Rapidan River rose over 5 m in 1 hour near Culpeper, Virginia, and the flood magnitude is estimated to have a recurrence interval of 500 years. The hydrologic context of the flooding is evident in the daily mean discharge shown in Figure 9.11. The flash floods in the watershed were responsible for three deaths and property losses exceeding $200 million (Pontrelli *et al.*, 1999).

The synoptic and mesoscale conditions responsible for initiating and maintaining the Rapidan storm resemble those reported for the Big Thompson and Rapid City floods on the eastern slopes of the Rocky Mountains (see Section 9.5.3) and Black Hills (see Section 9.5.1), respectively (Smith *et al.*, 1996; Pontrelli *et al.*, 1999). However, the Rapidan event consisted of two storm systems producing intense rainfall. A large horizontal dimension storm passed through the Rapidan River Basin in the morning and produced the first precipitation peak. A smaller storm moved slowly over the upper Rapidan watershed in the

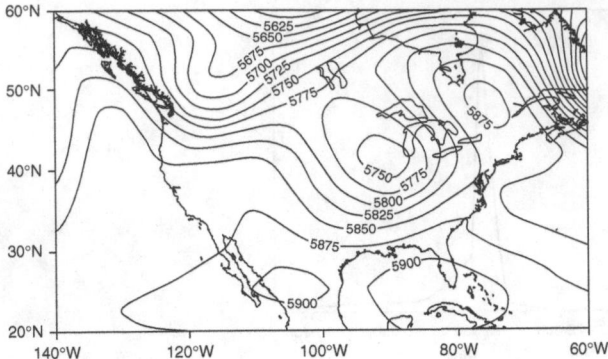

Fig. 9.12. Upper-air conditions indicated by 500 hPa geopotential heights on 27 June 1995 over North America when flash flooding occurred in the Rapidan River, Virginia. Heights are in m. (Data courtesy of NOAA and the Cooperative Institute for Research in Environmental Sciences Climate Diagnostics Center from their website at http://www.cdc.noaa.gov/.)

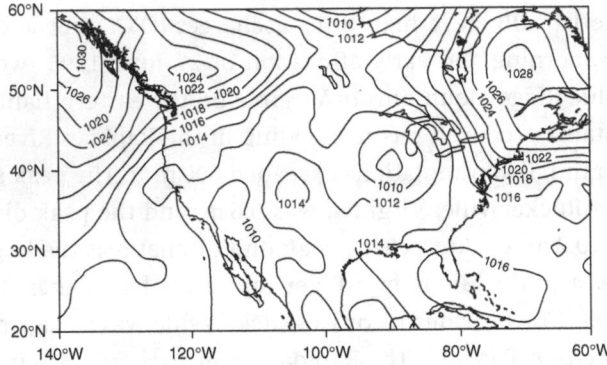

Fig. 9.13. Sea-level pressure over North America on 27 June 1995 when flash flooding occurred in the Rapidan River, Virginia. Units are in hPa. (Data courtesy of NOAA and the Cooperative Institute for Research in Environmental Sciences Climate Diagnostics Center from their website at http://www.cdc.noaa.gov/.)

afternoon and produced the second precipitation peak and the record flood peak (Smith *et al.*, 1996; Giannoni *et al.*, 2003).

For the Rapidan storm, an upper-level high-pressure ridge axis extended over New England and a weak shortwave trough was entering southwestern Virginia (Fig. 9.12). At the surface, high pressure from the system over New England pressed southward and funneled cool, dry air into a surface low positioned almost directly under the upper-level trough (Fig. 9.13). A weak frontal zone over Virginia marked the southern boundary of the cool, dry air (Pontrelli *et al.*,

1999). These features combined to produce strong low-level flow upslope toward the Virginia Blue Ridge Mountains (Smith *et al.*, 1996).

Upper-air soundings north and east of the front revealed high dew point temperatures and a high moisture content throughout a deep atmospheric layer. Such conditions indicate instability and the need for little lifting to initiate convection, and they promote warm rain processes (Smith *et al.*, 1996). In addition, high-speed low-level easterly winds in the lowest 1-2 km of the atmosphere were revealed by the soundings. The low-level easterly winds were several hundred kilometers wide and persisted for the entire storm period. Local terrain features determined where along the ridges convection would be focused. The first mountain elevations high enough to break the capping inversion were the Blue Ridge Mountains in the Rapidan River watershed. Individual convective cells were continually initiated, matured, and dissipated forming an overall quasi-stationary convective entity. The slow movement of the moist air inflow behind the front allowed convection to remain over the same area for several hours producing intense rain and flash floods (Pontrelli *et al.*, 1999).

9.6   General floods

Regional or general floods in a single drainage basin are produced by different types of precipitation events related to the season of the year and the geographic location of the watershed (Cunderlik *et al.*, 2004). Nearly all of the largest rainfall floods (see Table 9.1) occur in basins south of 40° N, but snowmelt and ice jams contribute to large floods at higher latitudes (O'Connor and Costa, 2004b). Equatorial sites dominated by the ITCZ may be subject to flooding year-round, or they may have a distinct seasonal characteristic if near the poleward limit of the annual ITCZ migration zone. The annual flooding of the Nile River is related to ITCZ summer rainfall in the Ethiopian highlands, and summer flooding in India follows the abundant precipitation delivered by the Asian monsoon. Flooding in tropical latitudes is often related to tropical cyclones that are frequent in summer and fall. During the 1990s, tropical cyclones were responsible for severe flooding in the Philippines, China, Vietnam, and Central America. However, extratropical circulation features can invade the tropics and cause extreme rainfall and flooding as occurred in northern Venezuela in 1999 (Lyon, 2003). In the middle latitudes, floods occur in all seasons but result from different mechanisms in different seasons depending on specific site characteristics (Cunderlik *et al.*, 2004). In general, individual convective storms and MCSs produce summer floods, and tropical cyclones produce infrequent summer and fall floods. Frontal systems related to extratropical cyclones can generate winter, spring, or fall floods (Hirschboeck *et al.*, 2000), and the MJO is related to early winter floods in the Pacific Northwest region of the United States (Bond and Vecchi, 2003). High latitude areas with a substantial seasonal snow-cover experience an annual flood related to spring snowmelt (Hayden, 1988). The Dartmouth Flood Observatory (http://www.dartmouth.edu/~floods/) main-tains an archive of large floods with descriptions of their individual causality.

Specific flood occurrences are commonly related to anomalous configura-tions in the general circulation, SST anomalies related to ENSO, and long-period circulation adjustments evident in upper-atmosphere pressure fields. These influences combine with seasonal and geographic factors to form a complex of flood occurrences resulting from several different types of meteorological circulation systems (Hirschboeck, 1988). The common feature in each case is that the anomalous atmospheric pattern produces excessive amounts of pre-cipitation. In some cases, it is a combination of several mechanisms operating simultaneously that produce an anomalous pattern and the flooding episode. A large-scale perspective is required to place these events in their appropriate regional and global framework, but it is often difficult to attribute flooding to a single factor and to account for natural and human variability influences (Robson *et al.*, 1998; Milly *et al.*, 2002).

ENSO influences illustrate the difficulty in attributing flood occurrence to a specific causal agent. Global examples of severe flooding resulting from ENSO-related rainfall include California in 1983, Somalia in 1997, and Peru in 1998. However, not all ENSO events are associated with flooding in a particular river basin because a combination of specific meteorological and hydrological factors is necessary to promote flooding. From a hydroclimatological perspective, one must search for modifications initiated by ENSO on large-scale atmospheric transport systems to identify associations between ENSO and regional floods (Loáiciga *et al.*, 1993). Furthermore, only a relatively small number of regions globally display a coherent ENSO-related precipitation response (Kane, 1997). Hirschboeck *et al.* (2000) conclude that a relationship between ENSO and flood-ing is geographically variable and inconsistent since most areas lack an identifi-able episode-to-episode relationship between ENSO and precipitation.

9.7   The 2002 central Europe flood

Europe has experienced an unusual frequency of severe floods since the winter floods along the Rhine and Moselle rivers in 1993 and 1995. Summer floods struck the Oder River in 1997 and the Rhine and Danube rivers in 1999. Extensive autumn floods in 2000 impacted southern England, Wales, and western Europe, and summer floods in 2001 on the Vistula River covered a large area in southern Poland and neighboring countries (Ulbrich et al., 2003b). The 2002 summer was one of the wettest on record for many European countries, and it produced widespread flooding over an area affecting 13 countries. The most severe flooding was in August. Many of the same countries were flooded again in August 2005 with Romania being the most severely impacted.

The 2002 summer floods were notable because they affected an uncommonly broad area over several months. The first event was heavy June rainfall causing extensive flooding in parts of southern Russia between the Black Sea and Caspian Sea. Intense rainfall events and flooding continued in different areas of Europe throughout the summer and ended with the flash flood in the Gard region of France discussed in Section 9.4.1 and a slow-moving storm system producing severe flooding in Albania in late September.

August emerged as the most disastrous month for flooding in the summer of 2002. Rainfall totaling 100–200 mm during the first 10 days of August initiated some of the worst flooding in more than a century across an area extending from England to the Black Sea. In the Czech Republic, the Vltava River reached its highest levels since records began in 1896, and many parts of Prague were flooded. The Danube River at Budapest, Hungary, had a maximum stage of 8.6 m exceeding the previous record set in 1965. The Elbe River in Germany flooded Dresden when the river stage reached 9.4 m, which was its highest level since 1275. The previous maximum level at Dresden was 8.77 m in 1845 (Ulbrich et al., 2003a). Regional flooding in August caused 100 deaths, and estimated damage totaled $25 billion making this the most expensive weather-related event in Europe in recent decades (Pal et al., 2004).

The meteorological and climatological factors leading to the August 2002 Europe flood became established in late July as an upper-level trough developed west of Ireland. A surface low evolved below the upper-level trough, and the atmospheric flow aloft and at the surface responding to the prevailing pressure gradient drew cool maritime air from off the North Atlantic toward western Europe. A cold front developed as the eastward flowing cool maritime air displaced warm, humid air. A series of surface low-pressure systems and cold fronts evolved over the next several days that delivered extreme rainfall at different times to a broad area of Europe. The extreme precipitation was associated primarily with two early August rainfall episodes.

The initial thrust of cold air at the beginning of August triggered widespread thunderstorms and locally extreme rainfall in northern Germany that saturated the soil. A cold front stretched from the North Sea to the Bay of Biscay on 6 August with a surface low over The Netherlands. A weak secondary surface low formed in the Gulf of Genoa on 6 August and moved across northern Italy

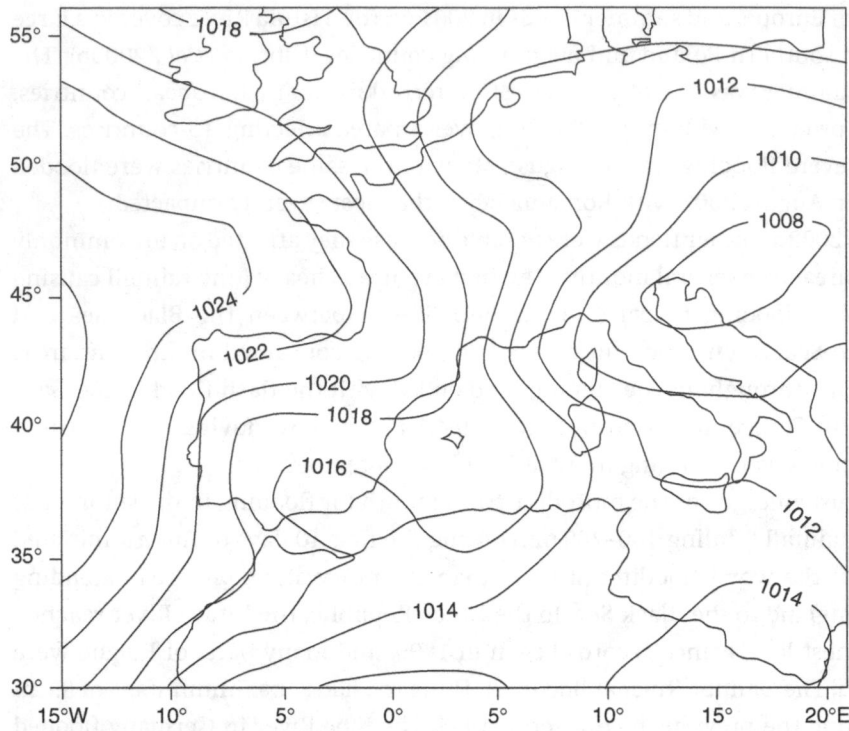

Fig. 9.14. Sea-level pressure over Western Europe and North Africa on 7 August 2002 when widespread flooding occurred in central Europe. Units are in hPa. (Data courtesy of NOAA and the Cooperative Institute for Research in Environmental Sciences Climate Diagnostics Center from their website at http://www.cdc.noaa.gov/.)

and the Adriatic Sea (Fig. 9.14) producing heavy rainfall in Austria, the southwestern Czech Republic, and southeastern Germany on 6 and 7 August (Ulbrich *et al.*, 2003b). This first large-scale rainfall event delivered more than 100 mm of rain to the affected area, but flooding did not occur in all areas. In eastern Bavaria, the rain replenished unsaturated soils and antecedent river flows were low. For Lower Austria, intense rainfall in small watersheds resulted in flooding. The Kemp River peak discharge was 800 m^3 s^{-1}, and this discharge has an estimated return period of several thousand years (Ulbrich *et al.*, 2003a).

A new cold air outbreak developed on 8 August with an upper-air low over northern France (Fig. 9.15). A surface low formed west of Ireland and moved to southern England where it filled on 10 August. However, the cut-off upper-air low continued southward and formed a secondary surface low over the Gulf of Genoa. This low drew in warm, moist Mediterranean air from the southeast to combine with the cold maritime air from the northwest. The upper-level low was displaced northeastward to a position southeast of

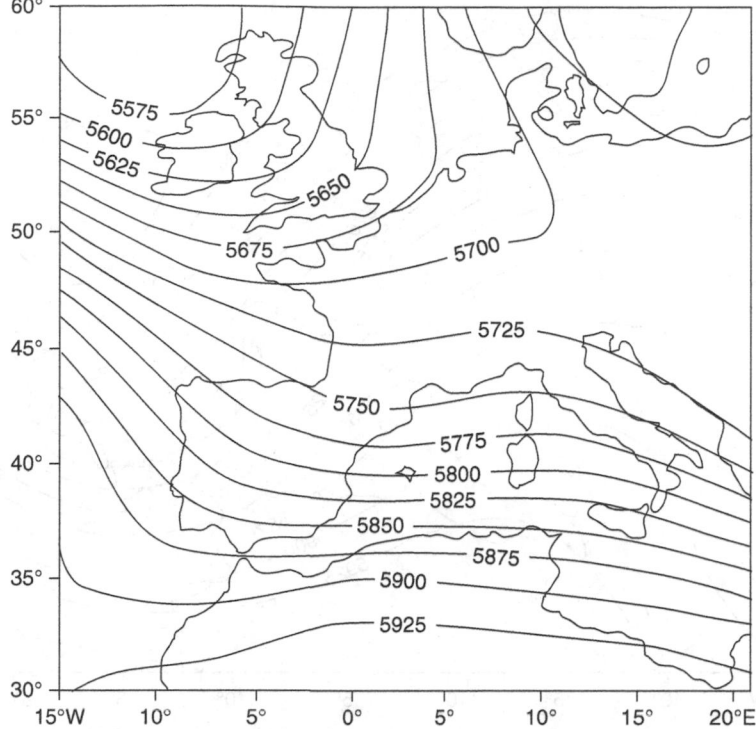

Fig. 9.15. Upper-air conditions indicated by 500 hPa geopotential heights on 8 August 2002 over Western Europe and North Africa when widespread flooding occurred in central Europe. Heights are in m. (Data courtesy of NOAA and the Cooperative Institute for Research in Environmental Sciences Climate Diagnostics Center from their website at http://www.cdc.noaa.gov/.)

Germany and the Czech Republic where it remained in a quasi-stationary position and induced upper-level divergence (Fig. 9.16). As the surface low moved northward in response to the repositioning of the upper-air low (Fig. 9.17), the frontal system stalled and frontal cloud bands with heavy rainfall and thunderstorm cells moved over the Alps. Intense rainfall developed east of Lake Constance and progressed through the Salzburg area, the western Czech Republic, and the Erzgebirge Mountains where orographic lifting reinforced frontal lifting. Rainfall totals for 10–12 August of 300 mm occurred in the Erzgebirge, a record precipitation total of 357.7 mm was measured at one station, and rainfall intensities were 10 mm hr^{-1} or greater for 18 hours out of one 24-hour period. Lower elevation areas received 100–150 mm in 24 hours which is 2–3.5 times the normal total rainfall for the month of August (Rudolf and Rapp, 2003; Unbrich et al., 2003b).

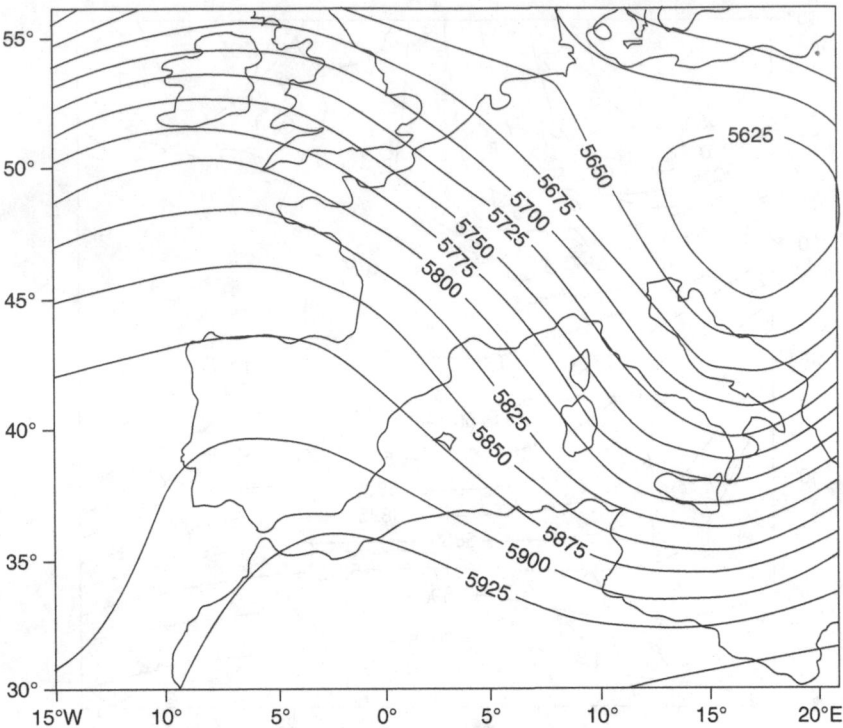

Fig. 9.16. Upper-air conditions indicated by 500 hPa geopotential heights on 12 August 2002 over Western Europe and North Africa when widespread flooding occurred in central Europe. Heights are in m. (Data courtesy of NOAA and the Cooperative Institute for Research in Environmental Sciences Climate Diagnostics Center from their website at http://www.cdc.noaa.gov/.)

The area of exceptionally high rainfall in the Erzgebirge occurred in the Elbe River watershed. Flash floods developed on tributary streams on the northern slopes of the Erzgebirge as stream levels rose 0.5–1.0 m hr^{-1}. The extreme flow rates in many tributary streams subsequently led to major flooding along the Elbe River. At Usti, near the Czech–German border, streamflow increased between 9 and 11 August forming a first flood wave with a peak of 1100 m^3 s^{-1}. Local and then more remote sources contributed to the flow at Usti, which reached a final flood crest of 5000 m^3 s^{-1} during the evening of 16 August. The flood crest reached Dresden on 17 August and was attenuated to 4500 m^3 s^{-1} due to water retention in the floodplain. Further downstream, the flood crest increased at some locations due to tributary stream contributions and decreased at other locations from breaches in the river embankment and controlled flooding (Ulbrich *et al.*, 2003a).

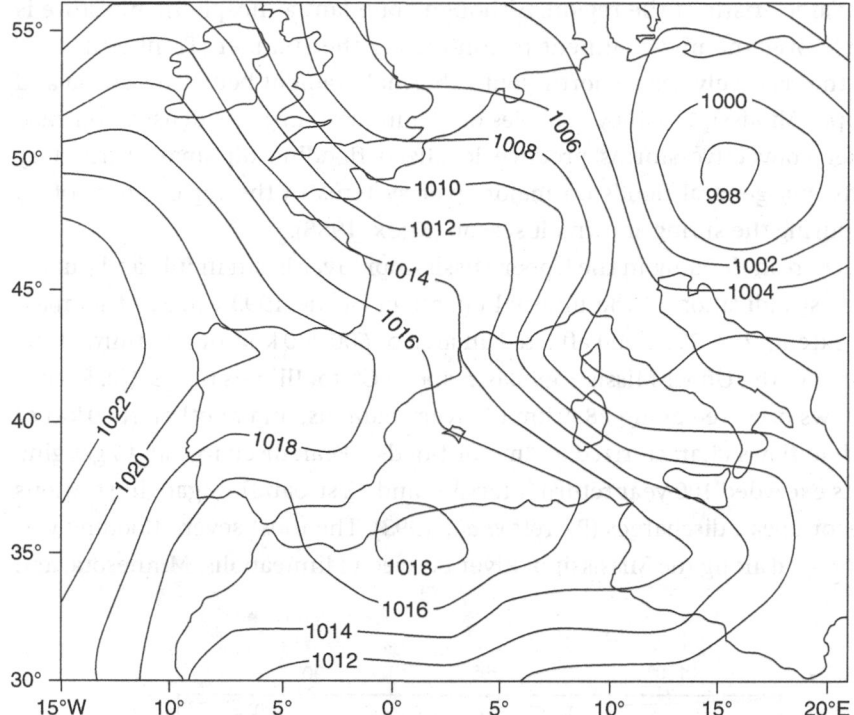

Fig. 9.17. Sea-level pressure over Western Europe and North Africa on 12 August 2002 when widespread flooding occurred in central Europe. Units are in hPa. (Data courtesy of NOAA and the Cooperative Institute for Research in Environmental Sciences Climate Diagnostics Center from their website at http://www.cdc.noaa.gov/.)

Heavy rainfall in the western Czech Republic, southeastern Germany, and northeastern Austria produced flooding in left bank tributaries of the Danube River. Intense rainfall in the northeastern Austrian Alps triggered flooding in the Salzach River, a right bank tributary of the Danube. At Salzburg, the Salzach River rose from 4.5 to 8.0 m in 12 hours and reached a peak flow of $2300\,\mathrm{m}^3\,\mathrm{s}^{-1}$, which equaled a value last observed in 1899 (Ulbrich *et al.*, 2003a).

9.8   The 1993 Midwestern United States flood

Extreme meteorological, climatological, and hydrological conditions led to widespread flooding in the upper Mississippi River and lower Missouri River basins during the summer of 1993. Summer flooding of major river systems seldom occurs in the Midwest because of the high space and time variability of summer convective rainfall and the high evapotranspiration rates characteristic of the region. Although abundant atmospheric moisture is available over the mid-continent region during the summer (Trenberth *et al.*, 2003), the relatively gentle north–south thermal gradient between Canada and the Upper Mississippi Valley provides only a modest air mass density contrast. Summer convective storms produce localized flooding on smaller tributary systems, but general floods on major river systems in the region most often occur during the spring snowmelt season (Knox, 1988).

Widespread flooding in the Upper Mississippi River Basin in July and August was a first indication of the unusual character of the 1993 event. At its maximum extent, the 1993 flood affected an area of 600 000 km^2 or one-third of the total area of the Upper Mississippi Basin above Cairo, Illinois (Fig. 9.18). Record high flows occurred along 2880 km of river channels, and another 2000 km of rivers had flows characterized as "major floods". Peak discharge at 46 gauging stations exceeded 100-year return intervals, and 42 streamflow gauging stations had record peak discharges (Parrett *et al.*, 1993). The most severe flooding was concentrated along the Mississippi River between Minneapolis, Minnesota, and

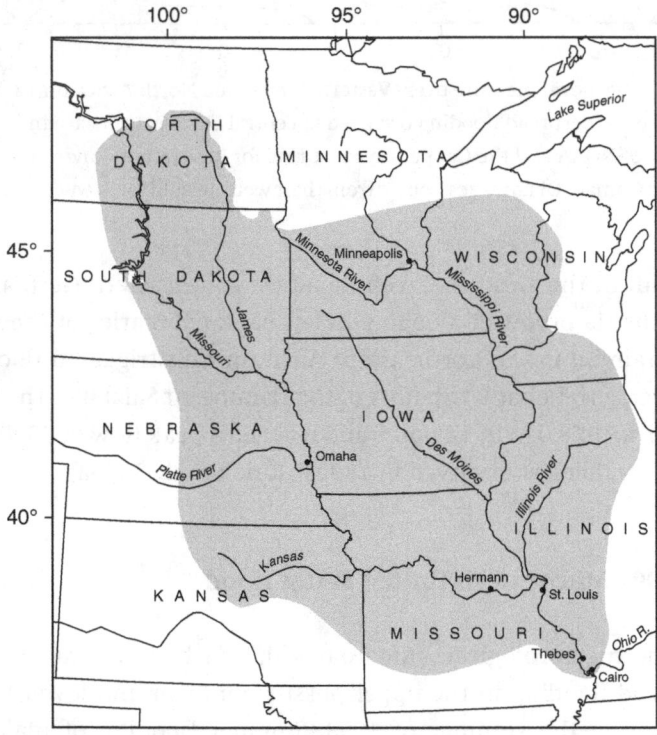

Fig. 9.18. Area of the Upper Mississippi River Basin flooded during the summer of 1993. (After NOAA, 1994, Figure 1–1.)

Cairo, Illinois, and along the Missouri River from Omaha, Nebraska, to St. Louis, Missouri (Lott, 1994). When compared to other historic floods in the Mississippi River Basin, the 1993 flood is larger in areal extent and magnitude, longer in duration, greater in volume of floodwater, and greater in amount of damage (Koellner, 1996). Compared to floods in other global regions during 1900–93, eastern Australia had two floods and southern Africa had three floods comparable to the 1993 United States flood (Dai *et al.*, 1997). The unprecedented extent and severity of the Mississippi River Basin flooding prompted the event to be called the Great Flood of 1993. However, the flooding in June to August was the most devastating, and only the hydroclimatology of this extreme summer flooding is addressed here. NOAA (1994) and Changnon (1996) provide comprehensive reviews of the 1993 flood.

The genesis of major summer flooding was established in July to November 1992 by relatively heavy rainfall throughout the Upper Mississippi Basin. Above normal soil moisture began to develop in many areas during August 1992 (Bell and Janowiak, 1995), and near normal winter precipitation maintained soils at saturated levels going into spring. Some early flooding occurred along the Upper Mississippi River as snowmelt augmented runoff from the especially wet spring. Above normal rainfall displayed a progressive increase from April 1993 to a peak in July (Fig. 9.19), but above average rainfall continued through September. Between January and July, almost all of the flooded area received 1.5 times the normal rainfall, and some areas received twice their normal rainfall. Flooding in early May was especially prevalent in Kansas, Missouri, and Minnesota. Eight consecutive days of rain in early June

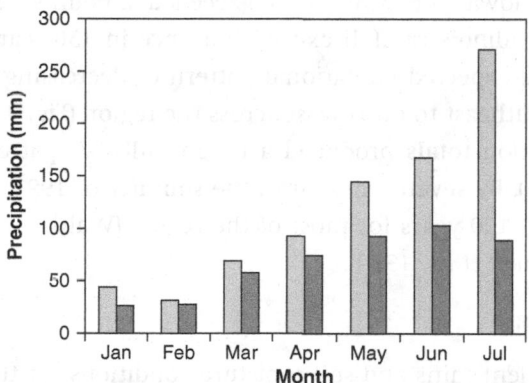

Fig. 9.19. Monthly precipitation in the Upper Mississippi River Basin for January to July 1993. Stippled columns are 1993 precipitation, and shaded columns are the 1961–90 average for the month. (After Kunkel *et al.*, 1994, Fig. 3a. Used with the permission of the American Meteorological Society.)

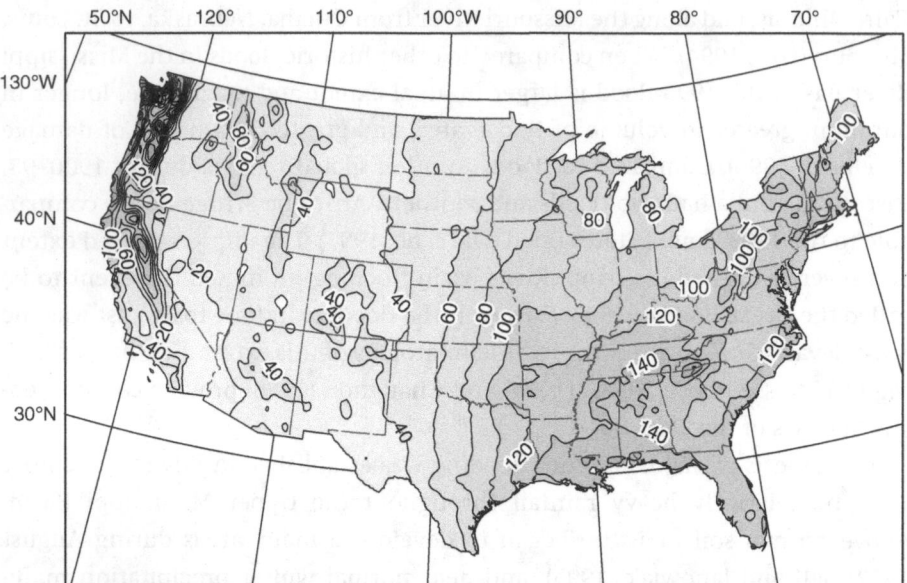

Fig. 9.20. Annual mean precipitation for the United States. Units are in cm. (Data courtesy of the USDA Natural Resources Conservation Service and the Prism Group from their website at http://www.ncgc.nrcs.usda.gov/.)

produced flooding across much of South Dakota, Iowa, Minnesota, and Wisconsin. Numerous precipitation records of various durations were set throughout the area, and some stations recorded in excess of 900 mm of rainfall during June to August, or 3 times the normal amount. June to August precipitation in Iowa and Minnesota exceeded amounts expected once in 200 years, while Illinois rainfall exceeded a once in 150-year return period. In contrast to the expected gradational pattern of decreasing annual precipitation from the southeast to northwest across the region (Fig. 9.20), the June to August precipitation totals produced a rough bulls-eye pattern centered over Iowa (Fig. 9.21). By several measures, the summer of 1993 was the wettest period in the past 100 years for most of the region (Wahl *et al.*, 1993; Guttman *et al.*, 1994; Kunkel *et al.*, 1994).

9.8.1   The atmospheric setting

Although antecedent rains and soil moisture conditions set the stage for the summer flood, much of the major river flooding is traced to six synoptic-scale rainfall events of up to nine days duration during mid June through mid August that are related to upper-level and surface atmospheric conditions (Koellner, 1996). An anomalously intense trough in the 500 hPa

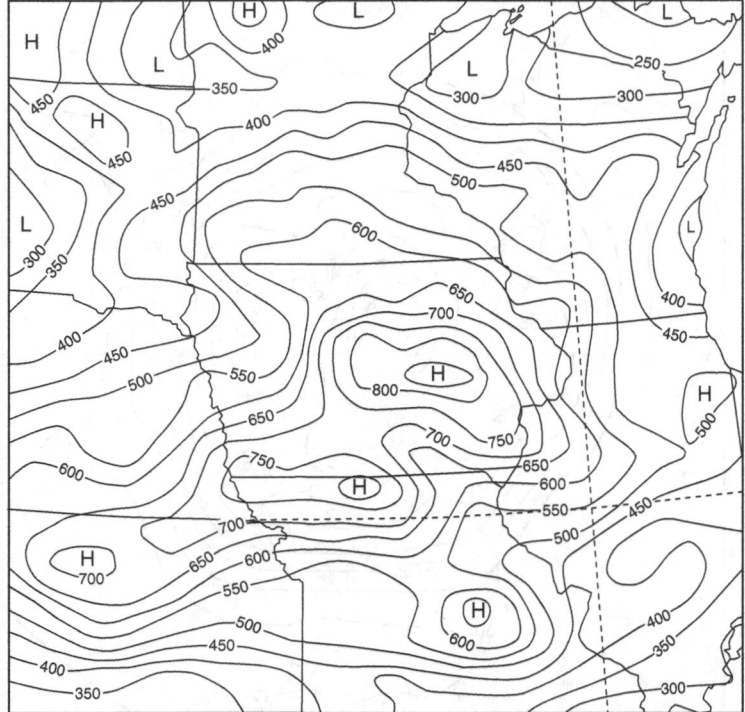

Fig. 9.21. June to August precipitation in the Upper Mississippi River Basin. Units are in mm. (From Kunkel *et al.*, 1994, Fig. 4. Used with the permission of the American Meteorological Society.)

circulation was positioned over the western United States (Fig. 9.22), and the entire Upper Mississippi River Basin was situated downstream of the large-scale trough axis and within a region of anomalously strong southwesterly geostrophic flow (Bell and Janowiak, 1995). The unusual character of the upper-air circulation is evident in the 500 hPa anomaly field for June–July 1993, which shows strong negative height anomalies over the Great Plains and Midwest and weak positive anomalies over the Gulf of Alaska and the eastern United States (Fig. 9.23). The persistence of the anomalous upper-level circulation pattern is revealed by the high negative daily percentages over the central United States and by high negative and positive values distributed around the Northern Hemisphere (Fig. 9.24). The anomalous upper-level circulation prompted displacement of the mean position of the jet stream southward of its usual summer position and its firm establishment over the northern portion of the Mississippi River Basin. In this configuration, the jet stream closely resembled its expected spring location with a southwest–northeast orientation.

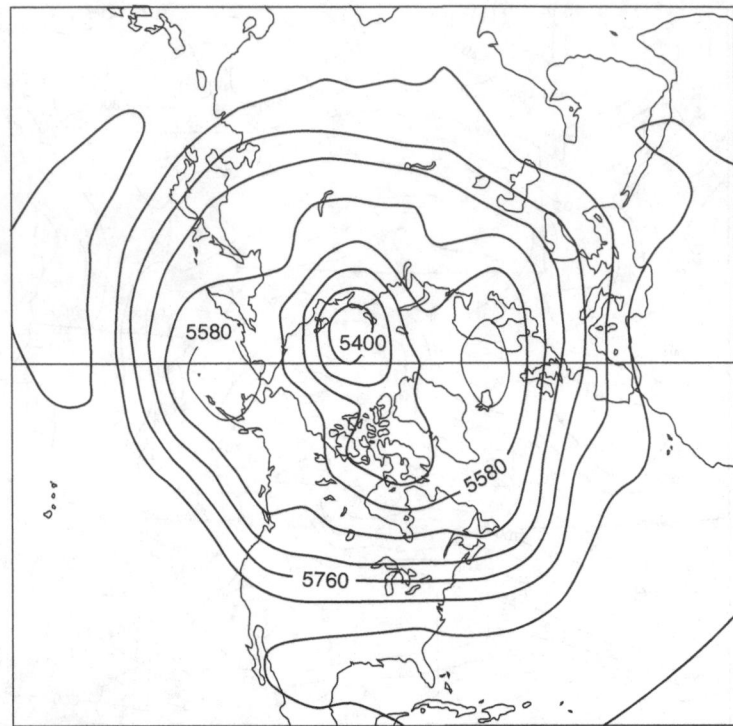

Fig. 9.22. Northern Hemisphere 500 hPa geopotential heights averaged for June and July 1993. Heights are in m. (From NOAA, 1994, Figure 3–5a.)

Bell and Janowiak (1995) and Trenberth and Guillemot (1996) relate the upper-level circulation anomalies producing the 1993 flood to a highly anomalous PNA teleconnection pattern and a mature El Niño circulation that dominated much of the Northern Hemisphere. However, Mo *et al.* (1995, 1997), and Liu *et al.* (1998) suggest an alternative explanation that strong diabatic heating over the North American continent largely compensated for the effect of cooling over the North Pacific. They propose that local vorticity forcing related to transient eddies upstream of the Rocky Mountains accounted for the strengthening of the jet stream leading to the zonal atmospheric flow across the Upper Mississippi River Basin.

Anomalous surface pressure patterns were coupled to the anomalous upper-level circulation features. A below normal sea-level pressure pattern was established in the central and western North Pacific in April 1993, and it extended westward during June and July to cover most of the United States (Fig. 9.25). Weak positive sea-level pressure anomalies dominated the eastern United States. The largest negative anomalies exceeded 2 hPa and were centered over the southern Great Plains, Texas, and New Mexico. This surface pressure pattern

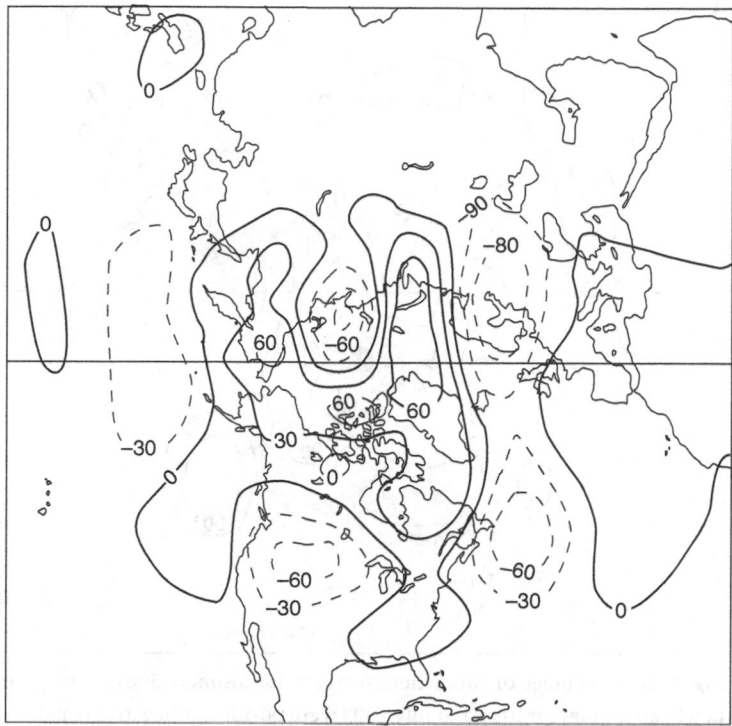

Fig. 9.23. Northern Hemisphere 500 hPa geopotential height anomalies for June and July 1993. Solid lines are positive anomalies, and dashed lines are negative anomalies. Contours are in m. (From NOAA, 1994, Figure 3–5b.)

was associated with enhanced southerly wind with speeds ranging from 6 to 10 m s $^{-1}$. The enhanced southerly flow was accompanied by strong and persistent moisture transport from the Gulf of Mexico into the Midwest (Bell and Janowiak, 1995; Paegle *et al.*, 1996).

The quasi-stationary jet stream aloft (Fig. 9.26) was associated with a stationary surface front that allowed frequent and nearly continuous overrunning of the cooler air to the north by moisture-laden air from the south (Lott, 1994; Bell and Janowiak, 1995; Trenberth and Guillemot, 1996). The flow of moist air was driven by a low-level jet related to the North Atlantic subtropical anticyclone and the strong upper-level jet stream (Mo *et al.*, 1995, 1997; Paegle *et al.*, 1996; Hu *et al.*, 1998). The front also served as a preferred location for unusually strong and frequent cyclones spawned by the combination of the unseasonably vigorous jet stream and the relatively strong frontal boundary at the surface. Many of the large rain areas during these months suggested the presence of MCSs or the larger MCCs (Kunkel *et al.*, 1994; Doswell *et al.*, 1996). By late July and early

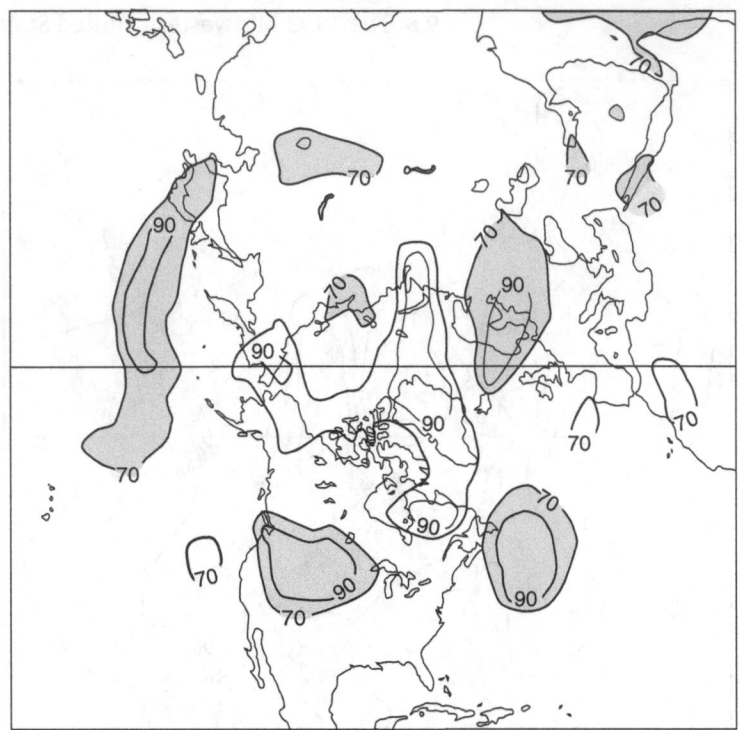

Fig. 9.24. Percentage of days when Northern Hemisphere 500 hPa geopotential height anomalies for June and July 1993 were positive or negative (shaded contours). (From NOAA, 1994, Figure 3–5c.)

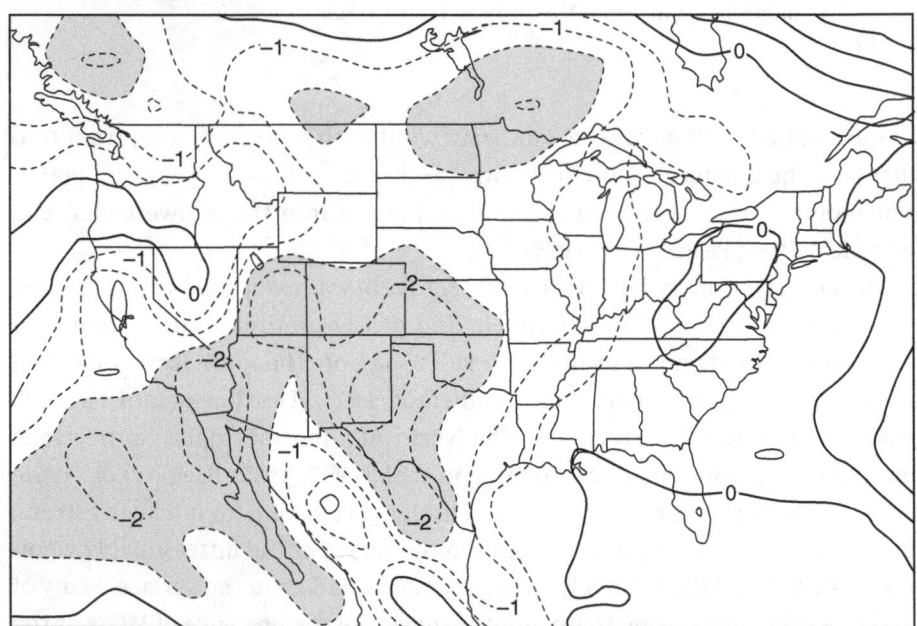

Fig. 9.25. Sea-level pressure anomalies for the United States during June and July 1993 based on the 1979–88 base period. Dashed lines are negative anomalies. Units are in hPa. (From Bell and Janowick, 1995, Fig. 9. Used with the permission of the American Meteorological Society.)

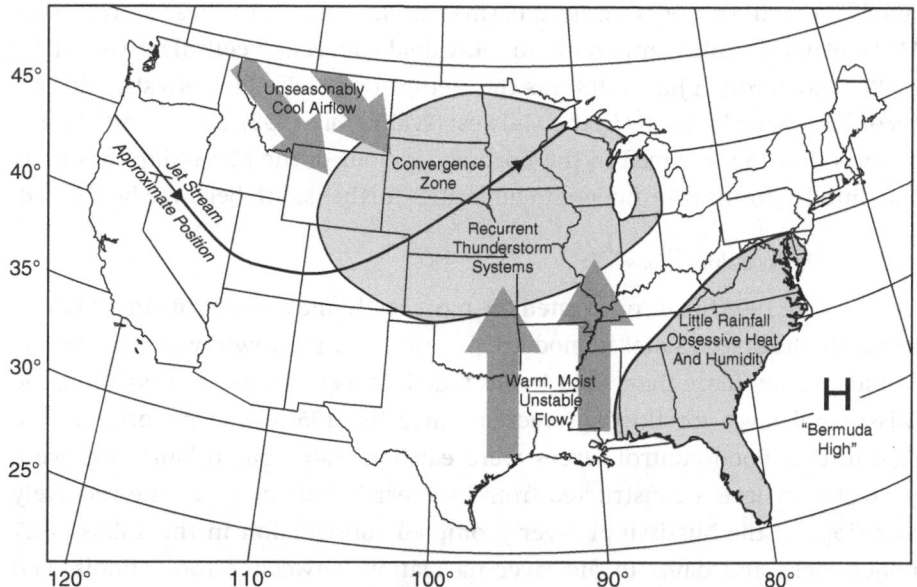

Fig. 9.26. Sketch of June and July 1993 dominant weather patterns over the United States. (From NOAA, 1994, Figure 3–6a.)

August, a change in the upper-air circulation pattern brought drier conditions to the Midwest as the trough shifted eastward.

The meteorological role in producing the flooding was particularly evident in the daily pattern of thunderstorms. In mid June, thunderstorms began to form almost daily along a persistent frontal boundary that stretched across the upper Midwest. Much of the rainfall occurred from the late afternoon to early morning hours with a pronounced midday minimum (Paegle *et al.*, 1996). The front remained in place for nearly 40 days in June and July and was positioned beneath the polar jet stream that was situated much farther south than usual for this time of year. The jet stream provided pockets of upper-level divergence for the development of weak surface waves that rippled along the front (Kunkel *et al.*, 1994). Converging surface air associated with large-scale transport of moisture from the Gulf of Mexico driven by the low-level jet and/or the subtropical anticyclone centered in the North Atlantic (Hu *et al.*, 1998) produced unstable conditions that promoted uplift for thunderstorm growth. Thunderstorms almost daily rolled through an area that stretched from South Dakota into Minnesota and southward into Missouri and Illinois. Unofficial reports indicate that 175 extreme thunderstorms producing 150 mm or more of rain occurred in the basin between late May and August 1993 (Kunkel, 1996). These intense, short-duration rainstorms were significant in producing flash floods on small to moderate-sized basins, and in some cases these storms were MCSs playing a role comparable to individual convective cells (Doswell, *et al.*, 1996). May through July 1993 was the wettest or nearly the wettest period on record for many locations in the Midwest (Wahl *et al.*, 1993). By the end of June, general flooding occurred in the northern regions of the Mississippi Basin. As the thunderstorms continued into July, cities further south began to be flooded.

9.8.2   *The flood dimensions*

The 1993 flood constituted the most costly and devastating inland flood to ravage the United States in modern history. At least 75 towns were completely inundated and more than 1000 levees failed. Levees have been used along the Mississippi River for flood protection since the 1850s. The majority of the agricultural flood control levees were earthen walls, but urban areas were protected by levees constructed from both earth and concrete. The relatively low slope of the Mississippi River prompted construction in the 1930s of 27 major locks and dams to aid river navigation between Alton, Illinois, and Minneapolis, Minnesota. The pools behind each lock and flood control reservoirs on tributary streams throughout the basin helped reduce flood levels and damage during the 1993 flood (Perry, 1994), but positive and negative effects of levees are debated (Kunkel, 1996).

The immense size of the Upper Mississippi Basin dictates that uniform flooding throughout the basin was unlikely. However, the uncommonly persistent and excessive summer rainfall in 1993 over much of the region resulted in a gradual coalescing and expansion of flooded areas and record-setting destructive flows on the main stem of the Mississippi and Missouri rivers. While the timing of flooding varied on tributary stream systems, flow for the Mississippi River at St. Louis, Missouri (39° N), illustrates the general nature of the flooding on the main stem at a location south of the confluence of the Mississippi, Missouri, and Illinois rivers. The Mississippi River at St. Louis stage hydrograph (Fig. 9.27) shows that the summer flood began on 26 June, crested on 1 August, and the river fell below flood stage on 13 September. Flood stage for the Mississippi River is 9.1 m at St. Louis, and the river reached a peak stage of 15.11 m. On 1 August, the Missouri contributed about 70% of the total discharge or about twice the discharge carried by the Mississippi River above St. Louis (Moody, 1995). The previous record flood stage of 13.2 m at St. Louis was exceeded continuously for three weeks. Another indicator of the severity of the flood is that the communities of Hannibal, Louisiana, and Clarksville, Missouri, all north of St. Louis on the Mississippi River between the Des Moines River and the Illinois River, were flooded for 153 consecutive days.

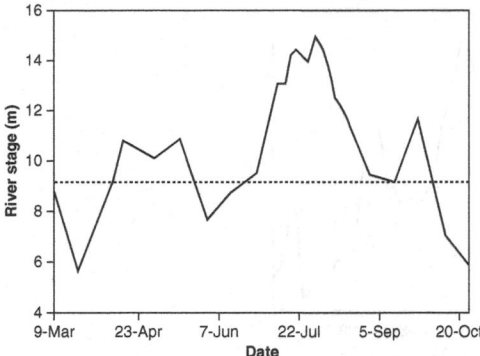

Fig. 9.27. Daily river stage for the Mississippi River at St. Louis, Missouri, for 9 March 1993 to 25 October 1993. Record stage reached on 1 August. Flood stage is indicated by the horizontal line. River stage is in m. (Data courtesy of the U.S. Geological Survey from their website at http://waterdata.usgs.gov/nwis/.)

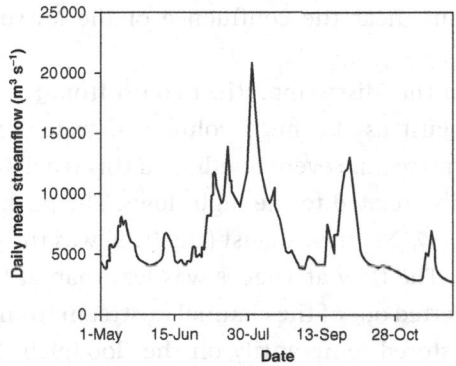

Fig. 9.28. Daily mean streamflow for the Missouri River at Hermann, Missouri, for 1 May 1993 to 1 December 1993. Flood crest on 31 July. (Data courtesy of the U.S. Geological Survey from their website at http://waterdata.usgs.gov/nwis/.)

The 1993 flooding on the Missouri River between Omaha, Nebraska, and St. Louis, Missouri, was the worst since 1952 and rated as 100-year to 500-year events at most locations (Koellner, 1996). The Missouri River at Hermann, Missouri (39° N), (Fig. 9.28) recorded four flood crests as runoff surges from various upstream tributaries arrived at this main stem location 104 km west of St. Louis. Floods in mid May and early June preceded major floods from 2 July to 25 August and from 13 September to 5 October. The Missouri River crested at Hermann on 31 July and at St. Louis on 2 August. Levee failures along the Missouri River resulted in especially severe flooding in Jefferson City, Missouri,

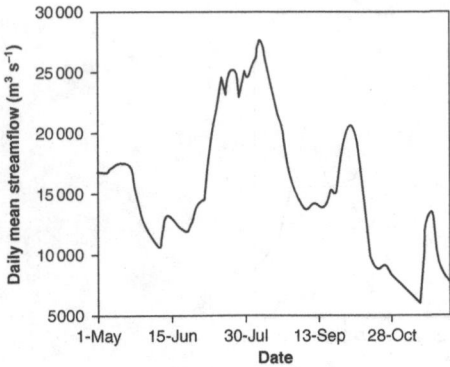

Fig. 9.29. Daily mean streamflow for the Mississippi River at Thebes, Illinois, for 1 May 1993 to 1 December 1993. Flow on 7 August was the greatest ever measured at this site. (Data courtesy of the U.S. Geological Survey from their website at http:// waterdata.usgs.gov/nwis/.)

and St. Charles County, Missouri, near the confluence of the Missouri and Mississippi rivers.

Flooding along the 288 km of the Mississippi River south from St. Louis to Cairo occurred in July and August as the huge volume of water from the combined rivers continued downstream. Severe flooding in this reach followed levee failures due to rapid erosion related to the high flows. The peak flow of 29 250 m^3 s^{-1} at Thebes, Illinois (37° N), on 7 August (Fig. 9.29) was the greatest flow ever measured at this site. The flow at Thebes was less than at St. Louis because water was naturally diverted out of the channel upstream from Thebes through failed levees and was stored temporarily on the floodplain (Moody, 1995). The Mississippi River flood dissipated south of Cairo where the river channel widened and provided a greater channel capacity for conveying the greater water volume (Parrett *et al.*, 1993). Also, it was fortunate that the Ohio River, which joins the Mississippi at Cairo, was at normal flow at this time.

Post-flood assessments aggregated the enormity of the flood impact. Parts of nine states totaling an area larger than Texas were flooded, but Iowa, Missouri, Minnesota, and Illinois sustained the most damage. Floods through November 1993 claimed 52 lives, but this is a relatively small number considering the expanse of the flooded area and that more than 54 000 people were evacuated and 71 000 homes were damaged or destroyed. Flooded land and soil losses had a tremendous impact on agriculture and accounted for financial losses of $8.5 billion that exceeded losses by all other sectors. Estimated crop losses were $6.5 billion, and an estimated 600 billion tons of topsoil were lost to erosion. The estimated total sediment moving past St. Louis was 55 million metric tons, but 35 million metric tons of sediment were deposited on the Missouri River floodplain between Hermann and the mouth of the Missouri River (Holmes, 1996). The flood was concentrated in the center of the United States rail and highway system and struck the heart of the Mississippi Waterway navigation system. The area and duration of the flood had a severe impact on transportation systems for up to six weeks. Railroad, barge, and surface transportation damages and lost revenue exceeded $1.9 billion. Total economic losses related to the 1993 flood were $23.1 billion or 4 times the mean annual United States flood loss (O'Connor and Costa, 2004a). This ranks the 1993 flood as the fourth most costly natural disaster during the last two decades of the twentieth century.

Review questions

9.1   What are the criteria used to characterize floods?
9.2   How is the role of precipitation different in producing a flash flood compared to a general flood?
9.3   What are the characteristics of the atmospheric conditions associated with most flash floods?
9.4   How do topography and atmospheric influences contribute to flash flood occurrences in the Mediterranean Basin?
9.5   What atmospheric conditions are shared by many of the most severe flash floods in the United States?
9.6   In what ways are floods commonly associated with atmospheric patterns characterized as being anomalous?
9.7   How did an upper-air trough contribute to the 2002 flood in central Europe?
9.8   What role did soil moisture play in the 1993 Mississippi Basin flood?
9.9   What were the atmospheric circulation features associated with the 1993 Mississippi Basin flood?

10

Drought: the hydroclimatic extreme of deficient moisture

10.1 Negative moisture anomalies

Drought is a naturally occurring phenomenon throughout most of the world, and it has tremendous damaging consequences for physical, economic, social, and political sectors of society (Woodhouse and Overpeck, 1998). The frequent and irregular occurrence of drought is a major reason for construction and management of water supply facilities in drought-prone areas because drought exacerbates natural limitations on water supplies. Unfortunately, drought duration, magnitude, severity, and recurrence remain difficult to forecast with precision (Loáiciga, 2005). Drought duration is the temporal component of the moisture deficiency, magnitude expresses the average deficiency, severity is the cumulative moisture deficiency, and recurrence is a probabilistic estimate of the next appearance of the moisture deficiency (Dracup and Kendall, 1990; Byun and Wilhite, 1999).

Drought is generally associated with low-frequency components of the climate system commonly related to SSTs or solar variability (Fye and Cleaveland, 2001; Hidalgo, 2004). Consequently, drought is a synthesis of two sets of processes. The first set is clearly related to the delivery of precipitation through climate of the first kind and the atmospheric branch of the hydrologic cycle. The second set of processes involves how precipitation is treated at the Earth–atmosphere interface, and this is the realm of climate of the second kind and the terrestrial branch of the hydrologic cycle.

Drought occupies the opposite extreme of floods on the hydroclimatic continuum. Negative moisture anomalies or water deficit are the central concept in defining drought, while floods are associated with positive moisture anomalies. One of the earliest written records of drought is found in the biblical story of Joseph and the pharaoh of Egypt (Walker *et al.*, 1991), and a long-term precipitation anomaly that corresponds to the critical-period problem in reservoir storage design is known today as a "Joseph event" (Kilmartin, 1980). Drought is commonly perceived as an abnormally long period without precipitation, but drought is also associated with weather systems that produce minimal precipitation. In hydroclimatic terms, water deficit implies a departure from expected moisture conditions and establishes drought as a non-event in that drought occurs when expected moisture levels are not realized. This non-event characteristic exacerbates the difficulty identifying when a drought begins and ends and combined with the gradual intensification of drought emphasizes its complex nature. Hydroclimatology is important to the study of drought because of the emphasis it places on the drought event in the context of the general circulation features related to the precipitation deficiency.

Truly extraordinary droughts have occurred in almost all parts of the world in modern times. Drought during 1998–2002 over an extensive area of the Northern Hemisphere mid-latitudes resulted from some regions receiving as little as 50% of the climatological annual average precipitation (Hoerling and Kumar, 2003). Africa is a particularly drought-prone continent due to its diversity of climatic zones and the relatively large year-to-year rainfall variability. The extreme east–west spatial coherence of rainfall anomalies across the Sahel is particularly notable (Rasmusson, 1987), and the extended dry period and famine in Ethiopia is evidence that drought is not limited to the Sahel. Recurring drought in India is related to a failure of the monsoon rains. Drought in Europe occurs in all regions and in all seasons, and large areas of Europe were affected by drought during the twentieth century (Lloyd-Hughes and Saunders, 2002). Drought occurred over much of central Asia from 1999 to 2002 in an area of steppe terrain and semiarid to arid climate that exacerbates susceptibility to drought, and a long-term decrease in mean precipitation has increased the area of drought in China (Easterling *et al.*, 2000). Central Chile experienced repeated droughts in the late 1990s (LeQuesne *et al.*, 2006), and catastrophic drought in northeast Brazil is evidence that tropical settings do not escape drought impacts.

In 1901–2 and 1982–3, drought seriously affected the eastern half of the Australian continent and wildfires burned tropical as well as semiarid zones. Drought is most economically damaging in southeastern Australia which contains about 75% of Australia's population and much of its agriculture. Drought in 2002–3 was not quite as dry over most of eastern Australia as the droughts of 1901–2 or 1982–3, but the 2002–3 drought had a particularly severe impact. Record high average maximum temperatures accompanied the 2002–3 drought, and the drought affected virtually the entire continent while the earlier severe droughts had little or no effect on Western Australia.

In the United States, the Dust Bowl drought of the 1930s is probably the most memorable of the twentieth century droughts because of its extent and related social upheaval, but paleoclimate data indicate several droughts in the past 2000 years were more severe (Woodhouse and Overpeck, 1998). More recently, severe drought visited the Great Plains in 1956, the Northeast in the early 1960s, California in the mid 1970s and 1990s, and the Mississippi Basin in 1988. Record dryness in parts of the Northeast and Mid-Atlantic states occurred from October 1994 through August 1995. During these 11 months, most areas received less than 75% of normal rainfall, rivers were reduced to 30% of normal flow, and wildfires were the worst in 60 years. Texas and Oklahoma experienced widespread drought for the 12 months ending in July 1996. Drought was especially severe in Texas where very little precipitation fell in most of the state from January through June (Hayes *et al.*, 1999). An estimated 10 000 ranchers and farmers went bankrupt, and Texas alone lost over $5 billion in crop and livestock production. Drought in the Upper Colorado River Basin from 1999 to 2004 was the most severe in 80 years.

10.2   Drought hydroclimatology

Identifying drought requires specification of the component or components of the hydrologic cycle affected by the water deficit and the period associated with the deficit (McNab and Karl, 1991). Precipitation is commonly considered the carrier of the drought signal, and streamflow and groundwater are the last indicators of drought occurrence. Soil moisture occupies an intermediate position as a drought indicator (Hare, 1987; Klemeš, 1987). Nevertheless, drought is often viewed as being climatological, meteorological, hydrological, agricultural, or socioeconomic depending on the purpose for examining the moisture deficit, but a precipitation deficiency is common to all types of drought. The characteristics considered essential for understanding the various drought types are reviewed by Klemeš (1987), Wilhite and Glantz (1987), and Heim (2002).

Atmospheric circulation anomalies must account for abnormal dryness for a specific area since climate determines the long-term characteristics of the precipitation signal. However, atmospheric circulation tends to undergo continuing change that would limit the temporal character of dryness and limit drought duration. SST anomalies may provide the steady atmospheric forcing necessary for achieving multiyear persistence of atmospheric circulation anomalies, abnormal dryness, and drought (Hoerling and Kumar, 2003). The strongest drought signal is evident when substantial precipitation is expected but does not occur (Karl *et al.*, 1987), and this is the basis for recognizing that periods of normal dryness that are expressions of a location's climate do not necessarily constitute drought. Drought is a meaningless term in areas like the Sahara and

Atacama deserts where rain is a rare event (Hare, 1987). At the same time, this characteristic illustrates why a given amount of precipitation is more effective in ameliorating drought during a period of minimal normal precipitation than during a wet period.

Drought occurrence is complicated further by its frequent association with higher than normal temperatures and lower relative humidity, especially when the drought occurs during the summer. Moisture deficits during droughts in the Northern Hemisphere from 1998 to 2002 were aggravated by increased moisture demands resulting from above normal temperatures that reached record levels (Hoerling and Kumar, 2003). However, the relationship between dry winters and warmer temperatures may be more evident at the national scale than at the regional scale, and not all droughts are associated with higher than normal surface-air temperatures (McNab and Karl, 1991). Warmer surface temperatures during a drought are due to increased solar radiation resulting from the below normal cloud cover and less available moisture for evaporative cooling (Durre *et al.*, 2000). Lower observed relative humidity during a drought is commonly related to both warmer surface temperatures and less atmospheric water vapor (Hanson, 1991), but the weight of evidence favors a substantial decrease in atmospheric water vapor during droughts (McNab and Karl, 1991). Additional significance of the warmer than normal temperatures is related to the contemporary view that drought has both a supply component provided by precipitation and a demand or water use component. This perspective is particularly evident in the natural landscape where vegetation and hydrologic systems develop in response to an expected moisture supply and a customary evapotranspiration demand. Greater energy loadings during drought increase the evapotranspiration demand. A quantitative basis is needed for drought analysis and comparison that facilitates examination of precipitation and streamflow records to extract information about the probabilistic nature and recurrence of drought (Loáiciga, 2005).

10.3   Drought indices

Comparing drought characteristics from region-to-region or event-to-event is facilitated by the use of a numerical standard. Unfortunately, the complexity of drought and the multiple drought definitions make it nearly impossible to capture drought intensity and severity in a single index. A number of drought indices are in use throughout the world based on the length of the period since a specified precipitation amount, the percentage of normal rainfall, precipitation deciles, soil moisture storage capacity, the water balance, a variety of streamflow characteristics, satellite measurements of vegetation health, and other variables. Heim (2002) describes drought indices developed in the United

Table 10.1. *Palmer Drought Severity Index (PDSI) categories developed by Palmer (1965)*

PDSI value	Moisture category
≥4.00	Extremely wet
3.00 to 3.99	Very wet
2.00 to 2.99	Moderately wet
1.00 to 1.99	Slightly wet
0.50 to 0.99	Incipient wet spell
0.49 to −0.49	Near normal
−0.50 to −0.99	Incipient drought
−1.00 to −1.99	Mild drought
−2.00 to −2.99	Moderate drought
−3.00 to −3.99	Severe drought
≤ −4.00	Extreme drought

States, and many of these indices have been applied in other parts of the world. An overview of selected indices is presented here to illustrate the range of characteristics included in quantifying drought intensity and duration.

10.3.1   Palmer Drought Severity Index (PDSI)

The moisture supply and demand concept is the foundation for the PDSI (Palmer, 1965) which is a commonly used and widely recognized drought index (Karl *et al.*, 1987; Heim, 2002; Burke *et al.*, 2006). The PDSI is known operationally in the United States as the Palmer Drought Index (PDI). The PDSI is a hydroclimatic accounting system calculated on a regular basis as a means of providing a single measure of meteorological drought severity, and it has been calculated for the 344 climate divisions in the United States back to 1895 (Karl, 1986a). It is essentially an index of soil moisture calculated on a monthly basis and intended to measure the duration and intensity of long-term drought, but daily or weekly time scales can be used. Droughts lasting longer than about one month are commonly related to atmospheric circulation patterns resulting in precipitation deficits. Since long-term drought is cumulative, drought intensity during the current month is dependent on current weather patterns plus the cumulative patterns of previous months. The computational procedures are presented by Palmer (1965) and in abbreviated form in numerous publications (e.g. Karl, 1986a; Heim, 2002; Wells *et al.*, 2004). The PDSI produces dimensionless index values divided into 11 categories (Table 10.1) that allow comparisons across time and space.

The monthly PDSI time-series for central Kansas (Fig. 10.1) shows that the PDSI represents both drought conditions (negative values) and moist conditions (positive values). It is evident in Figure 10.1 that severe or extreme droughts were more

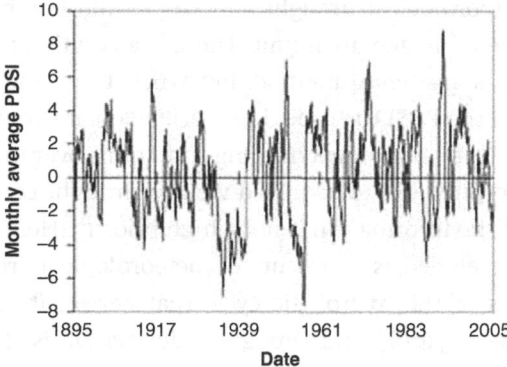

Fig. 10.1. Monthly average Palmer Drought Severity Index (PDSI) for the central Kansas climate division for January 1895 to December 2006. (Data courtesy of NOAA and the National Climatic Data Center from their website at http://www7.ncdc.noaa.gov/CDO/CDODivisionalSelect.jsp.)

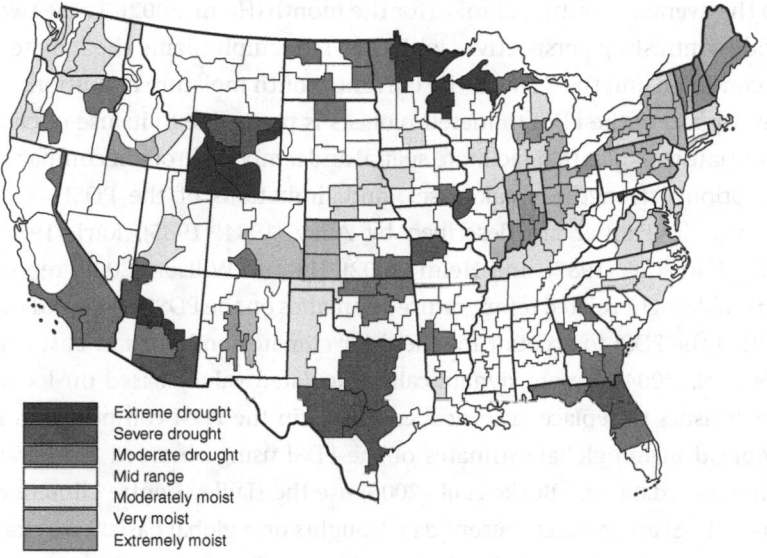

Fig. 10.2. Palmer Drought Severity Index (PDSI) for December 2006 for each climate division in the contiguous United States. Drought stages are shown by black tones (e.g. northern Minnesota and Wyoming). Wet stages are shown by gray tones (e.g. New England). (Data courtesy of NOAA and the National Climatic Data Center from their website at http://lwf.ncdc.noaa.gov/oa/climate/research/drought/palmer-maps/.)

common in the first three decades of the twentieth century than in the last three decades in central Kansas. Conversely, severe and extreme moist conditions were more common during the last three decades of the twentieth century.

Spatial drought characteristics are revealed by mapping PDSI values for all climate divisions in the area of interest for a specific time (Fig. 10.2). Mapped PDSI values show long-term, meteorological drought and wet conditions based on climate and soil data averaged for each areal unit. There is a certain amount of information sacrificed by this averaging method, but when the same climate period is used for calculating the PDSI the drought severity data are comparable and reveal drought distribution features and changing conditions over time. PDSI maps are commonly used to provide a generalized view of drought conditions, and the spatial resolution of the information is not a high priority (Heim, 2002).

Although the PDSI was developed as a measure of meteorological drought, it takes into account elements of the hydrologic cycle that permit its use as a hydrologic index. The Palmer Hydrological Drought Index (PHDI) is the index for the existing drought or wet spell in the PDSI computation. Therefore, the PHDI is the index for the long-term hydrologic moisture condition and it more accurately reflects groundwater conditions and reservoir levels. The Z Index is another element of the PDSI computation that serves as a stand-alone index. This moisture anomaly index quantifies the departure of the weather in a particular month from the average moisture climate for the month (Heim, 2002). These two indices provide contrasting perspectives with the PHDI emphasizing the long-term moisture condition and the Z Index the current month moisture condition.

The appeal of the PDSI to meteorologists is partly due to its use of climatically appropriate variables that facilitate spatial and temporal drought comparisons. The assumptions, strengths, weaknesses, and limitations of the PDSI as originally conceived by Palmer are described by Alley (1984, 1985), Karl (1983, 1986a, 1986b), Karl et al. (1987), and Heim (2002). Hu and Willson (2000) report on the effects of drought-related temperature anomalies on the PDSI. Ntale and Gan (2003) modified the PDSI to account for the drier climate conditions in East Africa, and Wells et al. (2004) present dynamically calculated values based on local climate characteristics to replace empirical constants in the PDSI computation. Dai et al. (2004b) calculated global estimates of the PDSI using observed temperature and precipitation data, and Burke et al. (2006) use the Hadley Centre climate model to estimate PDSI and predict present-day droughts on a global basis. Two examples of tree-ring reconstructions of the PDSI are their use for summer drought reconstructions for the continental United States over the past 200 years (Cook et al., 1999; Fye and Cleaveland, 2001) and to examine drought variability over the past 500 years in the western United States (Hidalgo, 2004).

10.3.2   The Standardized Precipitation Index (SPI)

The SPI introduced by McKee et al. (1993) quantifies the precipitation deficit over different averaging periods. The SPI was designed to be a consistent indicator of drought that recognizes the importance of time scales in the analysis of water availability and water use even during the winter (Hayes et al., 1999). In practical terms, it is a standardizing transform of the probability of observed

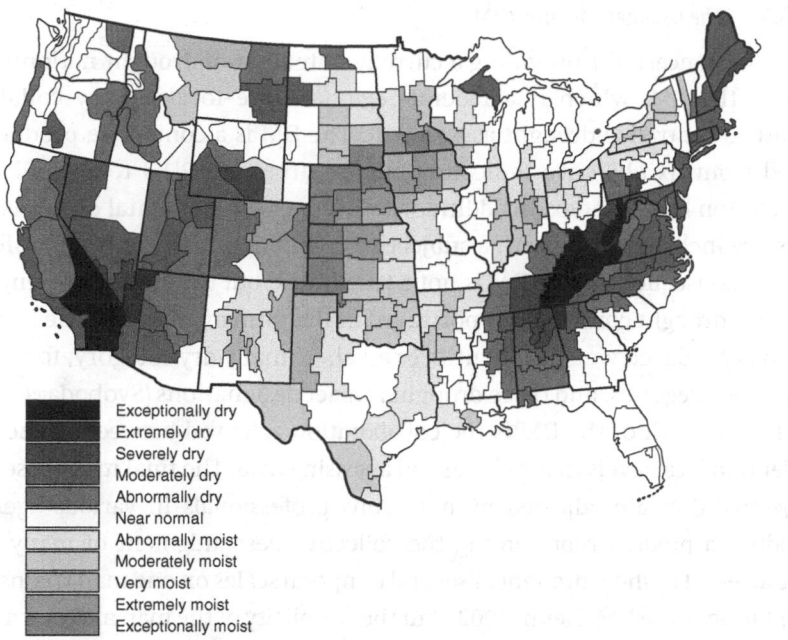

Fig. 10.3. Standardized Precipitation Index (SPI) for the three months of December 2006 through February 2007 for each climate division in the United States. Dry stages are shown by black tones (e.g. eastern Kentucky and Tennessee, western Virginia and North Carolina, and southeastern California). Moist stages are shown by gray tones (e.g. the area stretching from Wisconsin to western Texas). (Data courtesy of NOAA and the National Climatic Data Center from their website at http://lwf.ncdc.noaa.gov/oa/climate/research/prelim/drought/spi.html.)

precipitation over a specific averaging period (Guttman, 1998). Like the PDSI, the SPI is dimensionless and negative values indicate drought and positive values indicate moist conditions (Fig. 10.3). By computing the SPI for time scales of 1 to 24 months, various scales of both short-term and long-term drought are revealed with the shorter periods focused on agricultural and meteorological drought and the longer periods on climatological and hydrological drought (Heim, 2002).

The SPI was developed for use in Colorado, but it can be applied to any location (Heim, 2002). In addition, analysis of historical time-series of the PDSI and SPI indicates the SPI provides a better basis for comparing drought in one region with drought in another region (Guttman, 1998). Ntale and Gan (2003) found the flexible time scale of the SPI and its ease of interpretation to be particularly suitable for examining the 1949 East African drought. Lloyd-Hughes and Saunders (2002) employ the SPI to develop a high spatial resolution drought climatology of Europe for 1901–99. Their comparison of SPI and PDSI patterns found close correspondence, and analysis of extreme drought events showed SPI provided better spatial standardization.

10.3.3   The Drought Monitor (DM)

Concern for providing a current statement of drought nationwide produced the DM, which is a federal–regional–state–local agency collaborative effort to centralize drought assessment. The DM is a composite product developed from six objective indicators ranging from the PDSI to a satellite-based vegetation health index. In addition, a variety of supplemental objective indicators are included along with a subjective component representing professional input. Consequently, the DM is not a true index, but the spatial presentation of current drought conditions is portrayed in the form of an index. The DM in map form is produced weekly and utilizes an abnormally dry category, four drought severity categories, and three drought impact designations (Svoboda *et al.*, 2002).

The strength of the DM is the collaboration achieved between agencies from federal to local levels in providing and assessing data. The final maps based on the objective data are adjusted by numerous professionals in various agencies to produce a product representing the collective best judgment of many experts. The attempt to show drought at several temporal scales on one map is considered a limitation of the DM (Heim, 2002), but the simplicity of the map makes it attractive for the public, media, and a variety of applied areas (Svoboda *et al.*, 2002). The DM is available on the Internet at http://drought.unl.edu/dm/monitor.html.

10.3.4   Other drought indices

Byun and Wilhite (1999) suggest a series of indices based on the effective precipitation concept to improve drought monitoring on a daily basis. Effective precipitation is the summed value of daily precipitation with a time-dependent reduction function. Additional indices are computed from effective precipitation to incorporate climatological characteristics. In Africa, a National Rainfall Index is employed that allows comparison of precipitation patterns across time and space. National annual precipitation is weighted by the long-term average of all stations in the nation to produce a national-scale index. Australia employs the Australian Drought Watch System based on dividing the long-term monthly precipitation record into five deciles. The first and second deciles are grouped as precipitation much lower than normal indicating an occurrence less than 20% of the time. The system is relatively easy to calculate, but a long climatic record is needed to accurately define the deciles (Heim, 2002).

10.4   Proxy evidence of drought

Proxy data define long-term regional drought characteristics that are limited by sparse historical records, and they indicate that recent droughts in many regions have been exceeded in extent and severity in the last 2000 years.

Although all proxy data sources are utilized to reconstruct drought occurrences, tree rings offer the best temporal resolution and most accurate temporal control for examining drought (Woodhouse and Overpeck, 1998).

Annual tree-ring reconstructions provide improved quantitative information for examining drought frequency, severity, and duration over broad regions through an understanding of the relationships between hydroclimatic variables and the biological response of trees to hydroclimatic conditions (Loáiciga et al., 1993). The width of a tree ring in a particular year is proportional to the relative warmth and moisture availability of the ambient environment (Saltzman, 2002). Tree-ring widths have a particularly strong relationship with precipitation, and the annual tree-ring chronology conveys temporal hydroclimatic information not provided by other proxy data. The tree-ring width index expressing the departure of annual tree growth from normal is the most frequently used variable for examining drought (Hughes et al., 1981). The impact of negative precipitation anomalies on tree growth enhances the utility of tree rings for reconstructing a long-term perspective of year-to-year drought characteristics. Bradley and Jones (1995) and Bradley (1999) discuss the use of tree rings to reconstruct drought in different regions of the world.

Pfister (1995) reports evidence in documentary data of extraordinary dryness in Switzerland during the early decades of the 1800s, and Hughes (1995) found record dryness in Kashmir in the 1840s in precipitation reconstructed using tree rings. Woodhouse and Overpeck (1998) report that two droughts during the thirteenth to sixteenth centuries significantly exceeded the severity, duration, and spatial extent of twentieth century droughts throughout the western United States. A 1930s-magnitude Dust Bowl drought occurred once or twice a century over the past 400 years, and a decadal-length drought once every 500 years. Reconstructed precipitation time-series contain evidence that droughts more severe than twentieth century events occurred in the Pacific Northwest (Knapp et al., 2004), California (Fritts and Shao, 1995), the Great Basin (Hughes and Graumlich, 1996), the western Great Plains (Woodhouse et al., 2002), the Southwest deserts (Fritts and Shao, 1995), and the southeastern United States (Stahle and Cleaveland, 1992).

Drought across Europe in 2003 increased awareness for improved understanding of precipitation variability in the region which previously relied on documentary and historical data. The longest instrumental records for stations in Western Europe beginning in 1740 contain droughts exceeding those of the last century (Jones and Bradley, 1995), but relatively few tree-ring studies have been undertaken in the dominantly humid climate realm of central Europe to supplement other proxy data sources. The first dendroclimatic precipitation reconstruction for central Europe expressing centennial-scale hydroclimatic information is described by Wilson et al. (2005). They use a composite ring-width chronology from Norway spruce growing in the lower Bavarian Forest region of Germany and historical timbers to produce a 500-year spring–summer precipitation reconstruction. Their analysis indicates the twentieth century in central Europe was wetter than preceding centuries, and multiple-decade length dry periods occurred in each of the prior centuries back to AD 1500. In contrast, tree-ring chronologies from the Andean Cordillera of Chile from AD 1200 show that drought probability has increased dramatically during the late nineteenth and twentieth centuries in central Chile (LeQuesne et al., 2006).

10.5   Drought causes

The question of drought causation poses a potentially more difficult problem than specification of drought because drought is commonly related to multiple causes rather than a single physical cause, and it reflects an interaction between local-scale surface characteristics and regional-scale circulation phenomena (Knapp et al., 2004). Consequently, the actual ultimate causes of drought are often obscure. The difference between normal and deficient precipitation may depend on precipitation from just a few storms, and a precipitation deficiency may result from storms delivering only minimal precipitation (McNab and Karl, 1991). The task is made more difficult because hydrological processes are among the phenomena associated with the existence of drought but not the original causes (Klemeš, 1987).

10.5.1   Solar variability

At decadal and centennial time scales, it is common to regard solar irradiance as constant. However, solar radiation varies in response to sunspots, or magnetic disturbances that appear, migrate, and disappear from the Sun's surface, generally following an 11-year cycle or a 22-year double cycle. Although sunspots are areas of lower than normal temperature on the Sun's surface, they are associated with increased solar radiation of about 0.1% or $1.5\,\mathrm{W\,m^{-2}}$ from the maximum to the minimum in sunspot activity (Hartmann, 1994). The increased solar output during high sunspot numbers is related to the bright areas of higher than normal temperature that surround sunspots (Hoyt and Schatten, 1997). Nevertheless, the solar irradiance variability falls within a very narrow range during the sunspot cycle.

Connections between solar variability and climate first suggested in 1795 have been alternatively pronounced and dismissed (Haigh, 1996). A large body of work addresses the search for a causal solar influence on the variation of precipitation, lake levels, streamflow and drought (Hoyt and Schatten, 1997). Correlations between sunspots and precipitation and drought have received especially close examination. The emerging consensus is that the strongest cyclic variation for precipitation ranges from 18 to 22 years. This cycle implies that lunar tidal influences are more important than solar irradiance changes in explaining most of the co-variation. Drought, especially in the western United States, has a 20- to 22-year cycle and appears to be influenced by both lunar tidal influences and solar irradiance variations (Hoyt and Schatten, 1997). Empirical evidence linking solar variability and rainfall related to drought indicates that solar forcing amplifies internal oscillations of the climate system through feedback processes. One proposed amplification mechanism is related to changes in upper stratospheric ozone that penetrates into the troposphere and influences weather patterns (Shindell et al., 1999). Lean and Rind (2001) propose that complexity of this type may account for the varying strengths of sunspot and drought cycles.

10.5.2   Atmospheric circulation

Recurrent periods of deficient precipitation recognized as drought are best described as climate anomalies rather than weather anomalies. The basis for this assertion is the fact that rain-producing storm systems reach out to distances 3–5 times the radius of the precipitation region and draw in moisture over that area (Trenberth et al., 2003). Consequently, a hydroclimatic drought perspective is global, or at least hemispheric, and embraces interactions between the atmosphere, ocean, and land surfaces. A search for drought causes is not restricted to the atmosphere above the affected area, but looks for long-distance atmospheric circulation and surface anomalies revealed by time and space expressions consistent with the hydroclimatic perspective.

The immediate drought-producing mechanism involves persistent or persistently recurring atmospheric subsidence, but the subsidence is a manifestation of large-scale atmospheric circulation teleconnections and interactions with the underlying surface (Namias, 1983). Daily atmospheric circulation patterns associated with drought are not notably different from daily circulation patterns during non-drought periods, but synoptic-scale atmospheric circulation patterns averaged for a month or more during a drought display anomalies when compared to the climatological average. The long-term anomaly patterns capture the drought circulation patterns that persistently produce little or no precipitation (McNab and Karl, 1991). Mid-tropospheric ridge and trough patterns, blocking high-pressure systems (Knapp et al., 2004), storm tracks, and frontal passages are often associated with severe and widespread drought and are identified as drought precursor mechanisms. The persistence of these general circulation features largely determines the extent and severity of drought. A global, or at least a hemispheric, climatic and atmospheric circulation perspective is needed to understand the time and space features of drought resulting from anomalous atmospheric circulation patterns. However, even this perspective is unable to define the actual sequence of dry and wet days during a drought (McNab and Karl, 1991).

Drought in tropical climates illustrates the challenging task of addressing drought causality. SSTs and related oceanic circulations are initial physical features serving as precursors for tropical droughts. These oceanic changes trigger teleconnections that result in influences far from the area of origin. ENSO is the most widely known of these effects. ENSO produces drought in some regions and wet conditions in other areas. Some major summer droughts in India coincide with a major ENSO, but others have occurred in years without an ENSO, and drought does not always accompany a major ENSO. The complex pattern suggests that ENSO may be related to some other circulation change that also influences the Indian monsoon (Peixoto and Oort, 1992). The impact of ENSO on drought in Central and South America is more direct.

Namias (1978, 1983) argues that precursor circulation features are descriptions of the atmospheric circulation associated with below normal precipitation and not the physical cause of drought. The ultimate causes of the atmospheric circulation anomalies identified with drought are in reality the ultimate issue of drought causation. For example, causes of the SST anomalies that alter atmospheric circulation, storm tracks, and precipitation patterns are the root cause of drought. A comprehensive physical understanding of why SST anomalies form is still being sought. Consequently, the causes of drought commonly cited are in reality the immediate manifestations of atmospheric circulation anomalies rather than the fundamental cause of drought. While this may appear to be an excessively fine distinction, it is an important point relative to our understanding of the involved physical processes and our ability to anticipate the recurrence of these mechanisms and drought. The continuing search for the ultimate causes of drought includes both internal and external influences on the climate system (Namias, 1983; Trenberth and Branstator, 1992; Trenberth and Guillemot, 1996; Mo *et al.*, 1997).

The complex circumstances and characteristics of drought encourage the use of specific examples to illustrate representative features of selected droughts. This perspective acknowledges that an attempt to produce a comprehensive overview of the drought phenomenon is likely to be unsuccessful. The advantage of this approach is that specific details can be emphasized that are especially informative in understanding drought mechanics in different settings.

10.6   West Africa Sahel drought

The highly populated West Africa Sahel is a transition zone between the Sahara Desert and the humid savanna to the south. It extends from about

10° N to 18° N across Africa from Senegal and Mauritania on the west to the Sudan on the east, and it is one of the most climatically sensitive zones in the world because of the influence of a wide range of factors (Zeng, 2003). Rainfall is highly variable in space and time due to its convective origin, and any significant rainfall decrease sharply impacts the vegetation and the living conditions of the local population. The Sahel has been prone historically to severe droughts, but drought in the early 1970s was of unprecedented severity in recorded history and captured global attention. By the mid 1970s, millions of people from the region were dependent on external food sources, and an estimated 250 000 people and 12 million cattle amounting to 40% to 50% of the domestic livestock population died of starvation (Nicholson et al., 1998; Tarhule and Lamb, 2003).

10.6.1 Sahel rainfall deficits

Annual rainfall in the Sahel ranges between 200 and 800 mm along a north-to-south latitudinal gradient. Most of the annual rainfall occurs from April to October, but August contributes 32% to 40% of the annual rainfall total for different latitudinal bands. For most locations, the dry season is generally longer than the wet season (Nicholson et al., 2000). Year-to-year precipitation variability ranges between 20% and 80% of the long-term mean (Tarhule and Lamb, 2003), and the southern Indian Ocean is a significant source of water vapor that contributes to precipitation for the Sahel (Bosilovich and Schubert, 2002).

Since the late 1960s, precipitation in the Sahel and most of the semiarid region of West Africa has been markedly below normal (Fig. 10.4). The rainfall decline has been most evident during the months of August and September. These characteristics became dominant following a major atmospheric circulation shift in the early 1960s coinciding with a general reversal of the Hadley–Walker circulation, but the cause of this reversal remains unresolved (Long et al., 2000; Grist and Nicholson, 2001). During the 1970s and 1980s, mean decadal rainfall was generally 20% or more below the long-term mean, and the 1983-4 minimum followed the major 1982-3 ENSO event (Dai et al., 2004a). Western Sahel precipitation in 1988 was near the 1951 to 1980 average for the first time since 1975 (Ropelewski, 1988), and although rainfall improved during the 1990s it was still well below the long-term mean (Nicholson et al., 2000; Dai et al., 2004a). The decade of the 1980s was the driest of the twentieth century for West Africa, and earlier droughts of the twentieth century were minor events compared to the recent prolonged dryness which may have been equaled only once during each of the two preceding centuries (Nicholson, 1993; Tarhule and Lamb, 2003).

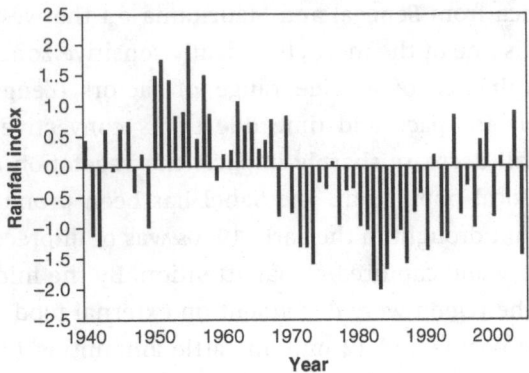

Fig. 10.4. Standardized Sahel precipitation index for 1940–2004. Annual values standardized to have a mean of zero and a standard deviation of one. (Data courtesy NOAA and the University of Washington Joint Institute for the Study of Atmosphere and Ocean from their website at http://jisao.washington.edu/data_sets/sahel/index2.html.)

10.6.2   Proposed causes for the Sahel drought

The succession of anomalously dry years in the Sahel has prompted numerous studies of large-scale atmospheric circulation processes and other physical mechanisms related to producing rainfall deficiencies, but an explanation for the Sahel drought has proven elusive. A consensus has not been achieved that satisfies the complex circumstances of the region and the observed Sahel rainfall variability. The drought has attributed to ENSO events, tropical Atlantic SST anomalies, Mediterranean Sea SST anomalies, the Indian monsoon, atmospheric wave disturbances, displacement of the ITCZ, the global increase in carbon dioxide, and human land-use change feedback mechanisms (Eltahir and Gong, 1996; Long *et al.*, 2000; Nicholson *et al.*, 2000; Raicich *et al.*, 2003; Rowell, 2003; Zeng, 2003; Dai *et al.*, 2004a; Thiaw and Mo, 2005). However, Taylor *et al.* (2002) modeled land-use changes and concluded that while the climate of the Sahel is sensitive to small changes in albedo and leaf area index the recent changes in land use are not large enough to be the principal cause of the drought. Additional factors contributing to the difficulty in identifying the underlying cause of the prolonged dryness are that interannual rainfall variability across the region is related to the ITCZ migration, the African monsoon, and ENSO, and interdecadal precipitation variability is related to SSTs in the North Pacific and the North Atlantic (Thiaw and Mo, 2005). A final consideration in accounting for Sahel rainfall variability is that different ocean basins influence precipitation on interannual and interdecadal time scales (Zeng, 2003).

The Sahel climate is dominated by the annual cycles of the quasi-permanent North Atlantic subtropical high-pressure system centered over the Azores, the South Atlantic subtropical high, and the intervening equatorial low-pressure trough. An upper-tropospheric tropical easterly jet (TEJ) occurs between 100 and 200 hPa levels that is sensitive to global SST changes. When the TEJ extends to the west it provides high vertical wind shear across the Gulf of Guinea and enables convection to move northward into the Sahel (Thiaw and Mo, 2005). Examination of reanalysis data suggests the TEJ has weakened and shrunk in areal extent over the past four to five decades reducing one component of convective activity (Sathiymoorthy, 2005). Upper-air flow at 500 hPa is westerly in the winter, but the westerlies are weaker and display larger deviations from zonality during the summer. The local Hadley cell and its vertical circulation display a dramatic shift between summer and winter producing a displacement of the convective area from 0°–5° N to 15°–20° N (Raicich et al., 2003). Surface flows related to these pressure systems produce dry northeasterly winds off the Sahara Desert for most of the year and humid southwesterly flow during April–October that support a rainy season in July–September. A narrow zone of contrasting humidity migrates meridionally in response to the annual insolation cycle, and areas approximately 200 km equatorward of the contrasting airflow boundary zone receive rainfall (Tarhule and Lamb, 2003).

The focus on atmospheric circulation features in attempting to account for the precipitation variability recognizes the influence of both interannual and interdecadal processes and the coupling of atmospheric and surface processes. A majority of studies have shown that dry years are characterized by a suite of circulation conditions identifiably different from those present during wet years, and Rowell (2003) and Raicich et al. (2003) demonstrate that dry years are related to cooler than average SSTs in the Mediterranean Sea, particularly on decadal time scales. The most prominent atmospheric circulation characteristics during prolonged dry years include weak anticyclonic and vertical shear over the Sahel, weaker development of the ITCZ rather than a difference in its surface position, and a shallower moist layer (Eltahir and Gong, 1996; Grist and Nicholson, 2001).

Atmospheric circulation differences during wet and dry years are evident in zonal wind fields at 600 hPa over West Africa. Horizontal wind shear increases during composite wet years compared to composite dry years (Fig. 10.5). In general, wind shear is an abrupt change in wind speed or direction or both. It is common to specify horizontal or vertical shear and to distinguish cyclonic (positive shear in the Northern Hemisphere) and anticyclonic (negative shear in the Northern Hemisphere) wind shear when addressing the horizontal wind field. The increased anticyclonic (negative) horizontal wind shear during wet

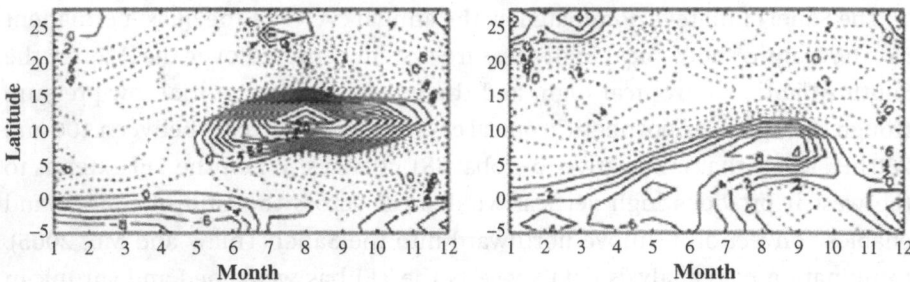

Fig. 10.5. Sahel region annual cycle of horizontal wind sheer. Left panel is a wet years (1958-61) composite. Right panel is a dry years (1982-5) composite. Units are $10^{-6}\,s^{-1}$, and dashed contours are positive values. (From Grist and Nicholson, 2001, Fig. 6. Used with the permission of the American Meteorological Society.)

years evident in Figure 10.5 is a possible instability mechanism that may enhance rainfall (Grist and Nicholson, 2001). During wet years, the horizontal shear is greater than $10^{-6}\,s^{-1}$ and negative from June through October over the Sahel. During dry years, the negative shear zone is weaker and about 6° of latitude equatorward, intense horizontal shear is limited to July to September, and the horizontal wind shear reverses sign over the Sahel.

The most consistently demonstrated atmospheric circulation difference between wet and dry years in the Sahel is in the location rather than the intensity of the southwesterly flowing African easterly jet (AEJ) (Eltahir and Gong, 1996; Grist and Nicholson, 2001; Thiaw and Mo, 2005). The AEJ is a mid-tropospheric wind maximum at 600–700 hPa that appears from late spring to summer and acts to maintain and provide energy to wave disturbances that produce Sahel rainfall (Thiaw and Mo, 2005). The AEJ is about 5° further pole-ward from June through September in wet years compared to dry years, but during October to May the jet is somewhat more poleward during dry years. While the jet is at the highest latitude during wet years in July and August, the most poleward advance of the jet in dry years is a spike in late July with generally a modest poleward advance in August and September. Since the region of maximum atmospheric moisture is just to the south of the jet core, the southward displacement of the jet is often associated with the decline in observed August and September rainfall in dry years (Thiaw and Mo, 2005).

Although the reduced seasonal migration of the AEJ in dry years is signifi-cant, differences in the strength of the jet are less pronounced. The jet is at maximum strength in August in wet years and is strong throughout August and September. The jet's maximum strength in dry years occurs in early June and it is strong from May through July and again in September. Differences in the intensity of the AEJ are greatest from May through July when the jet is stronger during dry years, and relatively small differences characterize the rest of the year (Grist and Nicholson, 2001).

Various measures of the atmospheric moisture field over West Africa have been examined for evidence of differences between wet and dry years. In most instances, differences in the occurrence of maxima or minima or the character of latitudinal shifts in moisture fields are observed. For example, moist layer heights are considerably greater at three selected latitudes for wet years than for the dry years. However, the role of the moist layer thickness and other moisture field variables in explaining interannual rainfall variability in West Africa is unclear (Grist and Nicholson, 2001).

10.7   Western United States drought

Drought in the western United States is initiated during the winter when most of the annual precipitation accumulates. Atmospheric circulation patterns deliver moisture from the Pacific Ocean and topography influences the spatial distribution of precipitation. Seasonal variations in atmospheric circulation produce contrasting distinct wet and dry seasons. The magnitude of abnormal dryness during the summer is limited by the meager precipitation received during this season. Nevertheless, summer aridity can contribute to drought duration and severity when drought is initiated in the preceding winter (Shelton, 1984).

10.7.1   California's 1975–1977 drought

The 1975 to 1977 California drought illustrates the severe dryness resulting from deficient winter rainfall. This drought and the 1928 to 1937 drought are severe statewide droughts (Fig. 10.6) often cited by different sources as the worst in the twentieth century (Hunrichs, 1991). October 1975 to September 1976 was the fourth driest 12 months in the twentieth century in California, and October 1976 to September 1977 was the driest 12 months. However, the spatial and temporal drought characteristics display significant differences across the state relative to the occurrence and duration of the most severe conditions. Moisture deficits occurred in September 1975 in southern California and in November throughout most of California (Fig. 10.7). The 2 years of abnormal moisture deficiency severely stressed the statewide water supply system that depends on northern California runoff for most of its developed water.

Long-term drought is a particular concern in the Sacramento River Basin in northern California because about one-third of the major surface reservoirs and nearly one-half of the total reservoir storage capacity in California are in this watershed (Shelton, 1984). The 1975 to 1977 streamflow deficiency in the Sacramento River Basin was the most severe in more than 100 years

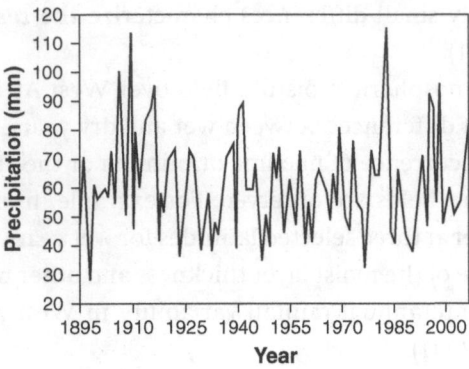

Fig. 10.6. Annual (July to June) precipitation for California, 1895–2005. (Data courtesy NOAA and the National Climatic Data Center from their website at http://lwf. ncdc.noaa.gov/oa/climate/climatedata.html.)

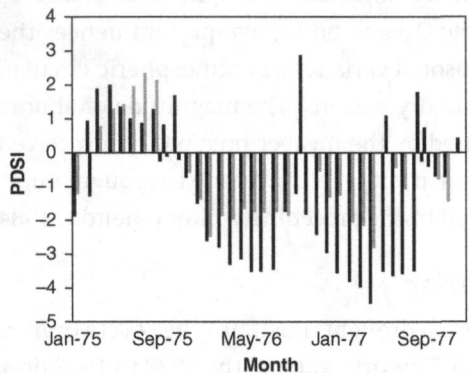

Fig. 10.7. Palmer Drought Severity Index (PDSI) for California's Sacramento Drainage climate division (solid column) and South Coast Drainage climate division (clear column) for January 1975 through December 1977. (Data courtesy of NOAA and the National Climatic Data Center from their website at http://www7.ncdc.noaa.gov/CDO/ CDODivisionalSelect.jsp.)

(Hunrichs, 1991). The Sacramento Drainage climate division nearly coincides with the Basin, and the PDSI for the Sacramento Drainage was negative for 22 months from November 1975 to August 1977 (see Fig. 10.7). Drought severity reached a maximum PDSI of −4.43 in April 1977. The historic maximum drought PDSI for this region was −4.45 in April 1898 during a 26-month drought. Shelton (1984) estimated that the areal average cumulative moisture deficiency for the watershed for the period May 1975 to November 1977 was 788 mm, or 87% of the average annual precipitation for the basin, and 26% of the basin had a cumulative moisture deficiency that exceeded 1000 mm.

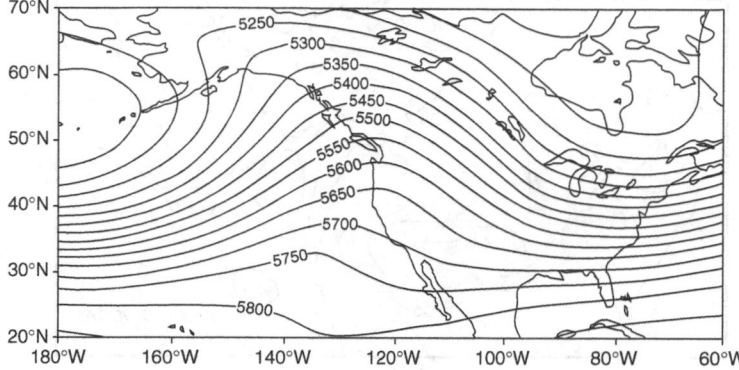

Fig. 10.8. Average 500 hPa geopotential heights for December 1976 through February 1977 over North America. This upper-air pattern promoted atmospheric blocking along the west coast of North America, and drought occurred in California. Heights are in m. (Data courtesy of NOAA and the Cooperative Institute for Research in Environmental Sciences Climate Diagnostics Center from their website at http://www.cdc.noaa.gov/.)

10.7.2   Ocean–atmosphere anomalies and drought

Ocean–atmosphere energy fluxes produce substantial variations in atmospheric heating that alters atmospheric pressure and circulation patterns influencing West Coast precipitation. The 1975 to 1977 California drought developed in response to anomalous SST patterns and related atmospheric teleconnections that amplified normal seasonal forcing influences. The immediate isobaric drought signature was anticyclonic vorticity associated with a high-pressure ridge (Fig. 10.8) that blocked cyclonic systems entering the region from the Pacific Ocean (Namias, 1983). Consequently, storm tracks were displaced north of the United States/Canadian border in a pattern that resembled summer conditions, and most of the United States West Coast was unusually dry. However, the physical causes for the presence and persistence of the high-pressure ridge occurred well outside the immediate drought area.

An extensive synoptic study by Namias (1978) identified multiple interrelated physical factors related to the anomalous atmospheric patterns resulting in abnormally low precipitation over the western United States. The Pacific Ocean north of 40° N was continuously colder than usual from spring 1975 to summer 1976 (Fig. 10.9). This vast Pacific pool of cold surface water was probably a result of the cold winters of the early 1970s in the Northern Hemisphere and the transition from a La Niña to El Niño event. Anomalously warm SSTs were present off the California coast, and the anomalous SST patterns helped maintain a southerly atmospheric flow via a PNA teleconnection resulting in a

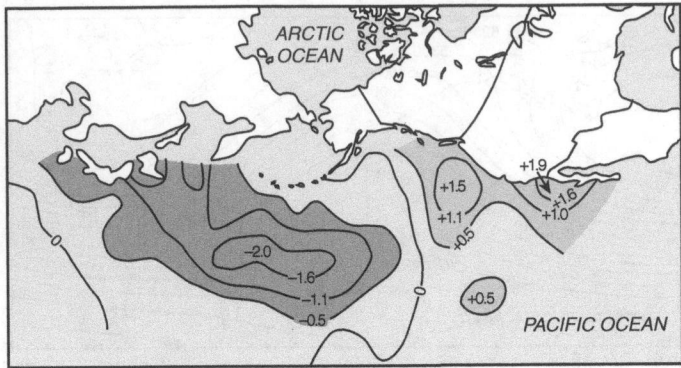

Fig. 10.9. North Pacific sea surface temperature anomalies for the winter of 1975-6. Cold water dominates the central North Pacific Ocean, but warm water occurs along the west coast of North America from southern Canada to Baja California. Units are in °C. (From McNab and Karl, 1991, Figure 5. Courtesy of the U.S. Geological Survey from their website at http://geochange.er.usgs.gov/sw/changes/natural/drought/.)

strong ridge over the West Coast (McNab and Karl, 1991). The jet stream was displaced between 5° and 10° of latitude northward and was persistent in this position with only short variations during the period. The long-wave pattern across North America was an amplified form of the climatological normal flow pattern resulting in an abnormal frequency of ridging over western North America and the eastern Pacific Ocean as part of the PNA teleconnection. Seasonal forcing influences operating on the abnormal wind pattern further stabilized the drought-producing ridge over the West Coast (Namias, 1978). A strong meridional wave pattern persisted from late October 1976 through mid February 1977. Northwesterly flow aloft supported repeated outbreaks of bitterly cold Arctic air into the eastern half of the United States and contributed to one of the coldest winters of the twentieth century for the region. An extensive snowcover over the East and Midwest assisted development of an East Coast trough which helped strengthen the western ridge. Southwesterly winds aloft brought unseasonably mild air to the far-western United States which amplified and stabilized the drought pattern (Namias, 1983).

10.8   Midwestern United States 1988 drought

Spring 1988 precipitation for the United States was the smallest spring precipitation in the twentieth century. Accumulated rainfall for May through June over much of the United States and southern Canada between 85° W and 110° W longitude was less than 60% of normal (Riebsame *et al.*, 1991). By late

Table 10.2 *Drought in world regions during 1900–93 comparable to the central United States 1988 drought*

Region	Number	Year
Central Europe 38°–57° N, 2° W–28° E	4	1904, 1921, 1929, 1948–9
Southeast China 22°–37° N, 107°–122° E	2	1917–18, 1963
India 8°–28° N, 74°–88° E	2	1909–10, 1968–9
Eastern Australia 40° S–5° N, 135°–160° E	9	1902, 1914, 1919, 1925–6, 1946, 1973, 1977, 1982–3, 1991–2
Southern Africa 18°–36° S, 17°–36° E	3	1972, 1972–3, 1982–3

After Dai *et al.*, 1997, Table 1. Used with permission of the American Meteorological Society.

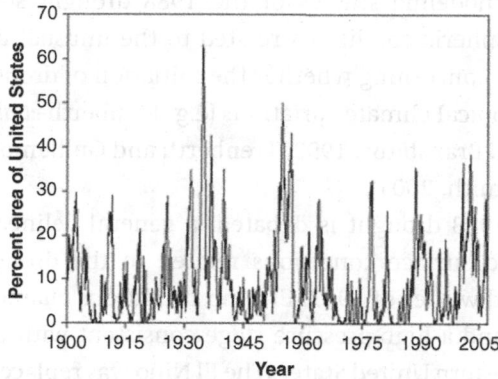

Fig. 10.10. Percentage area of the United States in severe or extreme drought as indicated by the Palmer Drought Severity Index (PDSI) for 1900–2006. (Data courtesy of NOAA and the National Climate Data Center from their website at http://www.ncdc.noaa.gov/oa/climate/research/2006/dec/us-drought.html.)

July, the PDSI categorized drought as either severe or extreme over 36% of the land area of the contiguous United States (Trenberth *et al.*, 1988). The 1988 event was the greatest expanse of severe or extreme drought in the United States since the 1930s and 1950s droughts (Fig. 10.10). In the United States, droughts in 2000 and 2002 have been comparable in severity to the 1988 drought, but regions on other continents have experienced more numerous occurrences of comparable droughts (Table 10.2). More than 1000 maximum temperature records were equaled or established during the summer of 1988, 10 000 deaths from heat stress were reported, and direct economic losses were estimated at $39 billion (Riebsame *et al.*, 1991; Trenberth and Branstator, 1992). Drought developed at different times in different parts of the country, but it was most severe in the Great Plains, the Midwest and the lower Mississippi Valley during the summer of 1988 (Trenberth and Branstator, 1992). Extreme temperature anomalies spanned the United States from the east to west coasts (Ropelewski, 1988).

The 1988 summer drought in the Great Plains and the Midwest shared many characteristics with the 1930s Dust Bowl drought. April to June in the Corn Belt was the driest for this period since 1895, and the PDSI categorized over 70% of this region as having severe or extreme drought. These conditions were exceeded only by the 1930s drought in the Corn Belt and by the 1950s drought in most of the Great Plains (Riebsame et al., 1991; Hu et al., 1998). The PDSI time-series for central Kansas (see Fig. 10.1) shows the longer duration of the 1930s drought and the greater severity of the 1950s drought for this climatic region.

10.8.1   Tropical or extratropical forcing

Observational and modeling studies of the 1988 drought suggest a variety of anomalous atmospheric conditions related to the unusual dryness. However, differences persist concerning whether the initiation of drought was rooted in tropical or extratropical climate variations (e.g. Trenberth et al., 1988; Namias, 1991; Trenberth and Branstator, 1992; Trenberth and Guillemot, 1996; Mo et al., 1997; Chang and Smith, 2001).

While the cause of the 1988 drought is debated, a general delineation of atmospheric circulation and surface conditions related to the drought has emerged. During the fall and winter of 1987, El Niño conditions characterized the tropical Pacific Ocean, and a high-pressure ridge consistent with this SST pattern persisted over the western United States. The El Niño was replaced in the late spring of 1988 by the rapid onset of a strong La Niña displaying very cold water along the equator and suppressed convective activity in this region (Ropelewski, 1988; Namias, 1991; Trenberth and Branstator, 1992). Warmer water, a residual from the previous El Niño, was pushed up to a region southeast of Hawaii. This SST pattern favored a northward displacement of the ITCZ by about 3° of latitude and a change of atmospheric heating patterns through the latent heat released in rainfall (Trenberth et al., 1988; Trenberth and Guillemot, 1996). Modeling studies indicate such atmospheric heating changes can disrupt the atmospheric flow by setting up anomalous waves in the atmosphere (Trenberth and Branstator, 1992). By late spring 1988, an altered atmospheric wave pattern was manifested as a change in the jet stream and associated storm tracks in the mid-latitudes and a shift in the position of the Pacific subtropical high to a mean position further west over the Pacific Ocean (Fig. 10.11). The westward shift of the Pacific high-pressure region promoted trough over the West Coast and an abnormally strong upper-level ridge over central North America with anticyclonic conditions favoring drought development and maintenance of the circulation anomalies (Liu et al., 1998). Positive 500 hPa

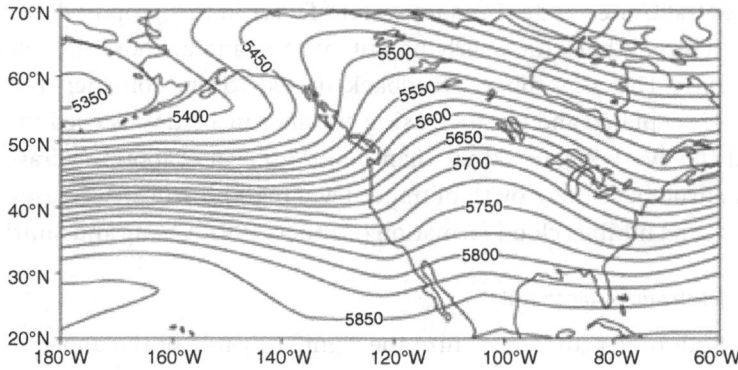

Fig. 10.11. Average 500 hPa geopotential heights for April–June 1988 over the North Pacific Ocean and North America. A strong ridge is present east of the Rocky Mountains while drought is widespread in the Midwest. Heights are in m. (Data courtesy of NOAA and the Cooperative Institute for Research in Environmental Sciences Climate Diagnostics Center from their website at http://www.cdc.noaa.gov/.)

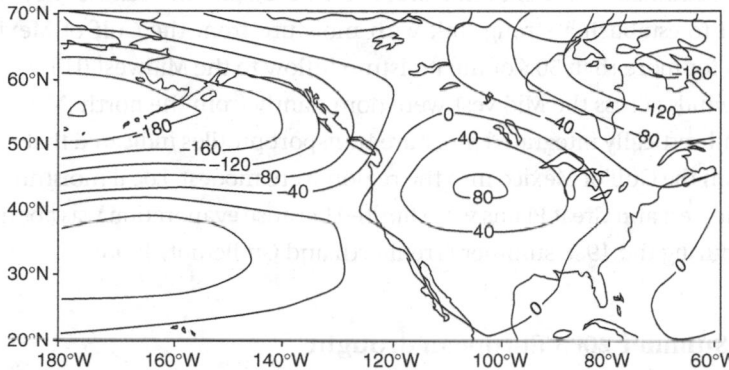

Fig. 10.12. Average 500 hPa geopotential height anomalies for April–June 1988 over the North Pacific Ocean and North America. Negative anomalies dominate the North Pacific, and positive anomalies dominate central North America where drought was widespread in the Midwest. Contours are in m. (Data courtesy of NOAA and the Cooperative Institute for Research in Environmental Sciences Climate Diagnostics Center from their website at http://www.cdc.noaa.gov/.)

geopotential height anomalies (Fig. 10.12) and positive temperature anomalies dominated the Midwest (Ropelewski, 1988).

An alternative view suggests that natural variability of the extratropical circulation accounts for the development of the wave train with upper-level anticyclonic anomalies over the North Pacific, North Atlantic, and continental United States in early summer (Namias, 1991; Chang and Smith, 2001). Dry soil in the central United States is an important factor in this proposed mechanism because it strengthens the overlying anticyclone and anchors the teleconnection pattern. The atmospheric feedback linked to dry soil increases as soil moisture is depleted and the increasing summer insolation is used to heat the soil and the overlying air rather than being used in evapotranspiration. This process encourages the growth of upper-level high pressure, decreases relative humidity, and inhibits cloud formation (Namias, 1991; Chang and Smith, 2001).

10.8.2   Jet stream displacement

Both perspectives recognize the significant role of the anomalous wave pattern in 1988. The West Coast trough turned the jet stream northeastward as it approached the continent so that moisture-bearing weather systems were diverted into central Canada and well north of their usual paths. The jet stream over North America east of the Rocky Mountains was farther north in 1988 than in any year from 1979 to 1993, and transient eddy activity was weak over North America (Trenberth and Guillemot, 1996). Disturbances in the westerlies over the continent encountered an environment unfavorable for development because they were too far north to establish a strong link with moisture from the Gulf of Mexico that accounts for more than 50% of the moisture inflow to the Midwest (Hu et al., 1998). Surface winds across the Midwest were dominantly from the north (McCabe et al., 1991), and vertically integrated moisture transport profiles indicated that moisture flows from the Gulf of Mexico into the region were modest. Local moisture sources in the Midwest and Great Plains were limited because evaporation was about 60% of normal during the 1988 summer (Trenberth and Guillemot, 1996).

10.9   Summer 2003 European drought

European summers in recent decades have been characterized by a drying trend and the occurrence of especially severe droughts and floods. Widespread flooding in 2002 (see Section 9.7) was followed by equally widespread drought in 2003. Although the summer 2003 drought was not among the most severe in central Europe, the drought impact was amplified by high evapotranspiration rates and precipitation deficits during the previous spring (Fink et al., 2004). The summer 2003 drought and related heatwave caused nearly 30 000 fatalities, and many countries experienced their worst harvest in decades (Pal et al., 2004). The combination of deficient water and excessive heat significantly lowered yields of grains, vegetables, fruits, and wines, and total crop losses were an estimated $12.3 billion (Zaitchik et al., 2006). Forests in western, central, and southern Europe were severely damaged due to an unprecedented reduction in plant primary productivity (Ciais et al., 2005), and wildfires burned

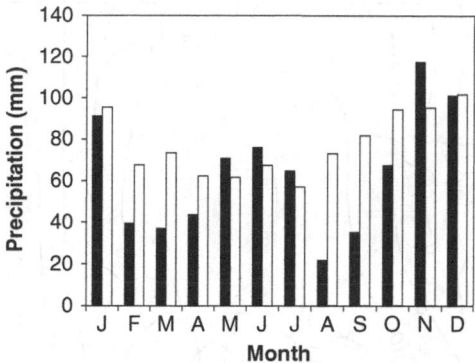

Fig. 10.13. Birmingham, England, 2003 monthly precipitation (solid column) and 1971–2000 monthly mean precipitation (clear column). (Data courtesy of NOAA and the National Climate Data Center from their website at http://www.ncdc.noaa.gov/oa/climate/ghcn-monthly/index.php.)

large areas of eight countries. Wildfires were most severe in Portugal where 450 000 ha burned (Trigo *et al.*, 2006).

10.9.1   Widespread meager precipitation

Precipitation was especially low in much of Europe in June and August 2003, but some areas had deficient rainfall from February thorough September. In the Bavarian Alpine region of Germany, precipitation deficits developed in late March and the deficit by mid June was the largest ever observed for that time of the year (Fink *et al.*, 2004). Annual precipitation deficits reached 50% of annual average precipitation for some areas of central Europe (Ciais *et al.*, 2005; Rebetez *et al.*, 2006). England (Fig. 10.13) and areas along the west coast of Europe benefited from slightly above average precipitation from May to July when most of continental Europe received precipitation below the long-term average. Even these small moisture pulses in England were enough to make this one of the relatively wetter areas of Europe with 2003 precipitation totaling 82% of the long-term average compared to 71% in Germany and 74% in Austria. Nevertheless, 2003 total precipitation was the smallest annual total for England since 1975 and the 25th lowest total in the record. August precipitation was only 29% of the expected amount for England, and it ranked as the sixth driest August since 1766 (Parker *et al.*, 2004). Meager August precipitation was the common condition throughout Europe.

10.9.2   Altered atmospheric circulation

European weather from May to August 2003 was dominated by anticyclonic flow related to the Azores subtropical high-pressure circulation. The

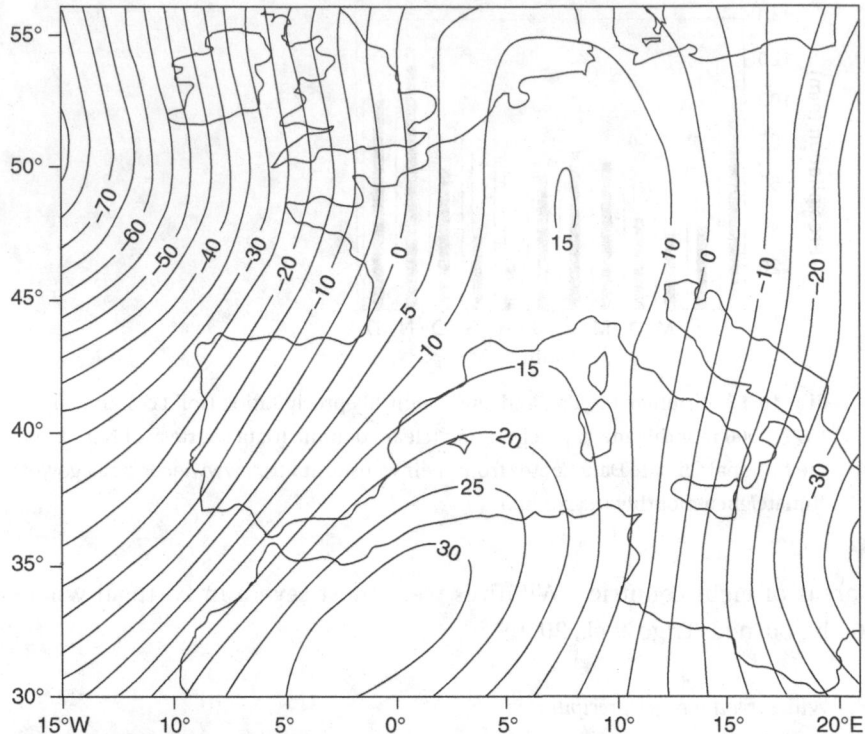

Fig. 10.14. Average 500 hPa geopotential height anomalies for June–August 2003 over Europe and North Africa coinciding with drought in central Europe. Positive anomalies characterize Europe east of 5° W. Contours are in m. (Data courtesy of NOAA and the Cooperative Institute for Research in Environmental Sciences Climate Diagnostics Center from their website at http://www.cdc.noaa.gov/.)

Azores high was displaced northward and extended from the mid Atlantic through eastern Europe in the form of an anomalous high-pressure ridge (Fig. 10.14). The upper-air ridge produced warm advection into western Europe that contrasted markedly with the usual zonal upper-air flow pattern delivering air from over the North Atlantic (Fig. 10.15). The displaced high-pressure ridge promoted a general strengthening of westerly flow on its poleward flank over the United Kingdom and southern Scandinavia as it advected warm air into western Europe (Black *et al.*, 2004).

The summer 2003 circulation varied with stronger and more cyclonic circulation in June and August and weaker cyclonic circulation in July. The details of these changes are evident in the 850 hPa level which is representative of the low tropospheric state, but it does not suffer from some of the problems that affect near-surface reanalysis variables (Trigo *et al.*, 2006). The cyclonic anomaly over

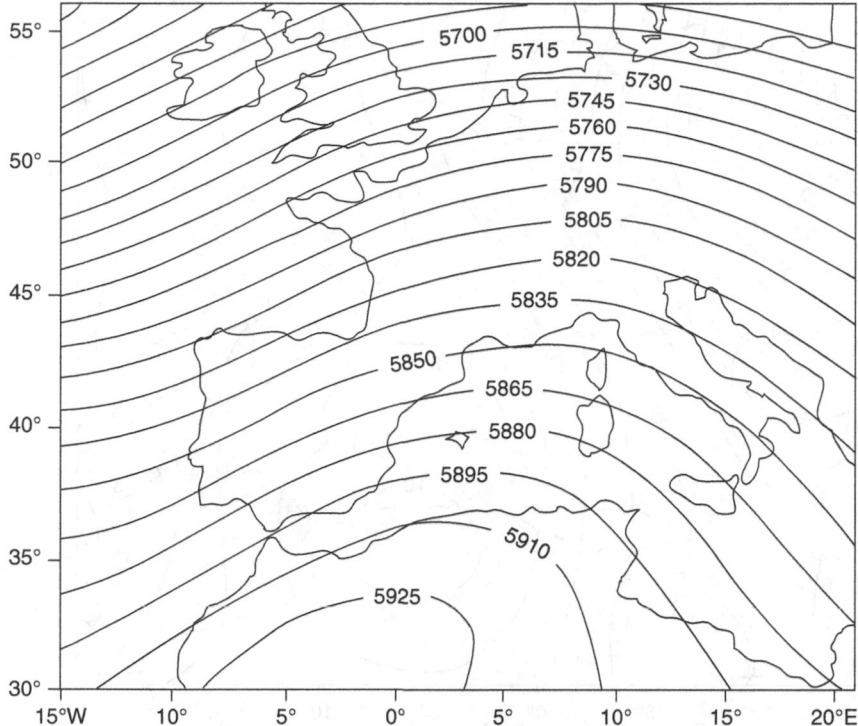

Fig. 10.15. Average 500 hPa geopotential heights for June–August 2003 over Europe and North Africa coinciding with drought in central Europe. The axis of a high-pressure ridge extends from North Africa to Germany. Heights are in m. (Data courtesy of NOAA and the Cooperative Institute for Research in Environmental Sciences Climate Diagnostics Center from their website at http://www.cdc.noaa.gov/.)

the Atlantic Ocean was centered west of the United Kingdom in June (Fig. 10.16) and July, and it moved further southwest in August (Fig. 10.17). The anticyclonic anomaly over North Africa remained relatively stationary, but its intensity increased and it expanded from June to August. The circulation resulting from the pressure anomalies produced southwesterly flow in May over the United Kingdom and coastal Europe that changed to southeasterly in August (Black et al., 2004). The southwesterly flow increased the tendency for advected moist air from the Atlantic over the United Kingdom and coastal Europe producing precipitation slightly above average in May to July for these regions (see Fig. 10.13), but precipitation was limited in these months over many inland areas. The southeasterly flow in August increased advection tendencies of warm, dry air from continental Europe and the Mediterranean region, and the United Kingdom and coastal Europe experienced relatively dry conditions during August and September similar to inland areas.

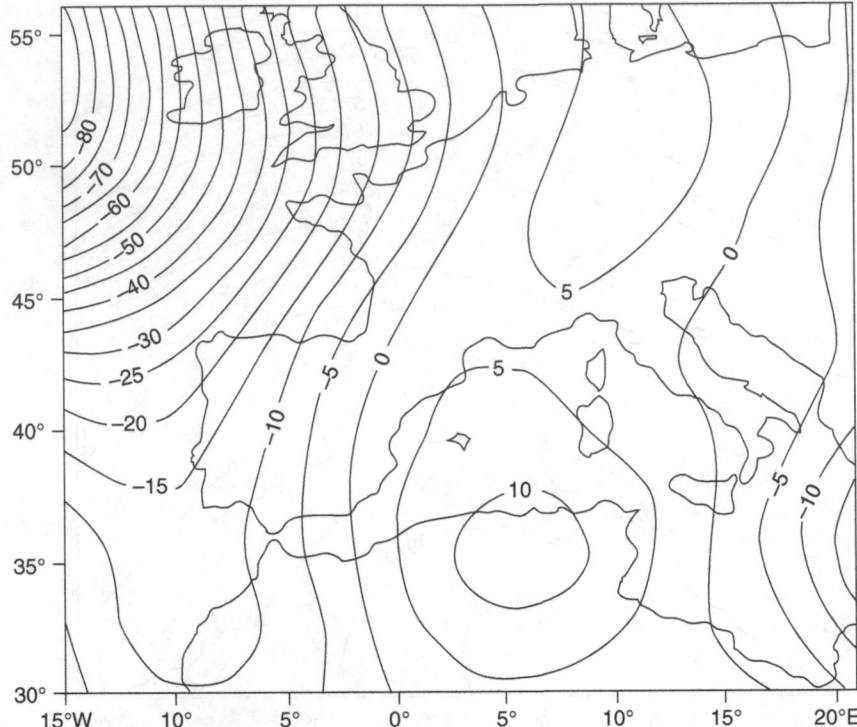

Fig. 10.16. Lower troposphere 850 hPa geopotential height anomalies for June 2003 over Europe and North Africa coinciding with drought in central Europe. Strong negative anomalies west of the United Kingdom suggest an active cyclonic zone, while positive anomalies over Europe east of 0° longitude inhibit air flow from the Atlantic Ocean. Contours are in m. (Data courtesy of NOAA and the Cooperative Institute for Research in Environmental Sciences Climate Diagnostics Center from their website at http://www.cdc.noaa.gov/.)

The increasing geopotential heights from June to August over central Europe shown in Figures 10.16 and 10.17 indicate the presence of a blocking feature inhibiting the inflow of cooler air from the Atlantic Ocean and the North Sea. This blocking pattern deflected storms northward along the western limb of the anomalous ridge over Europe permitting the intrusion of exceptionally few cyclonic storms from the end of June until the end of August (Pal *et al.*, 2004; Rebetez *et al.*, 2006). The dominating anticyclonic circulation promoted atmospheric stability, suppressed convection and precipitation, and enhanced clear skies, sunshine, and sensible heat flux. The interaction of these factors produced severe dry conditions, record high August temperatures, widespread wildfires, and record low stream levels (Black *et al.*, 2004; Fink *et al.*, 2004). Maximum temperatures of 35 °C to 40 °C were repeatedly recorded in July and

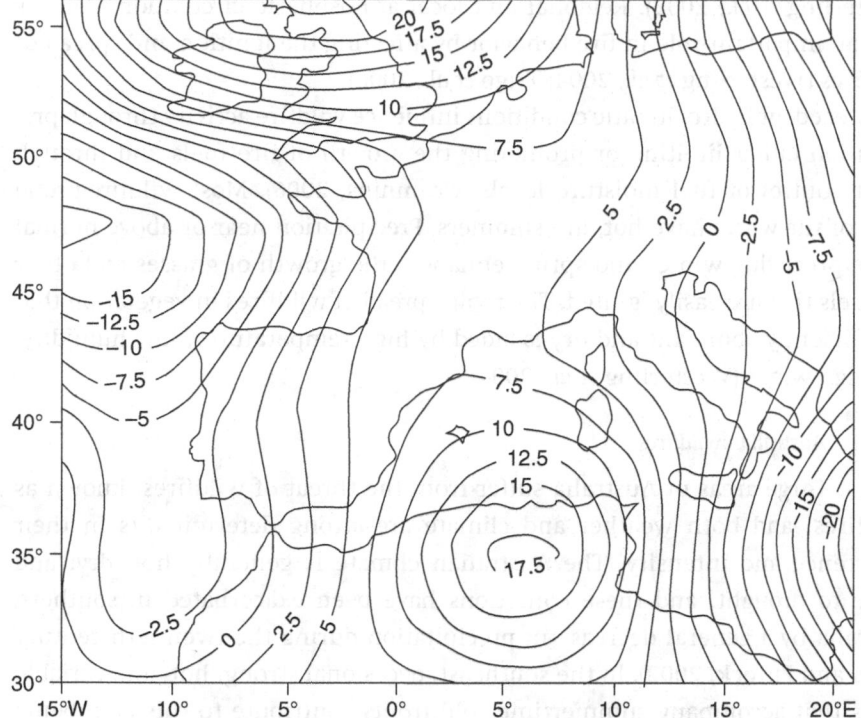

Fig. 10.17. Lower troposphere 850 hPa geopotential height anomalies for August 2003 over Europe and North Africa coinciding with drought in central Europe. Negative anomalies are weaker and positive anomalies are stronger than the conditions shown in Figure 10.16. Contours are in m. (Data courtesy of NOAA and the Cooperative Institute for Research in Environmental Sciences Climate Diagnostics Center from their website at http://www.cdc.noaa.gov/.)

August, and mean summer temperatures over a large area of continental Europe exceeded mean values for the period by about 3 °C, which exceeded normal distribution ranges for the data (Schär *et al.*, 2004).

10.10 Drought and wildfires

Drought and wildfires are related through their shared association with moisture deficiencies and enhanced energy loadings. Both short-term and long-term atmospheric states influence precipitation and fire-fuel production mechanisms, drought occurrence and severity, and daily variation in temperature, relative humidity, and wind speed (Crimmins, 2006). Large-scale atmospheric circulation patterns and anomalies affect the frequency and extent of drought and have been linked to wildfire occurrence and extent of burned area

(Westerling *et al.*, 2003). Regional and local atmospheric circulation features have an important role in fire behavior by affecting the ignition and spread of wildfires (Westerling *et al.*, 2004; Trigo *et al.*, 2006).

Antecedent hydroclimatic conditions influence wildfire activity through precipitation either limiting or promoting the growth of fire fuels and through direct control of fuel moisture levels (Crimmins, 2006). Most wildfire-prone areas of the world have hot, dry summers. Precipitation near or above normal in the preceding winter and spring enhances the growth of grasses and other fire fuels that are easily ignited. The rapid spread of wildfires in vegetation that is sufficiently abundant and dry is aided by high temperatures, low humidity, and high winds (Westerling *et al.*, 2004).

10.10.1 *Australian wildfires*

Large areas of Australia suffer from the threat of wildfires, known as bushfires, and both weather and climate are strong determinants in their occurrence and intensity. The Australian climate is generally hot, dry, and prone to drought, and these conditions have been exacerbated in southern Australia by a general decrease in precipitation during the twentieth century (Jones and Pittock, 2002). In the southeast, occasional strong, hot, and variable winds that accompany summertime cold fronts contribute to this area being recognized as having the greatest wildfire hazard in the world. The seasonal occurrence of bushfires varies in different parts of Australia ranging from the winter dry season in northern Australia to summer and early autumn in southeastern Australia. Large fires most commonly occur in Australia's arid zones in the months following an abnormally wet season when there is enough vegetation to provide fuel. The largest bushfires may burn for weeks after ignition until stopped naturally by rain or lack of fuel.

Many of Australia's native plants burn easily. Eucalyptus has a high oil content that makes it particularly fire prone, and vast areas of dry grass common in mid-to-late summer burn readily. Consequently, bushfires in Australia occur as either grass or forest fires. Grass fires occur mainly on grazing, farming, or remote scrub country, and they rarely result in heavy loss of human life. Fires in Australia's eucalyptus forests cannot be stopped and often destroy homes and settlements that border the forests. Huge amounts of flammable eucalyptus vapor, transpired from leaves, create fireballs that race ahead of the main firefront in the forest upper canopy. Most of Australia's most devastating bushfires have involved fires in dense eucalypt forests with the bushfire spreading to the suburban fringes of major cities.

Australian bushfires in 1939, 1967, and 1983 caused especially high human fatalities, and bushfires in 1992, 2003, and 2006–7 were among the most

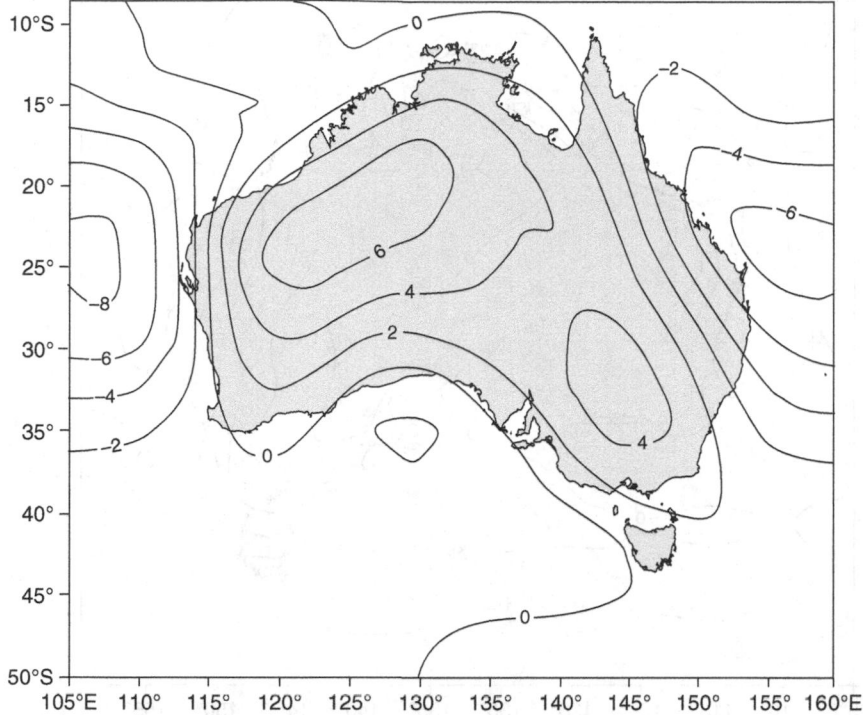

Fig. 10.18. Average 1000 hPa geopotential height temperature anomalies for October–December 2006 over Australia during widespread drought on the continent. Units are in °C. (Data courtesy of NOAA and the Cooperative Institute for Research in Environmental Sciences Climate Diagnostics Center from their website at http://www.cdc.noaa.gov/.)

protracted and extensive since European settlement of Australia. Bushfires ignited by lightning began in Victoria in late November 2006 and by mid January 2007 had burned over 1.1 million ha in Victoria and New South Wales. Unusually hot weather began in October 2006 that exacerbated the dryness resulting from five years of drought across a large part of Australia, and positive temperature anomalies for October to December 2006 (Fig. 10.18) covered a similarly large area. The high temperature anomalies were accompanied by high negative specific humidity anomalies except for coastal areas (Fig. 10.19).

10.10.2 European wildfires

Wildfires in Europe annually burn an average of 67 000 ha, but this average represents fires ranging in size from less than 1 ha to fires covering 2 000 000 ha in Spain in 1990. Greece, Italy, Portugal, Spain, and the Mediterranean parts of France account for the greatest number and largest wildfires and on average

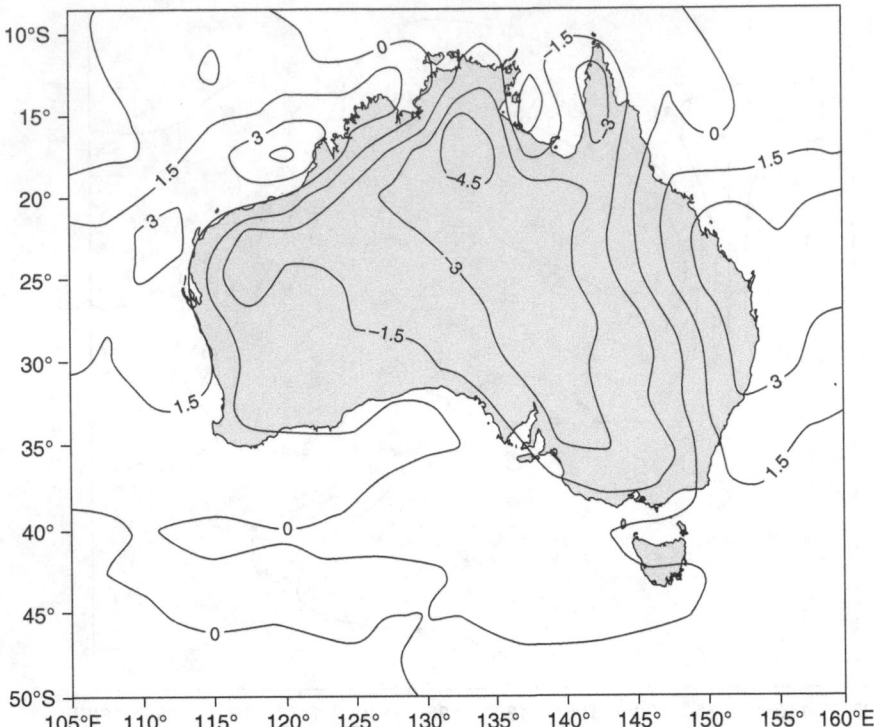

Fig. 10.19. Average 1000 hPa geopotential height specific humidity for October–December 2006 over Australia during widespread drought on the continent. Units are g kg^{-1}. (Data courtesy of NOAA and the Cooperative Institute for Research in Environmental Sciences Climate Diagnostics Center from their website at http://www.cdc.noaa.gov/.)

account for 94% of the annual burned area of Europe. Northern Europe wildfires are smaller and less frequent (Päätalo, 1998).

The largest wildfires in Portugal in modern times occurred during August 2003 when much of Europe was experiencing severe drought and a severe heatwave (see Section 10.9). A wet winter in Portugal provided moisture to produce fire fuel, and a dry May contributed to establishing summer moisture deficits that maximized antecedent conditions for August wildfires. Between 1 and 7 August wildfires burned 205 000 ha, which is nearly equal to the previous annual maximum burned area recorded in Portugal. The total burned area for 2003 was 450 000 ha or 5% of the total mainland area of the nation (Trigo *et al.*, 2006).

10.10.3 *United States wildfires*

The western United States and Canada have experienced significant increases since the 1980s in major wildfires and area burned. Throughout the region, links have been identified between antecedent moisture anomalies and anomalous wildfire activity that vary regionally and temporally. Wildfires in the western United States are strongly seasonal with a predominant occurrence between May and October, and the fire season severity has been related to the PDSI and the timing of spring snowmelt (Westerling *et al.*, 2003, 2006). Regional-scale winds in southern California and the Southwest play important roles in elevating temperatures and decreasing humidity which contribute to the winds' influence in spreading wildfires (Westerling *et al.*, 2004; Crimmins, 2006).

The average annual area burned by wildfires in the western United States is 1.73 million ha, but the 2005 burned area was more than twice the average. Over half of the wildfires in the United States are human-caused. The largest number of wildfires and the largest areas burned occur in July and August, but the length of the wildfires season has increased by 78 days since the 1980s (Westerling *et al.*, 2006). July and August have the warmest temperatures, and most areas of the western United States are characterized by summer dryness, which is aggravated when winter or spring precipitation is below normal. The seasonality of wildfires displays regional variability in the western United States as temperature and dryness increase at different rates with the progression of the seasons. Wildfires typically begin in May and June in Arizona and New Mexico, but Idaho and Montana experience their greatest wildfire activity in August (Westerling *et al.*, 2003). Although a small fraction of wildfires get very large, less than 5% of all wildfires account for more than 95% of the total annual burned area. Annual government agency expenditures to fight wildfires in the United States are $1.7 billion (Running, 2006).

Wildfires in southern California chaparral and adjacent woodlands illustrate the contributions of climatic and meteorological influences on the development and spread of wildfires in the western United States. In late October 2003, 15 major wildfires started in southern California and Baja California that burned over 300 000 ha and destroyed 3500 homes and 1100 other buildings (Westerling *et al.*, 2004). The antecedent climate provided positive precipitation anomalies in late winter and spring following a protracted drought extending over 53 consecutive months of negative PDSI values (Fig. 10.20). Monthly PDSIs of −4 and below indicating extreme drought occurred in 9 of the last 10 months prior to February 2003. Negative PDSIs returned in September and October after seven months of precipitation anomalies and positive PDSI values. The 2003 moisture regime was accompanied by a very warm winter, a cool spring, and a very warm summer and fall (Fig. 10.21). This pattern of moisture and temperature promoted growth of fire fuels over much of the area that ultimately burned (Westerling *et al.*, 2004).

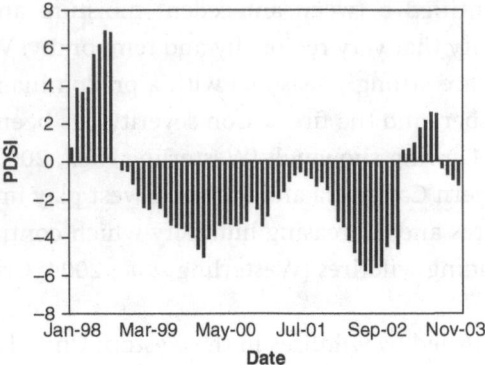

Fig. 10.20. Palmer Drought Severity Index (PDSI) for California's South Coast Drainage climate division for January 1998–December 2003. (Data courtesy of NOAA and the National Climatic Data Center from their website at http://www7.ncdc.noaa.gov/CDO/CDODivisionalSelect.jsp.)

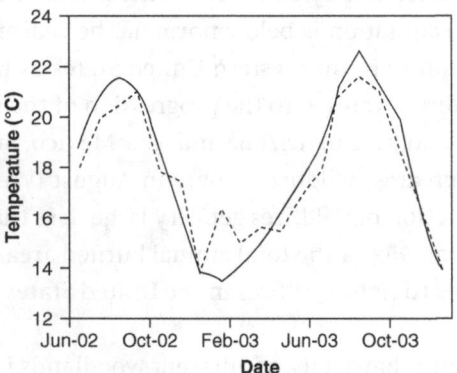

Fig. 10.21. San Diego, California, monthly temperature for June 2002–December 2003 (solid line) and the 1971–2000 monthly mean temperature (dashed line). (Data courtesy of NOAA's National Climate Data Center and the Oak Ridge National Laboratory, Carbon Dioxide Information Analysis Center from their website at http://cdiac.ornl.gov/epubs/ndp/ushcn/usa_monthly.html.)

The most evident meteorological role in the 2003 wildfires was provided by Santa Ana winds which promoted the ignition and rapid spread of wildfires. These foehn-like winds develop in fall, winter, and spring months from a regional meteorological setting like that shown in Figure 10.22. An anomalously deep Aleutian low occupies the Gulf of Alaska, and a high-pressure system with cool, dry air is centered over interior high elevations of the Pacific Northwest. The anticyclonic flow around the high-pressure feature directs air from interior basins toward the southern California coast. Lower pressure along the Pacific

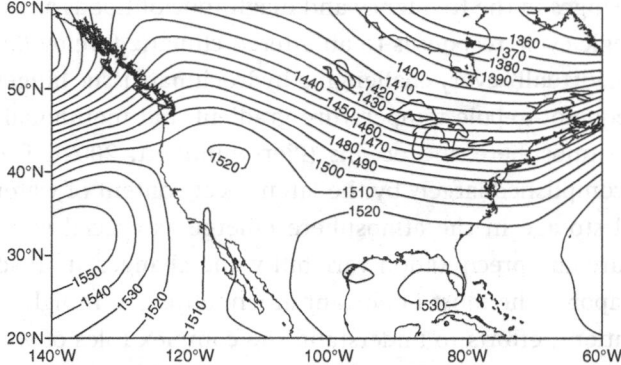

Fig. 10.22. Lower troposphere 850 hPa geopotential heights for October 2003 over North America and the eastern North Pacific Ocean coinciding with Santa Ana winds and wildfires in southern California. Heights are in m. (Data courtesy of NOAA and the Cooperative Institute for Research in Environmental Sciences Climate Diagnostics Center from their website at http://www.cdc.noaa.gov/.)

coast provides a pressure gradient producing easterly flow that is funneled through passes and canyons in the Coast Ranges and the Transverse Ranges. The air is warmed adiabatically as it descends to lower elevations and is compressed, and the resulting winds are hot and dry. Wind velocities of 40–60 km hr^{-1} are common as the air moves through the narrow mountain passes and canyons, and their influence affects the entire 400 km coastal region of southern California (Westerling *et al.*, 2004).

Autumn Santa Ana winds dry fuels already desiccated by a long dry summer. In October 2003, moderate Santa Ana winds in combination with anomalously warm temperatures reduced relative humidity to near 10% during the day and to only 20–30% at night. Abundant dry fuels and low relative humidity reinforced the influence of the moderate Santa Ana winds in extending the duration of the southern California wildfires (Westerling *et al.*, 2004).

10.11 Looking ahead

One of the most compelling global climate change manifestations is intensification of the global hydrologic cycle as more latent energy in the form of water vapor is stored in the atmosphere. Global temperatures are better constrained by climate observations than most other climate variables, but changes in rainfall distributions could have far more impact than global temperature increases (Allen and Ingram, 2002). Increased precipitation, evapotranspiration, and runoff, and exacerbations of extreme hydroclimatic events represented by changes in the frequency and magnitude of floods and drought are likely hydrologic cycle responses to an altered climatic forcing (Schlosser and Houser, 2007). Virtually every challenging hydroclimate issue hinges on the behavior of the equator-to-pole temperature gradient and meridional energy transport by the atmosphere and oceans (Pierrehumbert, 2002). The atmospheric role is accomplished largely by the latent heat content of water vapor. Water fluxes and storage in the atmosphere emerge as crucial variables in tracing temperature and precipitation variability and changes at all scales. In addition, water vapor is the most important greenhouse gas. In light of these multiple contributions, efforts to understand the complex roles of water provided by examination of the climate system and the hydrologic cycle are a promising pursuit.

The evidence that climate change is having a detectable influence on hydroclimate is inconclusive, but the evidence is accumulating. Trends in annual and seasonal precipitation and runoff are small relative to year-to-year variability, and change is difficult to detect under these circumstances. Development of global hydroclimatic metrics will advance as globally complete data for key hydrologic cycle rates and storages are achieved by on-site observations and satellite-based measurements for areas where observations are sparse or impractical (Schlosser and Houser, 2007). These global metrics must embrace the periodic fluctuations in the climate system due to modes like ENSO, PDO, and NAO that result in different regions displaying different temperature and moisture responses over time. At the global scale, uncertainties in assessing perturbations in the climate system are due largely to an incomplete understanding of non-linear feedback loops and threshold effects in hydrologic cycle processes that influence the climate system in numerous and complex ways. The terrestrial branch of the hydrologic cycle amplifies climate inputs, and regional models suggest that even modest changes in climate can lead to changes in water availability beyond the range of historical variability.

The persistence of severe drought in Africa since the early 1970s and the recurrence of drought in Australia, Europe, and North America are snapshots of what may occur in future decades if climate model estimates of global change prove to be real. One area of agreement among the models is that reduced rainfall and increased evapotranspiration will produce increased drought in mid-latitudes during summer (Arnell, 1996). Unfortunately, too few studies have focused on whether the occurrence of drought and other extreme weather conditions have already changed significantly (Zierl and Lischke, 2002). An increase in flood frequency is a concern in an altered climate regime because a relatively small change in hydroclimatic characteristics can produce large changes in flood risk (Arnell, 1996). Neither Third World nor Developed World countries are immune from the hydroclimatic changes suggested by the computer models. The climate models display some agreement on the character of future temperature changes, but precipitation estimates vary widely at the regional scale which is important hydroclimatically. Evapotranspiration, soil moisture, and runoff are even more difficult to predict when the precipitation input is uncertain. The contribution hydroclimatology provides toward understanding the ramifications of these predicated changes for agriculture, industry, forestry, and the water supply is its emphasis on the coupling of the climate system and the hydrologic cycle, but the relevant atmosphere, ocean, cryosphere, and land surface processes are inherently unpredictable beyond some period due to the non-linearity of the feedbacks.

Uncertainty makes most scientists uneasy, but the elimination of uncertainty inherent in natural phenomena like the climate system and the hydrologic cycle is impossible. Uncertainty related to many measurement factors exists in all forms of observational hydroclimatic data (Thorne et al., 2005; Changnon and Kunkel, 2006), and computer simulations of climate change involve huge uncertainties whose sources are attributed to complex causes ranging from insufficient understanding of natural mechanisms to weaknesses in representing complex processes (Petersen, 2000; Koutsoyiannis et al., 2007). However, uncertainty from a scientific perspective can be conceptualized, estimated, and addressed. Developing additional datasets, performing more sophisticated analyses, and conducting additional research are all used by scientists to reduce, quantify, and assess uncertainty (Morss et al., 2005). The uncertainty is not reduced, but it is acknowledged and incorporated in the result. Nevertheless, water managers and other practitioners often view scientific uncertainty as a hindrance to achieving the most cost-effective solution to a problem. Communication with practitioners on scientific uncertainty has been hindered by a lack of formal declarations concerning detection and attribution of climate change against a background of natural climate variability (Risbey and Kandlikar, 2002).

Past hydroclimatic data are unlikely to represent future hydroclimatic conditions if anthropogenic climate change is significant compared to natural climate variability. The chaotic dynamical features of the climate system and the long-term persistence of hydroclimatic phenomena present considerable barriers in our efforts to accurately represent future hydroclimatic states. Nevertheless, the hydroclimatic focus on the interactive processes among the atmosphere, oceans, and land surfaces over all time scales is a productive path for shaping advanced understanding of the complex natural processes that constitute the climate system and the hydrologic cycle.

Review questions

10.1   In what ways are droughts commonly associated with atmospheric patterns characterized as being anomalous?

10.2   What is the relationship between drought, temperature, and relative humidity?

10.3   How are drought and aridity different hydroclimatic phenomena?

10.4   What are the similarities and differences between the Palmer Drought Severity Index and the Standardized Precipitation Index?

10.5   Why are tree rings a favored proxy data source for studying drought?

10.6   Why has it proven difficult to specifically identify the atmospheric circulation conditions responsible for causing drought in the Sahel region of West Africa?

10.7   How did an upper-air ridge contribute to the 1975–7 drought in the western United States and the 2003 drought in Europe? What conditions accounted for the presence of these ridges?

10.8   What atmospheric conditions favored development of the 1988 drought in the Great Plains and Midwest of the United States?

10.9   What is the nature of the relationship between drought and wildfires?

References

Adam, D. P. and West, G. J. (1983). Temperature and precipitation estimates through the last glacial cycle from Clear Lake, California, pollen data. *Science*, **219**, 168–70.

Adler, R. F., Huffman, G. J., Chang, A. *et al.* (2003). The version-2 Global Precipitation Climatology Project (GPCP) monthly precipitation analysis (1979-present). *J. Hydrometeor.*, **4**, 1147–67.

Adler, R. F., Kidd, C., Petty, G., Morissey, M. and Goodman, H. M. (2001). Intercomparison of global precipitation products: the third Precipitation Intercomparison Project (PIP-3). *Bull. Amer. Meteor. Soc.*, **82**, 1377–96.

Akan, A. O. and Houghtalen, R. J. (2003). *Urban Hydrology, Hydraulics, and Stormwater Quality: Engineering Applications and Computer Modeling*. Hoboken, NJ: John Wiley and Sons.

Allen, M. R. and Ingram, W. J. (2002). Constraints on future changes in climate and the hydrologic cycle. *Nature*, **419**, 224–31.

Allen, R. G. (2000). Using the FAO-56 dual crop coefficient method over an irrigated region as part of an evapotranspiration intercomparison study. *J. Hydrol.*, **229**, 27–41.

Allen, R. G., Pereira, L. S., Raes, D. and Smith, M. (1998). *Crop Evapotranspiration: Guidelines for Computing Crop Water Requirements*. FAO Irrigation and Drainage Paper 56. Rome: United Nations Food and Agriculture Organization.

Alley, W. M. (1984). The Palmer drought severity index: limitations and assumptions. *J. Climate Appl. Meteor.*, **23**, 1001–9.

Alley, W. M. (1985). The Palmer drought severity index as a measure of hydrologic drought. *Water Resour. Bull.*, **2**, 105–14.

Ambaum, M. H. P., Hoskins, B. J. and Stephenson, D. B. (2001). Arctic oscillation or North Atlantic oscillation? *J. Climate*, **14**, 3495–507.

Anctil, F. and Coulibaly, P. (2004). Wavelet analysis of the interannual variability in southern Québec streamflow. *J. Climate*, **17**, 163–70.

Anderton, S. P., White, S. M. and Alvera, B. (2004). Evaluation of spatial variability in snow water equivalent for a high mountain catchment. *Hydrol. Process.*, **18**, 435–53.

Andrews, D. G. (2000). *An Introduction to Atmospheric Physics*. Cambridge: Cambridge University Press.

Angell, J. K. (2003). Effect of exclusion of anomalous tropical stations on temperature trends from a 63-station radiosonde network, and comparison with other analyses. *J. Climate*, **16**, 2288–95.

Angevine, W. M., Senff, C. J. and Westwater, E. R. (2003). Boundary layers: observational techniques – remote. In *Encyclopedia of Atmospheric Sciences*, ed. J. R. Holton, J. A. Curry and J. A. Pyle. London: Academic Press, pp. 271–9.

Appenzeller, C., Stocker, T. F. and Anklin, M. (1998). North Atlantic oscillation dynamics recorded in Greenland ice cores. *Science*, **282**, 446–9.

Argall, P. S. and Sica, R. J. (2003). Lidar: atmospheric sounding introduction. In *Encyclopedia of Atmospheric Sciences*, ed. J. R. Holton, J. A. Curry and J. A. Pyle. London: Academic Press, pp. 1169–76.

Arnell, N. (1996). *Global Warming, River Flows and Water Resources*. Chichester: John Wiley and Sons.

Arnold, J. G., Srinivasan, R., Muttiah, R. S. and Allen, P. M. (1999). Continental scale simulation of the hydrologic balance. *J. Amer. Water Resour. Assoc.*, **35**, 1037–51.

Ba, M. B., Ellingson, R. G. and Gruber, A. (2003). Validation of a technique for estimating OLR with the GOES Sounder. *J. Atmos. Oceanic Technol.*, **20**, 79–89.

Bailey, H. P. and Johnson, C. W. (1972). Potential evapotranspiration in relation to annual wave of temperature. *Public. Climatol.*, **25**, 4–12.

Bair, E. S. (1995). Hydrogeology. In *Environmental Hydrology*, ed. A. D. Ward and W. J. Elliot. New York, NY: Lewis Publishers, pp. 285–310.

Baldocchi, D. D. (2003). Assessing the eddy covariance technique for evaluating carbon dioxide exchange rates of ecosystems: past, present and future. *Glob. Change Biol.*, **9**, 479–92.

Barani, G. A. and Khanjani, M. J. (2002). A large electronic weighing lysimeter system: design and installation. *J. Amer. Water Resour. Assoc.*, **38**, 1053–60.

Barcelo, A., Robert, R. and Coudray, J. (1997). A major rainfall event: the 27 February–5 March 1993 rains on the southeastern slope of Piton de la Fournaise Massif (Reunion Island, southwest Indian Ocean). *Mon. Wea. Rev.*, **125**, 3341–6.

Barnston, A. G. and Livezey, R. E. (1987). Classification, seasonality and persistence of low-frequency atmospheric circulation patterns. *Mon. Wea. Rev.*, **115**, 1083–1126.

Barry, R. G. (1969). The world hydrological cycle. In *Water, Earth, and Man: A Synthesis of Hydrology, Geomorphology, and Socio-Economic Geography*, ed. R. J. Chorley. London: Methuen and Co. Ltd.

Basist, A., Bell, G. D. and Meentemeyer, V. (1994). Statistical relationships between topography and precipitation patterns. *J. Climate*, **7**, 1305–15.

Baumgartner, A. and Reichel, E. (1975). *The World Water Balance: Mean Annual Global, Continental and Maritime Precipitation, Evaporation and Run-Off*. R. Lee (translator). Amsterdam: Elsevier Scientific Publishers.

Beaubien, D. J., Bisberg, A. and Beaubien, A. F. (1998). Investigations in pyranometer design. *J. Atmos. Oceanic Technol.*, **15**, 677–86.

Becker, A. (1995). Problems and progress in macroscale hydrological modelling. In *Space and Time Scale Variability and Interdependencies in Hydrological Processes*, ed. R. A. Feddes. Cambridge: Cambridge University Press, pp. 135–43.

Bell, G. D. and Janowiak, J. E. (1995). Atmospheric circulation associated with the midwest floods of 1993. *Bull. Amer. Meteor. Soc.*, **76**, 681–95.

Beringer, J. and Tapper, N. (2002). Surface energy exchanges and interactions with thunderstorms during the Maritime Continent Thunderstorm Experiment (MCTEX). *J. Geophys. Res.*, **107**, 4552, doi:10.1029/2001JD001431.

Beven, K. (1997). Topmodel: a critique. *Hydrol. Process.*, **11**, 1069–85.

Beyrich, F., De Bruin, H. A. R., Meijninger, W. M. L., Schipper, J. W. and Lohse, H. (2002). Results from one-year continuous operation of a large aperture scintillometer over a heterogeneous land surface. *Bound.-Layer Meteor.*, **105**, 85–97.

Bidlake, W. R. (2000). Evapotranspiration from a bulrush-dominated wetland in the Klamath Basin, Oregon. *J. Amer. Water Resour. Assoc.*, **36**, 1309–20.

Bigg, G. R., Jickells, T. D., Liss, P. S. and Osborn, T. J. (2003). The role of the oceans in climate. *Int. J. Climatol.*, **23**, 1127–59.

Biondi, F., Gershunov, A. and Cayan, D. R. (2001). North Pacific decadal climate variability since 1661. *J. Climate*, **14**, 5–10.

Biswas, A. K. (1970). *History of Hydrology*. Amsterdam: North-Holland Publishing Co.

Bjerknes, J. (1969). Atmospheric teleconnections from the equatorial Pacific. *Mon. Wea. Rev.*, **97**, 163–72.

Black, E., Blackburn, M., Harrison, G., Hoskins, B. and Methven, J. (2004). Factors contributing to the summer 2003 European heatwave. *Weather*, **59**, 217–23.

Blaney, H. F. and Criddle, W. D. (1950). *Determining Water Requirements in Irrigated Areas From Climatological and Irrigation Data*. U.S. Department of Agriculture, Soil Conservation Service, SCS-TP 96.

Blunier, T. and Brook, E. J. (2001). Timing of millennial-scale climate change in Antarctica and Greenland during the last glacial period. *Science*, **291**, 109–12.

Bond, N. A. and Vecchi, G. A. (2003). The influence of the Madden-Julian oscillation on precipitation in Oregon and Washington. *Wea. Forecast.*, **18**, 600–13.

Bonsal, B. R. and Lawford, R. G. (1999). Teleconnections between El Niño and La Niña events and summer extended dry spells on the Canadian prairies. *Int. J. Climatol.*, **19**, 1445–58.

Bosart, L. F. and Sanders, F. (1981). The Johnstown flood of July 1977: a long-lived convective system. *J. Atmos. Sci.*, **38**, 1616–42.

Bosilovich, M. G. and Schubert, S. D. (2002). Water vapor tracers as diagnostics of the regional hydrologic cycle. *J. Hydrometeor.*, **3**, 149–65.

Boucher, R. J. and Wieler, J. G. (1985). Radar determination of snowfall rate and accumulation. *J. Climate Appl. Meteor.*, **24**, 68–73.

Bouwer, L. M., Vermaat, J. E. and Aerts, J. C. J. H. (2006). Winter atmospheric circulation and river discharge in northwest Europe. *Geophys. Res. Lett.*, **33**, L06403, doi:10.1029/2005GL025548.

Bradley, E. F. (2003). Boundary layers: observational techniques in situ. In *Encyclopedia of Atmospheric Sciences*, ed. J. R. Holton, J. A. Curry and J. A. Pyle. London: Academic Press, pp. 280–90.

Bradley, R. S. (1999). *Paleoclimatology: Reconstructing Climates of the Quaternary*, 2nd edn. San Diego, CA: Harcourt/Academic Press.

Bradley, R. S. and Jones, P. D. (1995). Climate since A.D. 1500: introduction. In *Climate Since A.D. 1500*, ed. R. S. Bradley and P. D. Jones. London: Routledge, pp. 1–16.

Bras, R. L. (1990). *Hydrology: An Introduction to Hydrologic Science*. Reading, MA: Addison-Wesley Publishing Company.

Braud, I., Haverkamp, R., Arrúe, J. L. and López, M. V. (2003). Spatial variability of soil surface properties and consequences for the annual and monthly water balance of a semiarid environment (EFEDA experiment). *J. Hydrometeor.*, **4**, 121–37.

Brooks, H. E. and Stensrud, D. J. (2000). Climatology of heavy rain events in the United States from hourly precipitation observations. *Mon. Wea. Rev.*, **128**, 1194–1201.

Brotzge, J. A. and Duchon, C. E. (2000). A field comparison among a domeless net radiometer, two four-component net radiometers, and a domed net radiometer. *J. Atmos. Oceanic Technol.*, **17**, 1569–82.

Bruce, J. P. and Clark, R. H. (1966). *Introduction to Hydrometeorology*. Oxford: Pergamon Press.

Bryson, R. A. (1997). The paradigm of climatology: an essay. *Bull. Amer. Meteor. Soc.*, **78**, 449–55.

Burke, E. J., Brown, S. J. and Christidis, N. (2006). Modeling the recent evolution of global drought and projections for the twenty-first century with the Hadley Centre climate model. *J. Hydrometeor.*, **7**, 1113–25.

Burns, S. P., Sun, J., Delany, A. C. *et al.* (2003). A field intercomparison technique to improve the relative accuracy of longwave radiation measurements and an evaluation of CASES-99 pyrgeometer data quality. *J. Atmos. Oceanic Technol.*, **20**, 348–61.

Burroughs, W. J. (2001). *Climate Change: A Multidisciplinary Approach*. Cambridge: Cambridge University Press.

Byun, H. R. and Wilhite, D. A. (1999). Objective quantification of drought severity and duration. *J. Climate*, **12**, 2747–56.

Caillault, K. and Lemaitre, Y. (1999). Retrieval of three-dimensional wind fields corrected for the time-induced advection problem. *J. Atmos. Oceanic Technol.*, **16**, 708–22.

Calder, I. R. (1993). Hydrologic effects of land-use change. In *Handbook of Hydrology*, ed. D. R. Maidment. New York, NY: McGraw-Hill, Inc., pp. 13.1–13.50.

Calder, I. R., Hall, R. L., Harding R. J. and Wright, I. R. (1984). The use of a wet-surface weighing lysimeter system in rainfall interception studies of heather (*Calluna vulgaris*). *J. Climate Appl. Meteor.*, **23**, 461–73.

Caracena, F., Maddox, R. A., Hoxit, L. R. and Chappell, C. F. (1979). Mesoanalysis of the Big Thompson storm. *Mon. Wea. Rev.*, **107**, 1–17.

Cashion, J., Lakshmi, V., Bosch, D. and Jackson, T. J. (2005). Microwave remote sensing of soil moisture: evaluation of the TRMM microwave imager (TMI) satellite for the Little River Watershed Tifton, Georgia. *J. Hydrol.*, **307**, 242–53.

Castello, A. F. and Shelton, M. L. (2004). Winter precipitation on the US Pacific Coast and El Niño-Southern Oscillation events. *Int. J. Climatol.*, **24**, 481–97.

Castellvi, F. (2004). Combining surface renewal analysis and similarity theory: a new approach for estimating sensible heat flux. *Water Resour. Res.*, **40**, W05201, doi:10.1029/2003WR002677.

Catriona, M., Gardner, K., Bell, J. P. *et al.* (1991). Soil water content. In *Soil Analysis: Physical Methods*, ed. K. A. Smith and C. E. Mullins. New York, NY: Marcel Dekker, Inc., pp. 1–73.

Cayan, D. R. and Webb, R. H. (1992). El Niño/Southern Oscillation and streamflow in the western United States. In *El Niño: Historical and Paleoclimatic Aspects of the Southern Oscillation*, ed. H. F. Diaz and V. Markgraf. Cambridge: Cambridge University Press, pp. 29–68.

Chahine, M. T. (1992). The hydrological cycle and its influence on climate. *Nature*, **359**, 373–79.

Champollion, C., Masson, F., Van Baelen, J. *et al.* (2004). GPS monitoring of the tropospheric water vapor distribution and variation during the 9 September 2002 torrential precipitation episode in the Cévennes (southern France). *J. Geophys. Res.*, **109**, D24102, doi:10.1029/2004JD004897.

Chang, A. T. C., Foster, J. L. and Hall, D. K. (1987). Nimbus-7 derived global snow cover parameters. *Ann. Glaciol.*, **9**, 39–44.

Chang, F-C. and Smith, E. A. (2001). Hydrological and dynamical characteristics of summertime droughts over U.S. great plains. *J. Climate*, **14**, 2296–316.

Changnon, S. A. (1996). *The Great Flood of 1993: Causes, Impacts, and Responses*. Boulder, CO: Westview Press.

Changnon, S. A. and Kunkel, K. E. (2006). Changes in instruments and sites affecting historical weather records: a case study. *J. Atmos. Oceanic Technol.*, **23**, 825–8.

Chao, Y., Ghil, M. and McWilliams, J. C. (2000). Pacific interdecadal variability in this century's sea surface temperatures. *Geophys. Res. Lett.*, **27**, 2261–4.

Chen, F. W. and Staelin, D. H. (2003). AIRS/AMSU/HSB precipitation estimates. *IEEE Trans. Geosci. Remote Sens.*, **41**, 410–7.

Chen, J. and Kumar, P. (2002). Role of terrestrial hydrologic memory in modulating ENSO impacts in North America. *J. Climate*, **15**, 3569–85.

Chen, M., Xie, P. and Janowiak, J. E. (2002). Global land precipitation: a 50-year monthly analysis based on gauge observations. *J. Hydrometeor.*, **3**, 249–66.

Ciais, P., Reichstein, M., Viovy, N. *et al.* (2005). Europe-wide reduction in primary productivity caused by the heat and drought in 2003. *Nature*, **437**, 529–33.

Clark, M. P. and Vrugt, J. A. (2006). Unraveling uncertainties in hydrologic model calibration: addressing the problem of compensatory parameters. *Geophys. Res. Lett.*, **33**, L06406, doi:10.1029/2005GL025604.

Clark, W. C. (1985). Scales of climate impacts. *Climatic Change*, **7**, 5–27.

Coe, M. T. (2000). Modeling terrestrial hydrological systems at the continental scale: testing the accuracy of an atmospheric GCM. *J. Climate*, **13**, 686–704.

Coleman, J. S. M. and Rogers, J. C. (2003). Ohio River Valley winter moisture conditions associated with the Pacific-North American teleconnection pattern. *J. Climate*, **16**, 969–81.

Contini, D., Mastrantonio, G., Viola, A. and Argentini, S. (2004). Mean vertical motions in the PBL measured by Doppler sodar: accuracy, ambiguities and possible improvements. *J. Atmos. Oceanic Technol.*, **21**, 1532–44.

Cook, E. R., Meko, D. M., Stahle, D. W. and Cleaveland, M. K. (1999). Drought reconstructions for the continental United States. *J. Climate*, **12**, 1145–62.

Cosh, M. H., Stedinger, J. R. and Brutsaert, W. (2004). Variability of surface soil
    moisture at the watershed scale. *Water Resour. Res.*, **40**, W12513, doi:10.1029/
    2004WR003487.

Costa, J. E. (1987). Hydraulics and basin morphometry of the largest flash floods in
    the conterminous United States. *J. Hydrol.*, **93**, 313–38.

Coulibaly, P. and Burn, D. H. (2005). Spatial and temporal variability of Canadian
    seasonal streamflows. *J. Climate*, **18**, 191–210.

Crescenti, G. H. (1997). A look back on two decades of Doppler sodar comparison
    studies. *Bull. Amer. Meteor. Soc.*, **78**, 651–73.

Crimmins, M. A. (2006). Synoptic climatology of extreme fire-weather conditions
    across the southwest United States. *Int. J. Climatol.*, **26**, 1001–16.

Crow, W. T., Wood, E. F. and Pan, M. (2003). Multiobjective calibration of land surface
    model evapotranspiration predictions using streamflow observations and
    spaceborne surface radiometric temperature retrievals. *J. Geophys. Res.*, **108**, 4725,
    doi:10.1029/2002JD003292.

Crum, T. D. and Alberty, R. L. (1993). The WSR-88D and the WSR-88D operational
    support facility. *Bull. Amer. Meteor. Soc.*, **74**, 1669–87.

Crum, T. D., Alberty, R. L. and Burgess, D. W. (1993). Recording, archiving, and using
    WSR-88D data. *Bull. Amer. Meteor. Soc.*, **74**, 645–53.

Crum, T. D., Saffle, R. E. and Wilson, J. W. (1998). An update on the NEXRAD program
    and future WSR-88D support to operations. *Wea. Forecast.*, **13**, 253–62.

Cunderlik, J. M., Ouarda, T. B. M. J. and Bobée, B. (2004). On the objective identification
    of flood seasons. *Water Resour. Res.*, **40**, W01520, doi:10.1029/2003WR002295.

Cushman, J. H. (1986). On measurement, scale, and scaling. *Water Resour. Res.*, **22**, 129–34.

Czikowsky, M. J. and Fitzjarrald, D. R. (2004). Evidence of seasonal changes in
    evapotranspiration in eastern U.S. hydrological records. *J. Hydrometeor.*, **5**, 974–88.

Dabberdt, W. F., Shellhorn, R., Cole, H. *et al.* (2003). Radiosondes. In *Encyclopedia of
    Atmospheric Sciences*, ed. J. R. Holton, J. A. Curry and J. A. Pyle. London: Academic
    Press, pp. 1900–13.

Dai, A. and Trenberth, K. E. (2002). Estimates of freshwater discharge from continents:
    latitudinal and seasonal variations. *J. Hydrometeor.*, **3**, 660–87.

Dai, A., Fung, I. Y. and Del Genio, A. D. (1997). Surface observed global land
    precipitation variations during 1900–88. *J. Climate*, **10**, 2943–62.

Dai, A., Lamb, P. J., Trenberth, K. E. *et al.* (2004a). The recent Sahel drought is real. *Int.
    J. Climatol.*, **24**, 1323–31.

Dai, A., Trenberth, K. E. and Qian, T. (2004b). A global dataset of Palmer Drought
    Severity Index for 1870–2002: relationship with soil moisture and effects of
    surface warming. *J. Hydrometeor.*, **5**, 1117–30.

Dai, A., Wang, J., Ware, R. H. and Van Hove, T. (2002). Diurnal variation in water vapor
    over North America and its implications for sampling errors in radiosonde
    humidity. *J. Geophys. Res.*, **107**, 4090, doi:10.1029/2001JD000642.

Daly, C., Gibson, W. P., Taylor, G. H., Johnson, G. L. and Pasteris, P. (2002). A
    knowledge-based approach to the statistical mapping of climate. *Climate Res.*, **22**,
    99–113.

Davi, N. K., Jacoby, G. C., Curtis, A. E., Baatarbileg, N. (2006). Extension of drought records for central Asia using tree rings: west-central Mongolia. *J. Climate*, **19**, 288–99.

De Bruin, H. (2002). Introduction: renaissance of scintillometry. *Bound.-Layer Meteor.*, **105**, 1–4.

DeFelice, T. P. (1998). *An Introduction to Meteorological Instrumentation and Measurement*. Upper Saddle River, NJ: Prentice Hall.

Delrieu, G., Ducrocq, V., Gaume, E. *et al.* (2005). The catastrophic flash-flood event of 8–9 September 2002 in the Gard region, France: a first case study for the Cévennes-Vivarais Mediterranean Hydrometeorological Observatory. *J. Hydrometeor.*, **6**, 34–52.

Deser, C. and Wallace, J. W. (1990). Large scale atmospheric circulation features of warm and cold episodes in the tropical Pacific. *J. Climate*, **3**, 1254–81.

Dettinger, M. D. and Diaz, H. F. (2000). Global characteristics of stream flow seasonality and variability. *J. Hydrometeor.*, **1**, 289–310.

Dingman, S. L. (1994). *Physical Hydrology*. Englewood Cliffs, NJ: Prentice Hall.

Dirmeyer, P. A. and Brubaker, K. L. (2006). Evidence for trends in the Northern Hemisphere water cycle. *Geophys. Res. Lett.*, **33**, L14712, doi:10.1029/2006GL026359.

Dirmeyer, P. A., Guo, Z. and Gao, X. (2004). Comparison, validation, and transferability of eight multiyear global soil wetness products. *J. Hydrometeor.*, **5**, 1011–33.

Dooge, J. C. I. (1995). Scale problems in surface fluxes. In *Space and Time Scale Variability and Interdependencies in Hydrological Processes*, ed. R. A. Feddes. Cambridge: Cambridge University Press, pp. 21–32.

Doorenbos, J. and Pruitt, W. O. (1977). *Guidelines for Predicting Crop Water Requirements*. FAO Irrigation and Drainage Paper 24. Rome: United Nations Food and Agriculture Organization.

Doswell III, C. A., Brooks, H. E. and Maddox, R. A. (1996). Flash flood forecasting: an ingredients-based methodology. *Wea. Forecast.*, **11**, 560–81.

Douglas, E. M. and Barros, A. P. (2003). Probable maximum precipitation estimation using multifractals: application in the eastern United States. *J. Hydrometeor.*, **4**, 1012–24.

Doviak, R. J. and Doviak, M. E. F. (2003). Radar: Doppler radar. In *Encyclopedia of Atmospheric Sciences*, ed. J. R. Holton, J. A. Curry and J. A. Pyle. London: Academic Press, pp. 1802–12.

Dracup, J. A. and Kendall, D. R. (1990). Floods and droughts. In *Climate Change and U.S. Water Resources*, ed. P. E. Waggoner. New York, NY: John Wiley and Sons, pp. 243–67.

Duchon, C. E. and Essenberg, G. R. (2001). Comparative rainfall observations from pit and aboveground rain gauges with and without wind shields. *Water Resour. Res.*, **37**, 3253–63.

Durre, I. and Wallace, J. M. (2001). Factors influencing the cold-season diurnal temperature range in the United States. *J. Climate*, **14**, 3263–78.

Durre, I., Wallace, J. M. and Lettenmaier, D. P. (2000). Dependence of extreme daily maximum temperature on antecedent soil moisture in the contiguous United States during summer. *J. Climate*, **13**, 2641–51.

Eagleson, P. S. (1978). Climate, soil, and vegetation. 6. Dynamics of the annual water balance. *Water Resour. Res.*, **14**, 749–64.

Eagleson, P. S. (1994). The evolution of modern hydrology (from watershed to continent in 30 years). *Adv. Water Resour.*, **17**, 3–18.

Easterling, D. R., Evans, J. L., Groisman, P. Y. *et al.* (2000). Observed variability and trends in extreme climate events: a brief review. *Bull. Amer. Meteor. Soc.*, **81**, 417–25.

Easterling, D. R., Horton, B., Jones, P. D. *et al.* (1997). Maximum and minimum temperature trends for the globe. *Science*, **277**, 364–7.

Ebbesmeyer, C. C., Cayan, D. R., McLain, D. R. *et al.* (1991). 1976 step in the Pacific climate: forty environmental changes between 1968–1975 and 1977–1984. In *Proceedings of the Seventh Annual Pacific Climate (PACLIM) Workshop*, ed. J. L. Betancourt and V. L. Tharp. Sacramento: California Department of Water Resources, Interagency Ecological Studies Program Technical Report 26, pp. 115–26.

Elliot, W. J. (1995). Precipitation. In *Environmental Hydrology*, ed. A. D. Ward and W. J. Elliot. New York, NY: Lewis Publishers, pp. 19–46.

Elliott, W. P., Ross, R. J. and Blackmore, W. H. (2002). Recent changes in NWS upper-air observations with emphasis on changes from VIZ to Vaisala radiosondes. *Bull. Amer. Meteor. Soc.*, **83**, 1003–17.

Eltahir, E. A. B. and Gong, C. (1996). Dynamics of wet and dry years in West Africa. *J. Climate*, **9**, 1030–42.

Engman, E. T. (1993). Remote sensing. In *Handbook of Hydrology*, ed. D. R. Maidment. New York, NY: McGraw-Hill, Inc., pp. 24.1–24.23.

Entekhabi, D., Asrar, G. R., Betts, A. K. *et al.* (1999). An agenda for land surface hydrology research and a call for the second international hydrological decade. *Bull. Amer. Meteor. Soc.*, **80**, 2043–58.

Eskridge, R. E., Alduchov, O. A., Chernykh, I. V. *et al.* (1995). A comprehensive aerological reference data set (CARDS): rough and systematic errors. *Bull. Amer. Meteor. Soc.*, **76**, 1759–75.

Eskridge, R. E., Luers, J. K. and Redder, C. R. (2003). Unexplained discontinuity in the U. S. radiosonde temperature data. Part I: Troposphere. *J. Climate*, **16**, 2385–95.

Esper, J., Cook, E. R. and Schweingruber, F. H. (2002). Low-frequency signals in long tree-ring chronologies for reconstructing past temperature variability. *Science*, **295**, 2250–3.

Fattorelli, S., Dalla Fontana, G. and Da Ros, D. (1999). Flood hazard assessment and mitigation. In *Floods and Landslides: Integrated Risk Assessment*, ed. R. Casale and C. Margottini. Berlin: Springer, pp. 19–38.

Feddes, R. A. (1995). Remote sensing–inverse modeling approach to determine large scale effective soil hydraulic properties in soil-vegetation-atmosphere systems. In *Space and Time Scale Variability and Interdependencies in Hydrological Processes*, ed. R. A. Feddes. Cambridge: Cambridge University Press, pp. 33–42.

Fekete, B. M., Vörösmarty, C. J., Roads, J. O. and Willmott, C. J. (2004). Uncertainties in precipitation and their impacts on runoff estimates. *J. Climate*, **17**, 294–304.

Feng, X. and Epstein, S. (1994). Climatic implications of the 8000-year hydrogen isotope time series from bristlecone pine trees. *Science*, **265**, 1079–81.

Ferraris, L., Rudari, R. and Siccardi, F. (2002). The uncertainty in the prediction of flash floods in the northern Mediterranean environment. *J. Hydrometeor.*, **3**, 714–27.

Ferraro, R. R. (1997). Special sensor microwave imager derived global rainfall estimates for climatological applications. *J. Geophys. Res.*, **102**, 16715–35.

Ferraro, R. R., Weng, F., Grody, N. C. and Basist, A. (1996). An eight-year (1987-1994) time series of rainfall, clouds, water vapor, snow cover, and sea ice derived from SSM/I measurements. *Bull. Amer. Meteor. Soc.*, **77**, 891–905.

Ferraro, R. R., Weng, F., Grody, N. C. *et al.* (2005). NOAA operational hydrological products derived from the advanced microwave sounding unit. *IEEE Trans. Geosci. Remote Sens.*, **43**, 1036–49.

Fink, A. H., Brücher, T., Krüger, A. *et al.* (2004). The 2003 European summer heatwaves and drought – synoptic diagnosis and impacts. *Weather*, **59**, 209–16.

Foster, J. L., Sun, C., Walker, J. P. *et al.* (2005). Quantifying the uncertainty in passive microwave snow water equivalent observations. *Remote Sens. Environ.*, **94**, 187–203.

Fowler, H. J. and Kilsby, C. G. (2003). A regional frequency analysis of United Kingdom extreme rainfall from 1961 to 2000. *Int. J. Climatol.*, **23**, 1313–34.

Francois, C., Quesney, A. and Ottlé, C. (2003). Sequential assimilation of ERS-1 SAR data into a coupled land surface-hydrological model using an extended Kalman filter. *J. Hydrometeor.*, **4**, 473–87.

Free, M., Durre, I., Aguilar, E. *et al.* (2002). Creating climate reference datasets: CARDS workshop on adjusting radiosonde temperature data for climate monitoring. *Bull. Amer. Meteor. Soc.*, **83**, 891–9.

Fritsch, J. M. and Carbone, R. E. (2004). Improving quantitative precipitation forecasts in the warm season: a USWRP research and development strategy. *Bull. Amer. Meteor. Soc.*, **85**, 955–65.

Fritts, H. C. (1976). *Tree Rings and Climate*. London: Academic Press.

Fritts, H. C. (1991). *Reconstructing Large-Scale Climatic Patterns from Tree-Ring Data: A Diagnostic Analysis*. Tucson, AZ: The University of Arizona Press.

Fritts, H. C. and Shao, X. M. (1995). Mapping climate using tree-rings from western North America. In *Climate Since A.D. 1500*, ed. R. S. Bradley and P. D. Jones. London: Routledge, pp. 269–95.

Fye, F. K. and Cleaveland, M. K. (2001). Paleoclimatic analyses of tree-ring reconstructed summer drought in the United States, 1700–1978. *Tree-Ring Res.*, **57**, 31–44.

Gay, L. W. (1979). Radiation budgets of desert, meadow, forest, and marsh sites. *Archiv. Meteor. Geophys. Bioklim. B*, **27**, 349–59.

Gedalof, Z., Peterson, D. L. and Mantua, N. J. (2004). Columbia River flow and drought since 1750. *J. Amer. Water Resour. Assoc.*, **40**, 1579–92.

Georgakakos, K. P., Tsintikidis, D., Attia, B. and Roskar, J. (2001). Estimation of pixel-scale daily rainfall over Nile River catchments using multi-spectral METEOSAT data. In *Remote Sensing and Hydrology 2000, IAHS Publication No. 267*, ed. M. Owe, K. Brubaker, J. Ritchie and A. Rango. Wallingford, UK: IAHS Press, pp. 11–5.

Gershunov, A. and Barnett, T. P. (1998). Interdecadal modulation of ENSO teleconnections. *Bull. Amer. Meteor. Soc.*, **79**, 2715–25.

Giannoni, F., Smith, J. A., Zhang, Y. and Roth, G. (2003). Hydrologic modeling of extreme floods using radar rainfall estimates. *Adv. Water Resour.*, **26**, 195–203.

Gillespie, A., Rokugawa, S., Matsunaga, T. *et al.* (1998). A temperature and emissivity separation algorithm for advanced spaceborne thermal emission and reflection radiometer (ASTER) images. *IEEE Trans. Geosci. Remote Sens.*, **36**, 1113–26.

Goody, R. M. and Yung, Y. L. (1989). *Atmospheric Radiation: Theoretical Basis*, 2nd edn. New York, NY: Oxford University Press.

Graham, N. E. and Diaz, H. F. (2001). Evidence for intensification of North Pacific winter cyclones since 1948. *Bull. Amer. Meteor. Soc.*, **82**, 1869–93.

Granger, R. J. (2000). Satellite-derived estimates of evapotranspiration in the Gediz basin. *J. Hydrol.*, **229**, 70–6.

Grayson, R. B., Blöschl, G., Western, A. W. and McMahon, T. A. (2002). Advances in the use of observed spatial patterns of catchment hydrological response. *Adv. Water Resour.*, **25**, 1313–34.

Green, A. E., Green, S. R., Astill, M. S. and Caspari, H. W. (2000). Estimating latent heat flux from a vineyard using scintillometry. *TAO*, **11**, 525–42.

Green, W. H. and Ampt, G. A. (1911). Studies in soil physics, part 1. The flow of air and water through soils. *J. Agric. Sci.*, **4**, 1–24.

Greene, J. S. and Morrissey, M. L. (2000). Validation and uncertainty analysis of satellite rainfall algorithms. *Prof. Geogr.*, **52**, 247–58.

Grist, J. P. and Nicholson, S. E. (2001). A study of the dynamic factors influencing the rainfall variability in the West Africa Sahel. *J. Climate*, **14**, 1337–59.

Groisman, P. Y. and Easterling, D. A. (1995). Variability and trends of precipitation and snowfall over North America. In *Natural Climate Variability on Decade-to-Century Time Scales*, ed. D. G. Martinson, K. Bryan, M. Ghil *et al.* Washington, D.C.: National Academy Press, pp. 67–79.

Groisman, P. Y. and Legates, D. R. (1994). The accuracy of United States precipitation data. *Bull. Amer. Meteor. Soc.*, **75**, 215–27.

Grotjahn, R. (2003). General circulation: mean characteristics. In *Encyclopedia of Atmospheric Sciences*, ed. J. R. Holton, J. A. Curry and J. A. Pyle. London: Academic Press, pp. 841–54.

Gruntfest, E. (2000). Nonstructural mitigation of flood hazards. In *Inland Flood Hazards: Human, Riparian, and Aquatic Communities*, ed. E. E. Wohl. New York, NY: Cambridge University Press, pp. 394–410.

Gualdi, S., Navarra, A. and Tinarelli, G. (1999). The interannual variability of the Madden-Julian oscillation in an ensemble of GCM simulations. *Climate Dyn.*, **15**, 643–58.

Gupta, S. K., Kratz, D. P., Wilber, A. C. and Nguyen, L. C. (2004). Validation of parameterized algorithms used to derive TRMM-CERES surface radiative fluxes. *J. Atmos. Oceanic Technol.*, **21**, 742–52.

Gutman, G., Csiszar, I. and Romanov, P. (2000). Using NOAA/AVHRR products to monitor El Niño impacts: focus on Indonesia in 1997–98. *Bull. Amer. Meteor. Soc.*, **81**, 1189–1205.

Guttman, N. B. (1998). Comparing the Palmer drought index and the standardized precipitation index. *J. Amer. Water Resour. Assoc.*, **34**, 113–21.

Guttman, N. B., Hosking, J. R. M. and Wallis, J. R. (1994). The 1993 Midwest extreme precipitation in historical and probabilistic perspective. *Bull. Amer. Meteor. Soc.*, **75**, 1785–92.

Guyot, G. (1998). *Physics of the Environment and Climate*. Chichester: John Wiley and Sons–Praxis Publishing.

Haigh, J. D. (1996). The impact of solar variability on climate. *Science*, **272**, 981–4.

Hall, D. K., Riggs, G. A., Salomonson, V. V. and Scharfen G. R. (2001). Earth Observing System (EOS) Moderate Resolution Imaging Spectroradiometer (MODIS) global snow-cover maps. In *Remote Sensing and Hydrology 2000 IAHS Publication No. 267*, ed. M. Owe, K. Brubaker, J. Ritchie and A. Rango. Wallingford, UK: IAHS Press, pp. 55–60.

Hannaford, J. and Marsh, T. (2006). An assessment of trends in UK runoff and low flows using a network of undisturbed catchments. *Int. J. Climatol.*, **26**, 1237–53.

Hanson, R. L. (1991). Evapotranspiration and droughts. In *National Water Summary 1988–89 – Hydrologic Events and Floods and Droughts*. U.S. Geological Survey Water-Supply Paper 2375. Washington, D.C.: U.S. Government Printing Office, pp. 99–104.

Hare, F. K. (1987). Drought and desiccation: twin hazards of a variable climate. In *Planning for Drought: Toward a Reduction of Societal Vulnerability*, ed. D. A. Wilhite, W. E. Easterling and D. A. Wood. Boulder, CO: Westview Press, pp. 3–9.

Hare, S. R. and Mantua, N. J. (2000). Empirical evidence for North Pacific regime shifts in 1977 and 1989. *Prog. Oceanogr.*, **47**, 103–45.

Harries, J. E. (2003). Satellite remote sensing: water vapor. In *Encyclopedia of Atmospheric Sciences*, ed. J. R. Holton, J. A. Curry and J. A. Pyle. London: Academic Press, pp. 2005–12.

Hartmann, D. L. (1994). *Global Physical Climatology*. San Diego, CA: Academic Press.

Haug, G. H., Hughen, K. A., Sigman, D. M., Peterson, L. C. and Röhl, U. (2001). Southward migration of the intertropical convergence zone through the holocene. *Science*, **293**, 1304–8.

Hayden, B. P. (1988). Flood climates. In *Flood Geomorphology*, ed. V. R. Baker, R. C. Kochel, and P. C. Patton. New York, NY: John Wiley and Sons, pp. 13–26.

Hayes, M. J., Svoboda, M. D., Wilhite, D. A. and Vanyarkho, O. V. (1999). Monitoring the 1996 drought using the standardized precipitation index. *Bull. Amer. Meteor. Soc.*, **80**, 429–38.

Hayes, P. S., Rasmussen, L. A. and Conway, H. (2002). Estimating precipitation in the central Cascades of Washington. *J. Hydrometeor.*, **3**, 335–46.

Heidinger, A. K. and Stephens, G. L. (2000). Molecular line absorption in a scattering atmosphere. Part II: application to remote sensing in the O_2 A band. *J. Atmos. Sci.*, **57**, 1615–34.

Heim, R. R. (2002). A review of twentieth-century drought indices used in the United States. *Bull. Amer. Meteor. Soc.*, **83**, 1149–65.

Herschy, R. W. (1997). Streamflow measurement for the 21st century. In *Land Surface Processes in Hydrology: Trials and Tribulations of Modeling and Measuring*, ed. S. Sorooshian, H. V. Gupta and J. C. Rodda. NATO ASI Series, I 46. Berlin: Springer-Verlag, pp. 389–421.

Hidalgo, H. G. (2004). Climate precursors of multidecadal drought variability in the western United States. *Water Resour. Res.*, **40**, W12504, doi:10.1029/2004WR003350.

Hidalgo, H. G. and Dracup, J. A. (2003). ENSO and PDO effects on hydroclimatic variations of the Upper Colorado River basin. *J. Hydrometeor.*, **4**, 5–23.

Higgins, R. W., Leetmaa, A. and Kousky, V. E. (2002). Relationships between climate variability and winter temperature extremes in the United States. *J. Climate*, **15**, 1555–72.

Hillel, D. (2004). *Introduction to Environmental Soil Physics*. San Diego, CA: Elsevier Academic Press.

Hirschboeck, K. K. (1988). Flood hydroclimatology. In *Flood Geomorphology*, ed. V. R. Baker, R. C. Kochel and P. C. Patton. New York, NY: John Wiley and Sons, pp. 27–49.

Hirschboeck, K. K. (1991). Climate and floods. In *National Water Summary 1988–89 – Hydrologic Events and Floods and Droughts*. U.S. Geological Survey Water-Supply Paper 2375. Washington, D.C.: U.S. Government Printing Office, pp. 67–88.

Hirschboeck, K. K., Ely, L. L. and Maddox, R. A. (2000). Hydroclimatology of meteorologic floods. In *Inland Flood Hazards: Human, Riparian, and Aquatic Communities*, ed. E. E. Wohl. New York, NY: Cambridge University Press, pp. 39–72.

Hisdal, H., Stahl, D., Tallaksen, L. M. and Demuth, S. (2001). Have streamflow droughts in Europe become more severe or frequent? *Int. J. Climatol.*, **21**, 317–33.

Hodgkins, G. A. and Dudley, R. W. (2006). Changes in the timing of winter-spring streamflows in eastern North America, 1913–2002. *Geophys. Res. Lett.*, **33**, L06402, doi:10.1029/2005GL025593.

Hoerling, M. and Kumar, A. (2003). The perfect ocean for drought. *Science*, **299**, 691–4.

Hoerling, M. P., Hurrell, J. W. and Xu, T. (2001). Tropical origins for recent North Atlantic climate change. *Science*, **292**, 90–2.

Holmes, R. R. (1996). *Sediment Transport in the Lower and Central Mississippi Rivers, June 26 Through September 14, 1993*. U.S. Geological Survey Circular 1120-I. Washington, D.C.: U.S. Government Printing Office.

Horton, R. E. (1940). An approach toward a physical interpretation of infiltration-capacity. *J. Soil Sci. Amer.*, **5**, 399–417.

Hoyt, D. V. and Schatten, K. H. (1997). *The Role of the Sun in Climate Change*. New York, NY: Oxford University Press.

Hu, Q. and Willson, G. D. (2000). Effects of temperature anomalies on the Palmer drought severity index in the central United States. *Int. J. Climatol.*, **20**, 1899–1911.

Hu, Q., Woodruff, C. M. and Mudrick, S. E. (1998). Interdecadal variations of annual precipitation in the central United States. *Bull. Amer. Meteor. Soc.*, **79**, 221–9.

Hudson, S. R., Town, M. S., Walden, V. P. and Warren, S. G. (2004). Temperature, humidity, and pressure response of radiosondes at low temperatures. *J. Atmos. Oceanic Technol.*, **21**, 825–36.

Huffman, G. J., Adler, R. F., Arkin, P. *et al.* (1997). The Global Precipitation Climatology Project (GPCP) combined precipitation dataset. *Bull. Amer. Meteor. Soc.*, **78**, 5–20.

Huffman, G. J., Adler, R. F., Morrissey, M. M. *et al.* (2001). Global precipitation at one-degree daily resolution from multisatellite observations. *J. Hydrometeor.*, **2**, 36–50.

Huffman, G. J., Adler, R. F., Rudolf, B., Schneider, U. and Keehn, P. R. (1995). Global precipitation estimates based on a technique for combining satellite-based estimates, rain gauge analysis, and NWP model precipitation information. *J. Climate*, **8**, 1284–95.

Hughes, M. (1995). Dendroclimatic evidence from the western Himalaya. In *Climate Since A.D. 1500*, ed. R. S. Bradley and P. D. Jones. London: Routledge, pp. 415–31.

Hughes, M. (1996). Tree-ring analysis. In *Encyclopedia of Climate and Weather*, ed. S. H. Schneider. New York, NY: Oxford University Press, pp. 773–5.

Hughes, M. and Graumlich, L. J. (1996). Multimillennial dendroclimatic studies from the western United States. In *Climatic Variations and Forcing Mechanisms of the Last 2000 Years*, ed. P. D. Jones, R. S. Bradley and J. Jouzel. Berlin: Springer-Verlag, pp. 109–24.

Hughes, M., Kelly, P. M., Pilcher, J. and LaMarche, V. C. Jr., ed. (1981). *Climate from Tree Rings*. Cambridge: Cambridge University Press.

Humes, K. S., Kustas, W. P. and Moran, M. S. (1994). Use of remote sensing and reference site measurements to estimate instantaneous surface energy balance components over a semiarid rangeland watershed. *Water Resour. Res.*, **30**, 1363–73.

Hunrichs, R. A. (1991). California floods and drought. In *National Water Summary 1988–89 – Hydrologic Events and Floods and Droughts*. U.S. Geological Survey Water-Supply Paper 2375. Washington, D.C.: U.S. Government Printing Office, pp. 197–206.

Hunt, B. G. (2001). A description of persistent climatic anomalies in a 1000-year climatic model simulation. *Climate Dyn.*, **17**, 717–33.

Hurrell, J. W. (1995). Decadal trends in the North Atlantic oscillation: regional temperatures and precipitation. *Science*, **269**, 676–9.

Hurrell, J. W. and van Loon, H. (1997). Decadal variations in climate associated with the North Atlantic oscillation. *Climatic Change*, **36**, 301–26.

Hurrell, J. W., Brown, S. J., Trenberth, K. E. and Christy, J. R. (2000). Comparison of tropospheric temperatures from radiosondes and satellites: 1979–98. *Bull. Amer. Meteor. Soc.*, **81**, 2165–77.

Hurrell, J. W., Kushnir, Y. and Visbeck, M. (2001). The North Atlantic oscillation. *Science*, **291**, 603–5.

Intergovernmental Panel on Climate Change (IPCC) (2001). *Climate Change 2001: The Scientific Basis. Contribution of Working Group I to the Third Assessment Report of the Intergovernmental Panel on Climate Change*, ed. J. T. Houghton, Y. Ding, D. J. Griggs, *et al.* Cambridge: Cambridge University Press.

International Commission on Large Dams (ICOLD). (2003). *World Registry of Dams*. Paris: ICOLD.

Jackson, T. J., Hsu, A. Y. and O'Neill, P. E. (2002). Surface soil moisture retrieval and mapping using high-frequency microwave satellite observations in the southern Great Plains. *J. Hydrometeor.*, **3**, 688–99.

Jacobowitz, H., Stowe, L. L., Ohring, G. *et al.* (2003). The advanced very high resolution radiometer Pathfinder atmosphere (PATMOS) climate dataset: a resource for climate research. *Bull. Amer. Meteor. Soc.*, **84**, 785–93.

Jacobs, J. M., Myers, D. A. and Whitfield, B. M. (2003). Improved rainfall/runoff estimates using remotely sensed soil moisture. *J. Amer. Water Resour. Assoc.*, **39**, 313–24.

Jain, S. and Lall, U. (2001). Floods in a changing climate: does the past represent the future? *Water Resour. Res.*, **37**, 3193–3205.

Jens, S. W. and McPherson, M. B. (1964). Hydrology and urban areas. In *Handbook of Applied Hydrology: A Compendium of Water-Resources Technology*, ed. V. T. Chow. New York, NY: McGraw-Hill, Section 20, pp. 1–45.

Jensen, M. E., Burman, R. D. and Allen, R. G. ed. (1990). *Evaporation and Irrigation Water Requirements*. New York, NY: American Society of Civil Engineers.

Jha, M., Pan, Z., Takle, E. S. and Gu, R. (2004). Impacts of climate change on streamflow in the Upper Mississippi River Basin: a regional climate model perspective. *J. Geophys. Res.*, **109**, D09105, doi:10.1029/2003JD003686.

Jin, M. (2004). Analysis of land skin temperature using AVHRR observations. *Bull. Amer. Meteor. Soc.*, **85**, 587–600.

Johnson, R. S., Williams, L. E., Ayars, J. E. and Trout, T. J. (2005). Weighing lysimeters aid study of water relations in tree and vine crops. *Calif. Agric.*, **59**, 133–6.

Jones, C. (2000). Occurrence of extreme precipitation events in California and relationships with the Madden-Julian oscillation. *J. Climate*, **13**, 3576–87.

Jones, J. A. A. (1997). *Global Hydrology: Processes, Resources and Environmental Management*. Harlow: Longman.

Jones, P. D. and Bradley, R. S. (1995). Climatic variations in the longest instrumental records. In *Climate Since A.D. 1500*, ed. R. S. Bradley and P. D. Jones. London: Routledge, pp. 246–68.

Jones, R. N. and Pittock, A. B. (2002). Climate change and water resources in an arid continent: managing uncertainty and risk in Australia. In *Climatic Change: Implications for the Hydrological Cycle and for Water Management*, ed. M. Beniston. Dordrecht: Kluwer Academic Publishers, pp. 465–501.

Justice, C. O., Townshend, J. R. G., Vermote, E. F. *et al.* (2002). An overview of MODIS land data processing and product status. *Remote Sens. Environ.*, **83**, 3–15.

Kalnay, E., Kanamitsu, M., Kistler, R. *et al.* (1996). The NCEP/NCAR 40-year reanalysis project. *Bull. Amer. Meteor. Soc.*, **77**, 437–71.

Kanamitsu, M., Ebisuzaki, W., Woollen, J. *et al.* (2002). NCEP-DOE AMIP-II reanalysis (R-2). *Bull. Amer. Meteor. Soc.*, **83**, 1631–43.

Kane, R. P. (1997). Relationship of El Niño-Southern oscillation and Pacific sea surface temperature with rainfall in various regions of the globe. *Mon. Wea. Rev.*, **125**, 1792–1800.

Karl, T. R. (1983). Some spatial characteristics of drought duration in the United States. *J. Climate Appl. Meteor.*, **22**, 1356–66.

Karl, T. R. (1986a). The sensitivity of the Palmer drought severity index and Palmer's z-index to their calibration coefficients including potential evapotranspiration. *J. Climate Appl. Meteor.*, **25**, 77–86.

Karl, T. R. (1986b). The relationship of soil moisture parameterizations to subsequent seasonal and monthly mean temperatures in the United States. *Mon. Wea. Rev.*, **114**, 675–86.

Karl, T. R., Jones, P. D., Knight, R. W. *et al.* (1993). A new perspective on recent global warming: asymmetric trends of daily maximum and minimum temperature. *Bull. Amer. Meteor. Soc.*, **74**, 1007–23.

Karl, T. R., Quinlan, F. and Ezell, D. S. (1987). Drought termination and amelioration: its climatological probability. *J. Climate Appl. Meteor.*, **26**, 1198–1209.

Katz, R. W., Parlange, M. B. and Naveau, P. (2002). Statistics of extremes in hydrology. *Adv. Water Resour.*, **25**, 1287–1304.

Kelly, R. E., Chang, A. T., Tsang, L. and Foster, J. L. (2003). A prototype AMSR-E global snow area and snow depth algorithm. *IEEE Trans. Geosci. Remote Sens.*, **41**, 230–42.

Kemball-Cook, S. R. and Weare, B. C. (2001). The onset of convection in the Madden-Julian oscillation. *J. Climate*, **14**, 780–93.

Kidd, C., Kniveton, D. R., Todd, M. C. and Bellerby, T. J. (2003). Satellite rainfall estimation using combined passive microwave and infrared algorithms. *J. Hydrometeor.*, **4**, 1088–1104.

Kidder, S. Q. (2003). Satellites: orbits. In *Encyclopedia of Atmospheric Sciences*, ed. J. R. Holton, J. A. Curry and J. A. Pyle. London: Academic Press, pp. 2024–38.

Kiehl, J. T. and Trenberth, K. E. (1997). Earth's annual global mean energy budget. *Bull. Amer. Meteor. Soc.*, **78**, 197–208.

Kiem, A. S. and Franks, S. W. (2001). On the identification of ENSO-induced rainfall and runoff variability: a comparison of methods and indices. *Hydrol. Sci. J.*, **46**, 715–27.

Kilmartin, R. F. (1980). Hydroclimatology – a needed cross-discipline. In *Improved Hydrologic Forecasting – Why and How*, ed. W. P. Henry. New York, NY: American Society of Civil Engineers, pp. 160–98.

Kite, G. W. and Droogers, P. (2000). Comparing evapotranspiration estimates from satellites, hydrological models and field data. *J. Hydrol.*, **229**, 3–18.

Klazura, G. E. and Imy, D. A. (1993). A description of the initial set of analysis products available from the NEXRAD WSR-88D system. *Bull. Amer. Meteor. Soc.*, **74**, 1293–1311.

Klemeš, V. (1987). Drought prediction: a hydrological perspective. In *Planning for Drought: Toward a Reduction of Societal Vulnerability*, ed. D. A. Wilhite, W. E. Easterling and D. A. Wood. Boulder, CO: Westview Press, pp. 81–94.

Knapp, P. A., Soulé, P. T. and Grissino-Mayer, H. D. (2004). Occurrence of sustained droughts in the interior Pacific Northwest (A.D. 1733–1980) inferred from tree-ring data. *J. Climate*, **17**, 140–50.

Knippertz, P., Ulbrich, U., Marques, F. and Corte-Real, J. (2003). Decadal changes in the link between El Niño and springtime North Atlantic oscillation and European-North African rainfall. *Int. J. Climatol.*, **23**, 1293–311.

Knox, J. C. (1988). Climatic influence on upper Mississippi valley floods. In *Flood Geomorphology*, ed. V. R. Baker, R. C. Kochel and P. C. Patton. New York, NY: John Wiley and Sons, pp. 279–300.

Koellner, W. H. (1996). The flood's hydrology. In *The Great Flood of 1993: Causes, Impacts, and Responses*, ed. S. A. Changnon. Boulder, CO: Westview Press, pp. 68–100.

Kohsiek, W., Meijninger, W. M. L., Moene, A. F. *et al.* (2002). An extra large aperture scintillometer for long range applications. *Bound.-Layer Meteor.*, **105**, 119–27.

Kongoli, C., Pellegrino, P., Ferraro, R. R., Grody, N. C. and Meng, H. (2003). A new snowfall detection algorithm over land using measurements from the Advanced Microwave Sounding Unit (AMSU). *Geophys. Res. Lett.*, **30**, 1756, doi:10.1029/2003GL017177.

Konrad, C. E. (1998). Intramonthly indices of the Pacific/North American teleconnection pattern and temperature regimes over the United States. *Theor. Appl. Climatol.*, **60**, 11–19.

Koutsoyiannis, D., Efstratiadis, A. and Georgakakos, K. P. (2007). Uncertainty assessment of future hydroclimatic predictions: a comparison of probabilistic and scenario-based approaches. *J. Hydrometeor.*, **8**, 261–81.

Krajewski, W. F. and Smith, J. A. (2002). Radar hydrology: rainfall estimation. *Adv. Water Resour.*, **25**, 1387–94.

Kummerow, C., Simpson, J., Thiele, O. *et al.* (2000). The status of the tropical rainfall measuring mission (TRMM) after two years in orbit. *J. Appl. Meteor.*, **39**, 1965–82.

Kump, L. R. (2002). Reducing uncertainty about carbon dioxide as a climate driver. *Nature*, **419**, 188–90.

Kundzewicz, Z. W. (2002). Floods in the context of climate change and variability. In *Climatic Change: Implications for the Hydrological Cycle and for Water Management*, ed. M. Beniston. Dordrecht: Kluwer Academic Publishers, pp. 225–47.

Kunkel, K. E. (1996). A hydroclimatological assessment of the rainfall. In *The Great Flood of 1993: Causes, Impacts, and Responses*, ed. S. A. Changnon. Boulder, CO: Westview Press, pp. 52–67.

Kunkel, K. E., Changnon, S. A. and Angel, J. R. (1994). Climatic aspects of the 1993 upper Mississippi River basin flood. *Bull. Amer. Meteor. Soc.*, **75**, 811–22.

Kustas, W. P. and Norman, J. M. (1996). Use of remote sensing for evapotranspiration monitoring over land surfaces. *Hydrol. Sci. J.*, **41**, 495–516.

Kustas, W. P., Albertson, J. D., Scanlon, T. M. and Cahill, A. T. (2001). Issues in monitoring evapotranspiration with radiometric temperature observations. In *Remote Sensing and Hydrology 2000, IAHS Publication No. 267*, ed. M. Owe, K. Brubaker, J. Ritchie and A. Rango. Wallingford, UK: IAHS Press, pp. 239–45.

Ladurie, E. L. and Baulant, M. (1980). Grape harvests from the fifteenth through the nineteenth centuries. *J. Interdis. Hist.*, **10**, 839–49.

Laing, A. G. (2003). Mesoscale meteorology: mesoscale convective systems. In *Encyclopedia of Atmospheric Sciences*, ed. J. R. Holton, J. A. Curry and J. A. Pyle. London: Academic Press, pp. 1251–61.

Lakshmi, V., Wood, E. F. and Choudhury, B. J. (1997a). A soil-canopy-atmosphere model for use in satellite microwave remote sensing. *J. Geophys. Res.*, **102**, 6911–27.

Lakshmi, V., Wood, E. F. and Choudhury, B. J. (1997b). Evaluation of special sensor microwave/imager satellite data for regional soil moisture estimation over the Red River Basin. *J. Appl. Meteor.*, **36**, 1309–28.

Lamb, H. H. (1982). *Climate, History and the Modern World*. London: Methuen and Co. Ltd.

Langbein, W. G. (1967). Hydroclimate. In *The Encyclopedia of Atmospheric Sciences and Astrogeology*, ed. R. W. Fairbridge. New York, NY: Reinhold, pp. 447–51.

Langenberg, H. (2002). Climate and water. *Nature*, **419**, 187.

LaPenta, K. D., McNaught, B. J., Capriola, S. J. *et al.* (1995). The challenge of forecasting heavy rain and flooding throughout the eastern region of the national weather service. Part 1: Characteristics and events. *Wea. Forecast.*, **10**, 78–90.

Larson, L. W., Ferral, R. L., Strem, E. T. *et al.* (1995). Operational responsibilities of the National Weather Service river and flood program. *Wea. Forecast.*, **10**, 465–76.

Lawford, R. G., Stewart, R., Roads, J. *et al.* (2004). Advancing global- and continental-scale hydrometeorology: contributions of GEWEX hydrometeorology panel. *Bull. Amer. Meteor. Soc.*, **85**, 1917–30.

Lean, J. and Rind, D. (2001). Earth's response to a variable sun. *Science*, **292**, 234–6.

Leathers, D. J. and Palecki, M. A. (1992). The Pacific/North American teleconnection pattern and United States climate. Part II: Temporal characteristics and index specification. *J. Climate*, **5**, 707–16.

Leathers, D. J., Yarnal, B. and Palecki, M. A. (1991). The Pacific/North American teleconnection pattern and United States climate. Part I: Regional temperature and precipitation associations. *J. Climate*, **4**, 517–28.

Leconte, R., Brissette, F., Galarneau, M. and Rousselle, J. (2004). Mapping near-surface soil moisture with RADARSAT-1 synthetic aperture radar data. *Water Resour. Res* **40**, W01515, doi:10.1029/2003WR002312.

Lee, S., Klein, A. and Over, T. (2004). Effects of the El Niño-Southern Oscillation on temperature, precipitation, snow water equivalent and resulting streamflow in the Upper Rio Grande river basin. *Hydrol. Process.*, **18**, 1053–71.

Leese, J., Williams, S., Jenne, R. and Ritchie, A. (2003). Data collection and management for Global Energy and Water Cycle Experiment (GEWEX) Continental-Scale International Project (GCIP). *J. Geophys. Res.*, **108**, 8620, doi:10.1029/2002JD003196.

Legates, D. R. (2000a). Real-time calibration of radar precipitation estimates. *Prof. Geogr.*, **52**, 235–46.

Legates, D. R. (2000b). Remote sensing in hydroclimatology: an introduction. *Prof. Geogr.*, **52**, 233–4.

Legates, D. R. and Willmott, C. J. (1990). Mean seasonal and spatial variability in gauge-corrected, global precipitation. *Int. J. Climatol.*, **10**, 111–27.

Lehner, B. and Döll, P. (2004). Development and validation of a global database of lakes, reservoirs and wetlands. *J. Hydrol.*, **296**, 1–22.

LeMone, M. A., Grossman, R. L., Coulter, R. L. *et al.* (2000). Land-atmosphere interaction research, early results, and opportunities in the Walnut River Watershed in southeast Kansas: CASES and ABLE. *Bull. Amer. Meteor. Soc.*, **81**, 757–79.

LeQuesne, C., Stahle, D. W., Cleaveland, M. K. *et al.* (2006). Ancient *Austrocedrus* tree-ring chronologies used to reconstruct central Chile precipitation variability from A.D. 1200 to 2000. *J. Climate*, **19**, 5731–44.

Li, J., Menzel, W. P., Sun, F., Schmit, T. J. and Gurka, J. (2004a). AIRS subpixel cloud characterization using MODIS cloud products. *J. Appl. Meteor.*, **43**, 1083–94.

Li, J., Menzel, W. P., Zhang, W. *et al.* (2004b). Synergistic use of MODIS and AIRS in a variational retrieval of cloud parameters. *J. Appl. Meteor.*, **43**, 1619–34.

Linacre, E. (1992). *Climate Data and Resources: A Reference and Guide*. London: Routledge.

Linacre, E. and Geerts, B. (1997). *Climates and Weather Explained*. London: Routledge.

Lins, H. F. and Slack, J. R. (1999). Streamflow trends in the United States. *Geophys. Res. Lett.*, **26**, 227–30.

Linsley, B. K., Wellington, G. M. and Schrag, D. P. (2000). Decadal sea surface temperature variability in the subtropical South Pacific from 1726 to 1997 A.D. *Science*, **290**, 1145–48.

Liou, Y-C. (2002). An explanation of the wind speed underestimation obtained from a least squares type single-Doppler radar velocity retrieval method. *J. Appl. Meteor.*, **41**, 811–23.

Liu, A. Z., Ting, M. and Wang, H. (1998). Maintenance of circulation anomalies during the 1988 drought and 1993 floods over the United States. *J. Atmos. Sci.*, **55**, 2810–32.

Liu, G. (2003). Satellite remote sensing: precipitation. In *Encyclopedia of Atmospheric Sciences*, ed. J. R. Holton, J. A. Curry and J. A. Pyle. London: Academic Press, pp. 1972–9.

Liu, J., Chen, J. M. and Cihlar, J. (2003). Mapping evapotranspiration based on remote sensing: an application to Canada's landmass. *Water Resour. Res.*, **39**, 1189, doi:10.1029/2002WR001680.

Liu, Q. and Weng, F. (2005). One-dimensional variational retrieval algorithm of temperature, water vapor, and cloud water profiles from advanced microwave sounding unit (AMSU). *IEEE Trans. Geosci. Remote Sens.*, **43**, 1087–95.

Liu, W. T. (2003). Satellite remote sensing: surface wind. In *Encyclopedia of Atmospheric Sciences*, ed. J. R. Holton, J. A. Curry and J. A. Pyle. London: Academic Press, pp. 1979–85.

Lloyd-Hughes, B. and Saunders, M. A. (2002). A drought climatology for Europe. *Int. J. Climatol.*, **22**, 1571–92.

Loáiciga, H. A. (2005). On the probability of droughts: the compound renewal model. *Water Resour. Res.*, **41**, W01009, doi:10.1029/2004WR003075.

Loáiciga, H. A., Haston, L. and Michaelsen, J. (1993). Dendrohydrology and long-term hydrologic phenomena. *Rev. Geophys.*, **31**, 151–71.

Lockwood, J. G. (2001). Abrupt and sudden climatic transitions and fluctuations: a review. *Int. J. Climatol.*, **21**, 1153–79.

Lohmann, D., Raschke, E., Nijssen, B. and Lettenmaier, D. P. (1998). Regional scale hydrology: I. Formulation of the VIC-2L model coupled to a routing model. *Hydrol. Sci. J.*, **43**, 131–41.

Long, M., Entekhabi, D. and Nicholson, S. E. (2000). Interannual variability in rainfall, water vapor flux, and vertical motion over West Africa. *J. Climate*, **13**, 3827–41.

Lorenz, D. J. and Hartmann, D. L. (2006). The effect of the MJO on the North American monsoon. *J. Climate*, **19**, 333–43.

Lott, J. N. (1994). The US summer of 1993: a sharp contrast in weather extremes. *Weather*, **49**, 370–83.

Luce, C. H. (1995). Forests and wetlands. In *Environmental Hydrology*, ed. A. D. Ward and W. J. Elliot. New York, NY: Lewis Publishers, pp. 253–83.

Luers, J. K. and Eskridge, R. E. (1998). Use of radiosonde temperature data in climate studies. *J. Climate*, **11**, 1002–19.

Lull, H. W. (1964). Ecological and silviculture aspects. In *Handbook of Applied Hydrology: A Compendium of Water-Resources Technology*, ed. V. T. Chow. New York, NY: McGraw-Hill, Section 6, pp. 1–30.

Lundberg, A., Eriksson, M., Halldin, I. S., Kellner, E. and Seibert, J. (1997). New approach to the measurement of interception evaporation. *J. Atmos. Oceanic Technol.*, **14**, 1023–35.

Lydolph, P. E. (1985). *The Climate of the Earth*. Totowa, NJ: Rowman and Allanheld Publishers.

Lyon, B. (2003). Enhanced seasonal rainfall in northern Venezuela and the extreme events of December 1999. *J. Climate*, **16**, 2302–6.

Madden, R. A. and Julian, P. R. (1994). Observations of the 40–60 day tropical oscillation – a review. *Mon. Wea. Rev.*, **122**, 814–35.

Maddox, R. A., Canova, F. and Hoxit, L. R. (1980). Meteorological characteristics of flash flood events over the western United States. *Mon. Wea. Rev.*, **108**, 1866–77.

Maddox, R. A., Chappell, C. F. and Hoxit, L. R. (1979). Synoptic and meso-scale aspects of flash flood events. *Bull. Amer. Meteor. Soc.*, **60**, 115–23.

Maddox, R. A., Hoxit, L. R., Chappell, C. F. and Caracena, F. (1978). Comparison of meteorological aspects of the Big Thompson and Rapid City flash floods. *Mon. Wea. Rev.*, **106**, 375–89.

Maidment, D. R., ed. (1993). *Handbook of Hydrology*. New York, NY: McGraw-Hill, Inc.

Mancini, M., Hoeben, R. and Troch, P. A. (1999). Multifrequency radar observations of bare surface soil moisture content: a laboratory experiment. *Water Resour. Res.*, **35**, 1827–38.

Mansell, M. G. (2003). *Rural and Urban Hydrology*. London: Thomas Telford Publishing.

Mantua, N. J., Hare, S. R., Zhang, Y., Wallace, J. M. and Francis, R. C. (1997). A Pacific interdecadal climate oscillation with impacts on salmon production. *Bull. Amer. Meteor. Soc.*, **78**, 1069–79.

Mao, L. M., Bergman, M. J. and Tai, C. C. (2002). Evapotranspiration measurement and estimation of three wetland environments in the upper St. Johns River Basin, Florida. *J. Amer. Water Resour. Assoc.*, **38**, 1271–85.

Marks, D. and Winstral, A. (2001). Comparison of snow deposition, the snow cover energy balance, and snowmelt at two sites in a semiarid mountain basin. *J. Hydrometeor.*, **2**, 213–27.

Mather, J. R. (1991). A history of hydroclimatology. *Phys. Geogr.*, **12**, 260–73.

Mather, J. R. and Ambroziak, R. A. (1986). A search for understanding potential evapotranspiration. *Geogr. Rev.*, **76**, 355–70.

Mather, J. R. and Sdasyuk, G. V., ed. (1991). *Global Change: Geographical Approaches.* Tucson, AZ: University of Arizona Press.

Matthai, H. F. (1990). Floods. In *Surface Water Hydrology*, ed. M. G. Wolman and H. C. Riggs. The Geology of North America, Vol. O-1. Boulder, CO: Geological Society of America, pp. 97–120.

Matrosov, S. Y. (1998). A dual-wavelength radar method to measure snowfall rate. *J. Appl. Meteor.*, **37**, 1510–21.

Mays, L. W. (2005). *Water Resources Engineering.* New York, NY: John Wiley and Sons.

McCabe, G. J. and Dettinger, M. D. (1999). Decadal variations in the strength of ENSO teleconnections with precipitation in the western United States. *Int. J. Climatol.*, **19**, 1399–1410.

McCabe, G. J. and Dettinger, M. D. (2002). Primary modes and predictability of year-to-year snowpack variations in the western United States from teleconnections with Pacific Ocean climate. *J. Hydrometeor.*, **3**, 13–25.

McCabe, G. J., Barker, J. L. and Chase, E. B. (1991). Review of water year 1988 hydrologic conditions and water-related events. In *National Water Summary 1988–89 – Hydrologic Events and Floods and Droughts*. U.S. Geological Survey Water-Supply Paper 2375. Washington, D.C.: U.S. Government Printing Office, pp. 14–18.

McClelland, J. W., Déry, S. J., Peterson, B. J., Holmes, R. M. and Wood, E. F. (2006). A pan-arctic evaluation of changes in river discharge during the latter half of the 20th century. *Geophys. Res. Lett.*, **33**, L06715, doi:10.1029/2006GL025753.

McCuen, R. H. (2005) *Hydrologic Analysis and Design.* 3rd edn. Upper Saddle River, NJ: Pearson Prentice Hall.

McKee, T. B., Doesken, N. J. and Kleist, J. (1993). The relationship of drought frequency and duration to times scales. In *Proceedings Eighth Conference on Applied Climatology.* Boston, MA: American Meteorological Society, pp. 179–84.

McKenney, M. S. and Rosenberg, N. J. (1993). Sensitivity of some potential evapotranspiration estimation methods to climate change. *Agric. Forest Meteor.*, **64**, 81–110.

McNab, A. L. and Karl, T. R. (1991). Climate and droughts. In *National Water Summary 1988–89 – Hydrologic Events and Floods and Droughts.* U.S. Geological Survey Water-Supply Paper 2375. Washington, D.C.: U.S. Government Printing Office, pp. 89–98.

Meehl, G. A., Karl, T., Easterling, D. R. *et al.* (2000). An introduction to trends in extreme weather and climate events: observations, socioeconomic impacts, terrestrial ecological impacts, and model projections. *Bull. Amer. Meteor. Soc.*, **81**, 413–6.

Meier, M. F. (1990). Snow and ice. In *Surface Water Hydrology*, ed. M. G. Wolman and H. C. Riggs. The Geology of North America, Vol. 0-1. Boulder, CO: Geological Society of America, pp. 131–58.

Meijninger, W. M. L. and De Bruin, H. A. R. (2000). The sensible heat fluxes over irrigated areas in western Turkey determined with a large aperture scintillometer. *J. Hydrol.*, **229**, 42–9.

Meijninger, W. M. L., Green, A. E., Hartogensis, O. K. *et al.* (2002). Determination of area-averaged water vapour fluxes with large aperture and radio wave scintillometers over a heterogeneous surface – Flevoland field experiment. *Bound.-Layer Meteor.*, **105**, 63–83.

Meko, D. M., Therrell, M. D., Baisan, C. H. and Hughes, M. K. (2001). Sacramento River flow reconstructed to A.D. 869 from tree rings. *J. Amer. Water Resour. Assoc.*, **37**, 1029–39.

Melack, J. M. (1992). Reciprocal interactions among lakes, large rivers, and climate. In *Global Climate Change and Freshwater Resources*, ed. P. Firth and S. G. Fisher. New York, NY: Springer-Verlag, pp. 68–87.

Mendoza, B., Velasco, V. and Jáuregui, E. (2006). A study of historical droughts in southeastern Mexico. *J. Climate*, **19**, 2916–34.

Menzel, L., Niehoff, D., Bürger, G. and Bronstert, A. (2002). Climate change impacts on river flooding: a modeling study of three meso-scale catchments. In *Climatic Change: Implications for the Hydrological Cycle and for Water Management*, ed. M. Beniston. Dordrecht: Kluwer Academic Publishers, pp. 249–69.

Menzel, W. P. (2001). Cloud tracking with satellite imagery: from the pioneering work of Ted Fujita to the present. *Bull. Amer. Meteor. Soc.*, **82**, 33–47.

Menzel, W. P. and Purdom, J. F. W. (1994). Introducing GOES-I: the first of a new generation of geostationary operational environmental satellites. *Bull. Amer. Meteor. Soc.*, **75**, 757–81.

Menzel, W. P., Holt, F. C., Schmit, T. J. *et al.* (1998). Application of GOES-8/9 soundings to weather forecasting and nowcasting. *Bull. Amer. Meteor. Soc.*, **79**, 2059–77.

Merz, R. and Blöschl, G. (2003). A process typology of regional floods. *Water Resour. Res.*, **39**, 1340, doi:10.1029/2002WR001952.

Michel, C., Andréassian, V. and Perrin, C. (2005). Soil conservation service curve number method: how to mend a wrong soil moisture accounting procedure? *Water Resour. Res.*, **41**, W02011, doi:10.1029/2004WR003191.

Miller, A. J. and Schneider, N. (2000). Interdecadal climate regime dynamics in the North Pacific Ocean: theories, observations and ecosystem impacts. *Progr. Oceanogr.*, **47**, 355–79.

Miller, A. J., Cayan, D. R., Barnett, T. P., Graham, N. E. and Oberhuber, J. M. (1994). The 1976–77 climate shift of the Pacific Ocean. *Oceanogr.*, **7**, 21–6.

Miller, D. H. (1977). *Water at the Surface of the Earth: An Introduction to Ecosystem Hydrodynamics*. New York, NY: Academic Press.

Miller, N. L., King, A. W., Miller, M. A. *et al.* (2005). The DOE Water Cycle Pilot Study. *Bull. Amer. Meteor. Soc.*, **86**, 359–74.

Milly, P. C. D., Wetherald, R. T., Dunne, K. A. and Delworth T. L. (2002). Increasing risk of great floods in a changing climate. *Nature*, **415**, 514–7.

Miloshevich, L. M., Paukkunen, A., Vömel, H. and Oltmans, S.J. (2004). Development and validation of a time-lag correction for Vaisala radiosonde humidity measurements. *J. Atmos. Oceanic Technol.*, **21**, 1305–27.

Mo, K. C., Nogues-Paegle, J. and Higgins, R. W. (1997). Atmospheric processes associated with summer floods and droughts in the central United States. *J. Climate*, **10**, 3028–46.

Mo, K. C., Nogues-Paegle, J. and Paegle, J. (1995). Physical mechanisms of the 1993 summer floods. *J. Atmos. Sci.*, **52**, 879–95.

Mo, R., Fyfe, J. and Derome, J. (1998). Phase-locked and asymmetric correlations of the wintertime atmospheric patterns with the ENSO. *Atmos.-Ocean*, **36**, 213–39.

Mo, X., Liu, S., Lin, Z. and Zhao, W. (2004). Simulating temporal and spatial variation of evapotranspiration over the Lushi basin. *J. Hydrol.*, **285**, 125–42.

Mokhov, I. I., Khvorostyanov, D. V. and Eliseev, A. V. (2004). Decadal and longer term changes in El Niño-Southern Oscillation characteristics. *Int. J. Climatol.*, **24**, 401–14.

Montaldo, N., Mancini, M. and Rosso, R. (2004). Flood hydrograph attenuation induced by a reservoir system: analysis with a distributed rainfall-runoff model. *Hydrol. Process.*, **18**, 545–63.

Monteith, J. L. (1965). Evaporation and environment. In *The State and Movement of Water in Living Organisms*, 19th Symposium of the Society for Experimental Biology. London: Cambridge University Press, pp. 205–34.

Moody, J. A. (1995). *Propagation and Composition of the Flood Wave on the Upper Mississippi River Basin, 1993*. U.S. Geological Survey Circular 1120-F. Washington, D.C.: U.S. Government Printing Office.

Moran, M. S., Kustas, W. P., Vidal, A. *et al.* (1994). Use of ground-based remotely sensed data for surface energy balance evaluation of a semiarid rangeland. *Water Resour. Res.*, **30**, 1339–49.

Morss, R. E., Wilhelmi, O. V., Downton, M. W. and Gruntfest, E. (2005). Flood risk, uncertainty, and scientific information for decision making: lessons from an interdisciplinary project. *Bull. Amer. Meteor. Soc.*, **86**, 1593–1601.

Mosley, M. P. and McKerchar, A. I. (1993). Streamflow. In *Handbook of Hydrology*, ed. D. R. Maidment. New York, NY: McGraw-Hill, Inc., pp. 8.1–8.39.

Mossin, L. and Ladekarl, U. L. (2004). Simple water balance modelling with few data – calibration and evaluation: investigations from a Danish Sitka spruce stand with a high interception loss. *Nordic Hydrol.*, **35**, 139–51.

Nace, R. L. (1974). General evolution of the concept of the hydrological cycle. In *Three Centuries of Scientific Hydrology*. Paris: UNESCO–WMO–IAHS, pp. 40–8.

Nair, U. S., Hjelmfelt, M. R. and Pielke Sr., R. A. (1997). Numerical simulation of the 9–10 June 1972 Black Hills storm using CSU RAMS. *Mon. Wea. Rev.*, **125**, 1753–66.

Namias, J. (1978). Multiple causes of the North American abnormal winter 1976–77. *Mon. Wea. Rev.*, **106**, 279–95.

Namias, J. (1983). Some causes of United States drought. *J. Climate Appl. Meteor.*, **22**, 30–9.

Namias, J. (1991). Spring and summer 1988 drought over the contiguous United States – causes and prediction. *J. Climate*, **4**, 54–65.

National Oceanic and Atmospheric Administration (NOAA). (1994). *The Great Flood of 1993. Natural Disaster Survey Report*. U.S. Department of Commerce. Washington, D.C.: U.S. Government Printing Office.

Neary, V. S., Habib, E. and Fleming, M. (2004). Hydrologic modeling with NEXRAD precipitation in middle Tennessee. *J. Hydrol. Eng.*, **9**, 339–49.

New, M., Todd, M., Hulme, M. and Jones, P. (2001). Precipitation measurements and trends in the twentieth century. *Int. J. Climatol.*, **21**, 1899–1922.

Nicholson, S. E. (1993). An overview of African rainfall fluctuations of the last decade. *J. Climate*, **6**, 1463–6.

Nicholson, S. E., Some, B. and Kone, B. (2000). An analysis of recent rainfall conditions in west Africa, including the rainy seasons of the 1997 El Niño and the 1998 La Niña years. *J. Climate*, **13**, 2628–40.

Nicholson, S. E., Tucker, C. J. and Ba, M. B. (1998). Desertification, drought, and surface vegetation: an example from the west African Sahel. *Bull. Amer. Meteor. Soc.*, **79**, 815–29.

Niemann, J. D. and Eltahir, E. A. B. (2004). Prediction of regional water balance components based on climate, soil, and vegetation parameters, with application to the Illinois River Basin. *Water Resour. Res.*, **40**, W03103, doi:10.1029/2003WR002806.

Nigam, S., Barlow, M. and Berbery, E. H. (1999). Analysis links Pacific decadal variability to drought and streamflow in United States. *Eos, Trans. Amer. Geophys. Un.*, **80**, 621.

Nissen, R., Hudak, D., Laroche, S. *et al.* (2001). 3D wind field retrieval applied to snow events using Doppler radar. *J. Atmos. Oceanic Technol.*, **18**, 348–62.

Njoku, E. G., Jackson, T. J., Lakshmi, V., Chan, T. K. and Nghiem, S. V. (2003). Soil moisture retrieval from AMSR-E. *IEEE Trans. Geosci. Remote Sens.*, **41**, 215–29.

Nokes, S. E. (1995). Evapotranspiration. In *Environmental Hydrology*, ed. A. D. Ward and W. J. Elliot. New York, NY: Lewis Publishers, pp. 91–131.

Norman, J. M., Anderson, M. C., Kustas, W. P. *et al.* (2003). Remote sensing of surface energy fluxes at 10^1-m pixel resolutions. *Water Resour. Res.*, **39**, 1221, doi:10.1029/2002WR001775.

Ntale, H. K. and Gan, T. Y. (2003). Drought indices and their application to East Africa. *Int. J. Climatol.*, **23**, 1335–57.

O'Connor, J. E. and Costa, J. E. (2004a). Spatial distribution of the largest rainfall-runoff floods from basins between 2.6 and 26,000 km² in the United States and Puerto Rico. *Water Resour. Res.*, **40**, W01107, doi:10.1029/2003WR002247.

O'Connor, J. E. and Costa, J. E. (2004b). *The World's Largest Floods, Past and Present: Their Causes and Magnitudes*. U.S. Geological Survey Circular 1254. Washington, D.C.: U.S. Government Printing Office.

Ogden, F. L., Sharif, H. O., Senarath, S. U. S. *et al.* (2000). Hydrologic analysis of the Fort Collins, Colorado, flash flood of 1997. *J. Hydrol.*, **228**, 82–100.

Ohmura, A. (2001). Physical basis for the temperature-based melt-index method. *J. Appl. Meteor.*, **40**, 753–61.

Ohmura, A., Dutton, E. G., Forgan, B. *et al.* (1998). Baseline surface radiation network (BSRN/WCRP): new precision radiometry for climate research. *Bull. Amer. Meteor. Soc.*, **79**, 2115–36.

Oke, T. R. (1987). *Boundary Layer Climates*. 2nd edn. London: Methuen and Company.

Oki, T. (1999). The global water cycle. In *Global Energy and Water Cycles*, ed. K. A. Browning and R. J. Gurney. Cambridge: Cambridge University Press, pp. 10–27.

Oudin, L., Hervieu, F., Michel, C., Perrin, C. *et al.* (2005). Which potential evapotranspiration input for a lumped rainfall-runoff model? Part 2 – Towards a simple and efficient potential evapotranspiration model for rainfall-runoff modelling. *J. Hydrol.*, **303**, 290–306.

Owe, M., Brubaker, K., Ritchie, J. and Rango, A., ed. (2001). *Remote Sensing and Hydrology 2000*. IAHS Publication No. 267. Wallingford, UK: International Association of Hydrological Sciences.

Päätalo, M-L. (1998). Factors influencing occurrence and impacts of fires in northern European forests. *Silva Fennica*, **32**, 185–202.

Paegle, J., Mo, K. C. and Nogues-Paegle, J. (1996). Dependence of simulated precipitation on surface evaporation during the 1993 United States summer floods. *Mon. Wea. Rev.*, **124**, 345–61.

Pal, J. S., Giorgi, F. and Bi, X. (2004). Consistency of recent European summer precipitation trends and extremes with future regional climate projections. *Geophys. Res. Lett.*, **31**, L13202, doi:10.1029/2004GLO19836.

Palmer, W. C. (1965). *Meteorological Drought: Its Measurement and Description*. U.S. Weather Bureau, Research Paper No. 45. Washington, D.C.: U.S. Department of Commerce.

Parker, D. E., Alexander, L. V. and Kennedy, J. (2004). Global and regional climate in 2003. *Weather*, **59**, 145–52.

Parker, D. E., Basnett, T. A., Brown, S. J. *et al.* (2000). Climate observations – the instrumental record. *Space Sci. Rev.*, **94**, 309–20.

Parkinson, C. L. (2003). Aqua: an earth-observing satellite mission to examine water and other climate variables. *IEEE Trans. Geosci. Remote Sens.*, **41**, 173–83.

Parlange, M. B., Eichinger, W. E. and Albertson, J. D. (1995). Regional scale evaporation and the atmospheric boundary layer. *Rev. Geophys.*, **33**, 99–124.

Parrett, C., Melcher, N. B. and James, R. W. (1993). *Flood Discharges in the Upper Mississippi River Basin, 1993*. U.S. Geological Survey Circular 1120-A. Washington, D.C.: U.S. Government Printing Office.

Paulson, R. W., Chase, E. B., Roberts, R. S. and Moody, D. W., comp. (1991). *National Water Summary 1988–89 – Hydrologic Events and Floods and Droughts*. U.S. Geological Survey Water-Supply Paper 2375. Washington, D.C.: U.S. Government Printing Office.

Paw U, K. T., Qiu, J., Su, H. B., Watanabe, T. and Brunet, Y. (1995). Surface renewal analysis: a new method to obtain scalar fluxes. *Agric. Forest Meteor.*, **74**, 119–37.

Paw U, K. T., Wharton, S. and Kochendorfer, J. (2004). Evapotranspiration: measuring and modeling. *Acta Hort.*, **664**, 537–54.

Pazwash, H. and Mavrigian, G. (1981). Millennial celebration of Karaji's hydrology. *Proc. ASCE, J. Hydraulics Div.*, **107**, 303–9.

Peck, E. L. (1997). Quality of hydrometeorological data in cold regions. *J. Amer. Water Resour. Assoc.*, **33**, 125–34.

Pederson, N., Jacoby, G. C., D'Arrigo, R. D. *et al.* (2001). Hydrometeorological reconstructions for northeastern Mongolia derived from tree rings: 1651–1995. *J. Climate*, **14**, 872–81.

Peel, M. C., McMahon, T. A. and Finlayson, B. L. (2002). Variability of annual precipitation and its relationship to the El Niño-Southern Oscillation. *J. Climate*, **15**, 545–51.

Peel, M. C., McMahon, T. A., Finlayson, B. L. and Watson, F. G. R. (2001). Identification and explanation of continental differences in the variability of annual runoff. *J. Hydrol.*, **250**, 224–40.

Peixoto, J. P. (1995). The role of the atmosphere in the water cycle. In *The Role of Water and the Hydrological Cycle in Global Change*, ed. H. R. Oliver and S. A. Oliver. Berlin: Springer-Verlag, pp. 199–252.

Peixoto, J. P. and Oort, A. H. (1992). *Physics of Climate*. New York, NY: American Institute of Physics.

Penman, H. L. (1948). Natural evaporation from open water, bare soil and grass. *Proc. Roy. Soc. A*, **193**, 120–45.

Pérez, I. A., García, M. A., Sánchez, M. L. and De Torre, B. (2004). Autocorrelation analysis of meteorological data from a RASS sodar. *J. Appl. Meteor.*, **43**, 1213–23.

Perry, C. A. (1994). *Effects of Reservoirs on Flood Discharges in the Kansas and the Missouri River Basins, 1993*. U.S. Geological Survey Circular 1120-E. Washington, D.C.: U.S. Government Printing Office.

Peters, G., Fischer, B. and Kirtzel, H. J. (1998). One-year operational measurements with a sonic anemometer-thermometer and a Doppler sodar. *J. Atmos. Oceanic Technol.*, **15**, 18–28.

Petersen, A. C. (2000). Philosophy of climate science. *Bull. Amer. Meteor. Soc.*, **81**, 265–71.

Petersen, W. A., Carey, L. D., Rutledge, S. A. *et al.* (1999). Mesoscale and radar observations of the Fort Collins flash flood of 28 July 1997. *Bull. Amer. Meteor. Soc.*, **80**, 191–216.

Peterson, B. J., Holmes, R. M., McClelland, J. W. *et al.* (2002). Increasing river discharge to the Arctic Ocean. *Science*, **298**, 2171–3.

Peterson, T., Daan, H. and Jones, P. (1997). Initial selection of a GCOS surface network. *Bull. Amer. Meteor. Soc.*, **78**, 2145–52.

Pfister, C. (1980). The little ice age: thermal and wetness indices for central Europe. *J. Interdis. Hist.*, **10**, 665–96.

Pfister, C. (1995). Monthly temperature and precipitation in central Europe 1525–1979: quantifying documentary evidence on weather and its effects. In *Climate Since A.D. 1500*, ed. R. S. Bradley and P. D. Jones. London: Routledge, pp. 118–42.

Philip, J. R. (1957). The theory of infiltration: 4. Sorptivity and algebraic infiltration equations. *Soil Sci.*, **84**, 257–64.

Philipona, R., Fröhlich, C., Dehne, K. *et al.* (1998). The baseline surface radiation network pyrgeometer round-robin calibration experiment. *J. Atmos. Oceanic Technol.*, **15**, 687–96.

Pierrehumbert, R. T. (2002). The hydrologic cycle in deep-time climate problems. *Nature*, **419**, 191–8.

Pilgrim, D. H. and Cordery, I. (1993). Flood runoff. In *Handbook of Hydrology*, ed. D. R. Maidment. New York, NY: McGraw-Hill, Inc., pp. 9.1–9.42.

Plüss, C. and Ohmura, A. (1997). Longwave radiation on snow-covered mountainous surfaces. *J. Appl. Meteor.*, **36**, 818–24.

Pomeroy, J. W., Toth, B., Granger, R. J., Hedstrom, N. R. and Essery, R. L. H. (2003). Variation in surface energetics during snowmelt in a subarctic mountain catchment. *J. Hydrometeor.*, **4**, 702–19.

Pontrelli, M. D., Bryan, G. and Fritsch, J. M. (1999). The Madison County, Virginia, flash flood of 27 June 1995. *Wea. Forecast.*, **14**, 384–404.

Pozo-Vázquez, D., Esteban-Parra, M. J., Rodrigo, F. S. and Castro-Díez, Y. (2001). The association between ENSO and winter atmospheric circulation and temperature in the North Atlantic region. *J. Climate*, **14**, 3408–20.

Priestley, C. H. B. and Taylor, R. J. (1972). On the assessment of surface heat flux and evaporation using large scale parameters. *Mon. Wea. Rev.*, **100**, 81–92.

Pumijumnong, N. (1999). Climate-growth relationships of teak (*Tectona grandis* L.) from northern Thailand. In *Tree-Ring Analysis: Biological, Methodological and Environmental Aspects*, ed. R. Wimmer and R. E. Vetter. Wallingford, UK: CABI Publishing, pp. 155–68.

Pypker, T. G., Bond, B. J., Link, T. E., Marks, D. and Unsworth, M. H. (2005). The importance of canopy structure in controlling the interception loss of rainfall: examples from a young and an old-growth Douglas-fir forest. *Agric. Forest Meteor.*, **130**, 113–29.

Quesney, A., Le Hégarat-Mascle, S., Taconet, O. *et al.* (2000). Estimation of watershed soil moisture index from ERS/SAR data. *Remote Sens. Environ.*, **72**, 290–303.

Quinn, W. H. and Neal, V. T. (1992). The historical record of El Niño events. In *Climate Since A.D. 1500*, ed. R. S. Bradley and P. D. Jones. London: Routledge, pp. 623–48.

Raicich, F., Pinardi, N. and Navarra, A. (2003). Teleconnections between Indian monsoon and Sahel rainfall and the Mediterranean. *Int. J. Climatol.*, **23**, 173–86.

Rajagopalan, B., Cook, E., Lall, U. and Ray, B. K. (2000). Spatiotemporal variability of ENSO and SST teleconnections to summer drought over the United States during the twentieth century. *J. Climate*, **13**, 4244–55.

Randall, D., Krueger, S., Bretherton, C. *et al.* (2003). Confronting models with data: the GEWEX cloud systems study. *Bull. Amer. Meteor. Soc.*, **84**, 455–69.

Randel, D. L., Vonder Haar, T. H., Ringerud, M. A. *et al.* (1996). A new global water vapor dataset. *Bull. Amer. Meteor. Soc.*, **77**, 1233–46.

Rango, A. and Martinec, J. (1995). Revisiting the degree-day method for snowmelt computations. *Water Resour. Bull.*, **31**, 657–69.

Rantz, S. E. (1972). *Runoff Characteristics of California Streams*. U.S. Geological Survey Water-Supply Paper 2009-A. Washington, D.C.: U.S. Government Printing Office.

Raschke, E., Meywerk, J., Warrach, K. *et al.* (2001). The Baltic Sea Experiment (BALTEX): a European contribution to the investigation of the energy and water cycle over a large drainage basin. *Bull. Amer. Meteor. Soc.*, **82**, 2389–413.

Raschke, E., Vonder Haar, T. H., Bandeen, W. R. and Pasternak, M. (1973). The annual radiation balance of the earth-atmosphere system during 1969–70 from Nimbus 3 measurements. *J. Atmos. Sci.*, **30**, 341–64.

Rasmusson, E. M. (1987). Global prospects for the prediction of drought: a meteorological perspective. In *Planning for Drought: Toward a Reduction of Societal Vulnerability*, ed. D. A. Wilhite, W. E. Easterling and D. A. Wood. Boulder, CO: Westview Press, pp. 31–43.

Rawls, W. J., Ahuja, L. R., Brakensiek, D. L. and Shirmohammadi, A. (1993). Infiltration and soil water movement. In *Handbook of Hydrology*, ed. D. R. Maidment. New York, NY: McGraw-Hill, Inc., pp. 5.1–5.51.

Rebetez, M., Mayer, H., Dupont, O. *et al.* (2006). Heat and drought 2003 in Europe: a climate synthesis. *Ann. For. Sci.*, **63**, 569–77.

Regonda, S. K., Rajagopalan, B., Clark, M. and Pitlick, J. (2005). Seasonal cycle shifts in hydroclimatology over the western United States. *J. Climate*, **18**, 372–84.

Reichle, R. H., Koster, R. D., Dong, J. and Berg, A. A. (2004). Global soil moisture from satellite observations, land surface models, and ground data: implications for data assimilation. *J. Hydrometeor.*, **5**, 430–42.

Renwick, J. A. and Wallace, J. M. (1996). Relationships between North Pacific wintertime blocking, El Niño, and the PNA pattern. *Mon. Wea. Rev.*, **124**, 2071–6.

Richards, L. A. (1931). Capillary conduction of liquids through porous mediums. *Physics*, **1**, 318–33.

Richner, H., Joss, J. and Ruppert, P. (1996). A water hypsometer utilizing high-precision thermocouples. *J. Atmos. Oceanic Technol.*, **13**, 175–82.

Riebsame, W. E., Changnon, Jr., S. A. and Karl, T. R. (1991). *Drought and Natural Resources Management in the United States: Impacts and Implications of the 1987-89 Drought*. Boulder, CO: Westview Press.

Risbey, J. S. and Kandlikar, M. (2002). Expert assessment of uncertainties in detection and attribution of climate change. *Bull. Amer. Meteor. Soc.*, **83**, 1317–26.

Robertson, A. W. and Mechoso, C. R. (1998). Interannual and decadal cycles in river flows of southeastern South America. *J. Climate*, **11**, 2570–81.

Robinson, D. A. and Frei, A. (2000). Seasonal variability of Northern Hemisphere snow extent using visible satellite data. *Prof. Geogr.*, **52**, 307–15.

Robock, A. (2003). Hydrology: soil moisture. In *Encyclopedia of Atmospheric Sciences*, ed. J. R. Holton, J. A. Curry and J. A. Pyle. London: Academic Press, pp. 987–93.

Robock, A., Konstantin, Y. V., Srinivasan, G. *et al.* (2000). The global soil moisture data bank. *Bull. Amer. Meteor. Soc.*, **81**, 1281–99.

Robson, A. J., Jones, T. K., Reed, D. W. and Bayliss, A. C. (1998). A study of national trend and variation in UK floods. *Int. J. Climatol.*, **18**, 165–82.

Roebber, P. J., Bruening, S. L., Schultz, D. M. and Cortinas Jr., J. V. (2003). Improving snowfall forecasting by diagnosing snow density. *Wea. Forecast.*, **18**, 264–87.

Romanov, P., Gutman, G. and Csiszar, I. (2000). Automated monitoring of snow cover over North America with multispectral satellite data. *J. Appl. Meteor.*, **39**, 1866–80.

Romero, R., Doswell III, C. A. and Riosalido, R. (2001). Observations and fine-grid simulations of a convective outbreak in northeastern Spain: importance of diurnal forcing and convective cold pools. *Mon. Wea. Rev.*, **129**, 2157–82.

Ropelewski, C. F. (1988). The global climate for June-August 1988: a swing to the positive phase of the Southern Oscillation, drought in the United States, and abundant rain in monsoon areas. *J. Climate*, **1**, 1153–74.

Rose, C. (2004). *An Introduction to the Environmental Physics of Soil, Water and Watersheds*. Cambridge: Cambridge University Press.

Rosen, R. D. (1999). The global energy cycle. In *Global Energy and Water Cycles*, ed. K. A. Browning and R. J. Gurney. Cambridge: Cambridge University Press, pp. 1–9.

Rosenberg, N. J., Blad, B. L. and Verma, S. B. (1983). *Microclimate: The Biological Environment*. 2nd edn. New York, NY: John Wiley and Sons.

Ross, R. J. and Elliott, W. P. (2001). Radiosonde-based Northern Hemisphere tropospheric water vapor trends. *J. Climate*, **14**, 1602–12.

Rossow, W. B., and Schiffer, R. A. (1999). Advances in understanding clouds from ISCCP. *Bull. Amer. Meteor. Soc.*, **80**, 2261–87.

Rouse, W. R., Blyth, E. M., Crawford, R. W. *et al.* (2003). Energy and water cycles in a high-latitude, north-flowing river system: summary of results from the Mackenzie GEWEX study – phase 1. *Bull. Amer. Meteor. Soc.*, **84**, 73–87.

Rouse, W. R., Oswald, C. J., Binyamin, J. *et al.* (2005). The role of northern lakes in a regional energy balance. *J. Hydrometeor.*, **6**, 291–305.

Rowell, D. P. (2003). The impact of Mediterranean SSTs on the Sahelian rainfall season. *J. Climate*, **16**, 849–62.

Rowntree, L. B. (1985). A crop-based rainfall chronology for pre-instrumental record southern California. *Climatic Change*, **7**, 327–41.

Rudolf, B. and Rapp, J. (2003). *The Century Flood of the River Elbe in August 2002: Synoptic Weather Development and Climatological Aspects*. Quarterly Report of the Operational NWP- Models of the Deutscher Wetterdienst, Special Topic July, pp. 7–22.

Running, S. W. (2006). Is global warming causing more, larger wildfires? *Science*, **313**, 927–8.

Rybski, D., Bunde, A., Havlin, S. and von Storch, H. (2006). Long-term persistence in climate and the detection problem. *Geophys. Res. Lett.*, **33**, L06718, doi:10.1029/2005GL025591.

Salby, M. L. (1996). *Fundamentals of Atmospheric Physics*. San Diego, CA: Academic Press.

Saltzman, B. (2002). *Dynamical Paleoclimatology: Generalized Theory of Global Climate Change*. San Diego, CA: Academic Press.

Salvucci, G. D., Saleem, J. A. and Kaufmann, R. (2002). Investigating soil moisture feedbacks on precipitation with tests of Granger causality. *Adv. Water Resour.*, **25**, 1305–12.

Sankarasubramanian, A. and Vogel, R. M. (2002). Annual hydroclimatology of the United States. *Water Resour. Res.*, **38**, 1083, doi:10.1029/2001WR000619.

Sankarasubramanian, A. and Vogel, R. M. (2003). Hydroclimatology of the continental United States. *Geophys. Res. Lett.*, **30**, 1363, doi:10.1029/2002GL015937.

Sathiyamoorthy, V. (2005). Large scale reduction in the size of the Tropical Easterly Jet. *Geophys. Res. Lett.*, **32**, L14802, doi:10.1029/2005GL022956.

Schär, C., Vidale, P. L., Lüthi, D. *et al.* (2004). The role of increasing temperature variability in European summer heatwaves. *Nature*, **427**, 332–6.

Schlosser, C. A. and Houser, P. R. (2007). Assessing a satellite-era perspective of the global water cycle. *J. Climate*, **20**, 1316–38.

Schmetz, J., Pili, P., Tjemkes, S. *et al.* (2002). An introduction to Meteosat Second Generation (MSG). *Bull. Amer. Meteor. Soc.*, **83**, 977–92.

Schmit, T. J., Feltz, W. F., Menzel, W. P. *et al.* (2002). Validation and use of GOES sounder moisture information. *Wea. Forecast.*, **17**, 139–54.

Schmugge, T. J., Kustas, W. P., Ritchie, J. C., Jackson, T. J. and Rango, A. (2002). Remote sensing in hydrology. *Adv. Water Resour.*, **25**, 1367–85.

Schneider, J. M., Fisher, D. K., Elliott, R. L., Brown, G. O. and Bahrmann, C. P. (2003). Spatiotemporal variations in soil water: first results from the ARM SGP CART network. *J. Hydrometeor.*, **4**, 106–20.

Scofield, R. A. and Kuligowski, R. J. (2003). Status and outlook of operational satellite precipitation algorithms for extreme-precipitation events. *Wea. Forecast.*, **18**, 1037–51.

Scott, N. A., Chédin, A., Armante, R. *et al.* (1999). Characteristics of the TOVS Pathfinder Path-B dataset. *Bull. Amer. Meteor. Soc.*, **80**, 2679–701.

Seidel, D. J., Angell, J. K., Christy, J. *et al.* (2004). Uncertainty in signals of large-scale climate variations in radiosonde and satellite upper-air temperature datasets. *J. Climate*, **17**, 2225–40.

Seidel, K. and Martinec, J. (2004). *Remote Sensing in Snow Hydrology: Runoff Modelling, Effect of Climate Change*. Chichester: Springer-Praxis Publishing.

Shelton, M. L. (1978). Calibrations for computing Thornthwaite's potential evapotranspiration in California. *Prof. Geogr.*, **30**, 389–96

Shelton, M. L. (1984). Hydroclimatic analysis of severe drought in the Sacramento River Basin, California. *Phys. Geogr.*, **5**, 262–86.

Shelton, M. L. (1988). *Climate and Weather: A Spatial Perspective*. Dubuque, IA: Kendall/Hunt Publishing Company.

Shelton, M. L. (1989). Spatial scale influences on modeled runoff for large watersheds. *Phys. Geogr.*, **10**, 368–83.

Shelton, M. L. (1995). Unimpaired and regulated discharge in the Sacramento River Basin, California. *Yearbook Assoc. Pacific Coast Geogr.*, **57**, 134–57.

Shelton, M. L. (2001). Mesoscale atmospheric 2XCO2 climate change simulation applied to an Oregon watershed. *J. Amer. Water Resour. Assoc.*, **37**, 1041–52.

Sherman, L. K. (1932). Stream flow from rainfall by unit hydrograph method. *Eng. News-Record*, **108**, 501–5.

Shindell, D., Rind, D., Balachandran, N., Lean, J. and Lonergan, P. (1999). Solar cycle variability, ozone, and climate. *Science*, **284**, 305–8.

Shuttleworth, W. J. (1993). Evaporation. In *Handbook of Hydrology*, ed. D. R. Maidment. New York, NY: McGraw-Hill, Inc., pp. 4.1–4.53.

Shuttleworth, W. J. (1995). Soil-vegetation-atmosphere relations: process and prospect. In *The Role of Water and the Hydrologic Cycle in Global Change*, ed. II. R. Oliver and S. A. Oliver. Berlin: Springer-Verlag, pp. 135–43.

Siccardi, F., Boni, G., Ferraris, L. and Rudari, R. (2005). A hydrometeorological approach for probabilistic flood forecast. *J. Geophys. Res.*, **110**, D05101, doi:10.1029/2004JD005314.

Slingo, J. M., Rowell, D. P., Sperber, K. R. and Nortley, F. (1999). On the predictability of the interannual behaviour of the Madden-Julian Oscillation and its relationship with El Niño. *Quart. J. Roy. Meteor. Soc.*, **125**, 583–609.

Smith, J. A., Baeck, M. L., Morrison, J. E. and Sturdevant-Rees, P. (2000). Catastrophic rainfall and flooding in Texas. *J. Hydrometeor.*, **1**, 5–25.

Smith, J. A., Baeck, M. L., Steiner, M. and Miller, A. J. (1996). Catastrophic rainfall from an upslope thunderstorm in the central Appalachians: the Rapidan storm of June 27, 1995. *Water Resour. Res.*, **32**, 3099–113.

Smith, T. M., Yin, X. and Gruber, A. (2006). Variations in annual global precipitation (1979–2004), based on the Global Precipitation Climatology Project 2.5° analysis. *Geophys. Res. Lett.*, **33**, L06705, doi:10.1029/2005GL025393.

Song, J., Wesely, M. L., Coulter, R. L. and Brandes, E. A. (2000a). Estimating watershed evapotranspiration with PASS. Part I: Inferring root-zone moisture conditions using satellite data. *J. Hydrometeor.*, **1**, 447–61.

Song, J., Wesely, M. L., LeMone, M. A. and Grossman, R. L. (2000b). Estimating watershed evapotranspiration with PASS. Part II: Moisture budgets during drydown periods. *J. Hydrometeor.*, **1**, 462–73.

Spano, D., Snyder, R. L., Duce, P. and Paw U, K. T. (2000). Estimating sensible and latent heat flux densities from grapevine canopies using surface renewal. *Agric. Forest Meteor.*, **104**, 171–83.

Speidel, D. H. and Agnew A. F. (1988). The world water balance. In *Perspectives on Water: Uses and Abuses*, ed. D. H. Speidel, L. C. Ruedisili and A. F. Agnew. New York, NY: Oxford University Press, pp. 27–36.

Stahle, D. W. and Cleaveland, M. K. (1992). Reconstruction and analysis of spring rainfall over the southeastern U.S. for the past 1000 years. *Bull. Amer. Meteor. Soc.*, **73**, 1947–61.

Stedinger, J. R., Vogel, R. M. and Foufoula-Georgiou, E. (1993). Frequency analysis of extreme events. *In Handbook of Hydrology*, ed. D. R. Maidment. New York, NY: McGraw-Hill, Inc., pp. 18.1–18.66.

Stewart, I. T., Cayan, D. R. and Dettinger, M. D. (2005). Changes toward earlier streamflow timing across western North America. *J. Climate*, **18**, 1136–55.

Stewart, R. E., Leighton, H. G., Marsh, P. *et al.* (1998). The Mackenzie GEWEX Study: the water and energy cycles of a major North American river basin. *Bull. Amer. Meteor. Soc.*, **79**, 2665–83.

Stockton, C. W. and Meko, D. M. (1990). Some aspects of the hydroclimatology of arid and semiarid lands. In *Human Intervention in the Climatology of Arid Lands*, ed. D. R. Haragan. Albuquerque, NM: University of New Mexico Press, pp. 1–26.

Street-Perrott, F. A. (1995). Natural variability of tropical climates on 10- to 100-year time scales: limnological and paleolimnological evidence. In *Natural Climate Variability on Decade-To-Century Time Scales*, ed. D. G. Martinson, K. Bryan, M. Ghil *et al.* Washington, D.C.: National Academy Press, pp. 506–11.

Stricker, J. N. M., Kim, C. P., Feddes, R. A. *et al.* (1993). The terrestrial hydrological cycle. In *Energy and Water Cycles in the Climate System*, ed. E. Raschke and D. Jacob. NATIO ASI Series, I 5. Berlin: Springer-Verlag, pp. 419–44.

Sturm, M. (2003). Snow (surface). In *Encyclopedia of Atmospheric Sciences*, ed. J. R. Holton, J. A. Curry and J. A. Pyle. London: Academic Press, pp. 2061–72.

Sturman, A. P., Bradley, S., Drummond, P. *et al.* (2003). The Lake Tekapo Experiment (LTEX): an investigation of atmospheric boundary layer processes in complex terrain. *Bull. Amer. Meteor. Soc.*, **84**, 371–80.

Sujono, J., Shikasho, S. and Hiramatsu, K. (2004). A comparison of techniques for hydrograph recession analysis. *Hydrol. Process.*, **18**, 403–13.

Susskind, J., Barnet, C. D. and Blaisdell, J. M. (2003). Retrieval of atmospheric and surface parameters from AIRS/AMSU/HSB data in the presence of clouds. *IEEE Trans. Geosci. Remote Sens.*, **41**, 390–409.

Susskind, J., Piraino, P., Rokke, L., Iredell, L. and Mehta, A. (1997). Characteristics of the TOVS Pathfinder Path A dataset. *Bull. Amer. Meteor. Soc.*, **78**, 1449–72.

Suzuki, K. and Ohta, T. (2003). Effect of larch forest density on snow surface energy balance. *J. Hydrometeor.*, **4**, 1181–93.

Sveinsson, O. G. B., Salas, J. D., Boes, D. C. and Pielke Sr., R. A. (2003). Modeling the dynamics of long-term variability of hydroclimatic processes. *J. Hydrometeor.*, **4**, 489–505.

Svoboda, M., LeComte, D., Hayes, M. *et al.* (2002). The drought monitor. *Bull. Amer. Meteor. Soc.*, **83**, 1181–90.

Syvitski, J. P. M., Vörösmarty, C. J., Kettner, A. J. and Green, P. (2005). Impact of humans on the flux of terrestrial sediment to the global coastal ocean. *Science*, **308**, 376–80.

Szilagyi, J. and Parlange, M. B. (1999). Defining watershed-scale evaporation using a normalized difference vegetation index. *J. Amer. Water Resour. Assoc.*, **35**, 1245–55.

Tarhule, A. and Lamb, P. J. (2003). Climate research and seasonal forecasting for West Africans: perceptions, dissemination, and use? *Bull. Amer. Meteor. Soc.*, **84**, 1741–59.

Taylor, C. M., Lambin, E. F., Stephenne, N., Harding, R. J. and Essery, R. L. H. (2002). The influence of land use change on climate in the Sahel. *J. Climate*, **15**, 3615–29.

Tessier, L., Guibal, F. and Schweingruber, F. H. (1997). Research strategies in dendroecology and dendroclimatology in mountain environments. *Clim. Change*, **36**, 499–517.

Thiaw, W. M. and Mo, K. C. (2005). Impact of sea surface temperature and soil moisture on seasonal rainfall prediction over the Sahel. *J. Climate*, **18**, 5330–43.

Thorne, P. W., Parker, D. E., Christy, J. R. and Mears, C. A. (2005). Uncertainties in climate trends. *Bull. Amer. Meteor. Soc.*, **86**, 1437–42.

Thornthwaite, C. W. (1948). An approach toward a rational classification of climate. *Geogr. Rev.*, **38**, 55–94.

Thornthwaite, C. W. with H. G. Wilm and others (1945). Report of the committee on transpiration and evaporation, 1943–1944. *Trans. Amer. Geophys. Union*, **25**, 686–93.

Tootle, G. A. and Piechota, T. C. (2006). Relationships between Pacific and Atlantic Ocean sea surface temperatures and U.S. streamflow variability. *Water Resour. Res.*, **42**, W07411, doi:10.1029/2005WR004184.

Trenberth, K. E. (1990). Recent observed interdecadal climate changes in the northern hemisphere. *Bull. Amer. Meteor. Soc.*, **71**, 988–93.

Trenberth, K. E. (1998). Atmospheric moisture residence times and cycling: implications for rainfall rates and climate change. *Climatic Change*, **39**, 667–94.

Trenberth, K. E. and Branstator, G. W. (1992). Issues in establishing causes of the 1988 drought over North America. *J. Climate*, **5**, 159–72.

Trenberth, K. E. and Caron, J. M. (2001). Estimates of meridional atmosphere and ocean heat transports. *J. Climate*, **14**, 3433–43.

Trenberth, K. E. and Guillemot, C. J. (1996). Physical processes involved in the 1988 drought and 1993 floods in North America. *J. Climate*, **9**, 1288–98.

Trenberth, K. E. and Hurrell, J. W. (1994). Decadal atmosphere-ocean variations in the Pacific. *Climate Dyn.*, **9**, 303–19.

Trenberth, K. E. and Smith, L. (2005). The mass of the atmosphere: a constraint on global analyses. *J. Climate*, **18**, 864–75.

Trenberth, K. E. and Solomon, A. (1994). The global heat balance: heat transports in the atmosphere and ocean. *Climate Dyn.*, **10**, 107–34.

Trenberth, K. E., Branstator, G. W. and Arkin, P. A. (1988). Origins of the 1988 North American drought. *Science*, **242**, 1640–5.

Trenberth, K. E., Caron, J. M., Stepaniak, D. P. and Worley, S. (2002). Evolution of El Niño-Southern Oscillation and global atmospheric surface temperatures. *J. Geophys. Res.*, **107**, 4065, doi:10.1029/2000JD000298.

Trenberth, K. E., Dai, A., Rasmussen, R. M. and Parsons, D. B. (2003). The changing character of precipitation. *Bull. Amer. Meteor. Soc.*, **84**, 1205–17.

Trigo, R. M., Pereira, J. M. C., Pereira, M. G. *et al.* (2006). Atmospheric conditions associated with the exceptional fire season of 2003 in Portugal. *Int. J. Climatol.*, **26**, 1741–57.

Trigo, R. M., Pozo-Vázquez, D., Osborn, T. J. *et al.* (2004). North Atlantic oscillation influence on precipitation, river flow and water resources in the Iberian Penninsula. *Int. J. Climatol.*, **24**, 925–44.

Turc, L. (1961). Estimate of irrigation water requirements, potential evapotranspiration: a simple climatic formula evolved up to date. *Ann. Agron.*, **12**, 13–49.

Twine, T. E., Kucharik, C. J. and Foley, J. A. (2005). Effects of El Niño-Southern Oscillation on the climate, water balance, and streamflow of the Mississippi River basin. *J. Climate*, **18**, 4840–61.

Ulbrich, U., Brücher, T., Fink, A. H. *et al.* (2003a). The central European floods of August 2002: Part 1 – Rainfall periods and flood development. *Weather*, **58**, 371–77.

Ulbrich, U., Brücher, T., Fink, A. H. *et al.* (2003b). The central European floods of August 2002: Part 2 – Synoptic causes and considerations with respect to climatic change. *Weather*, **58**, 434–42.

Uppenbrink, J. (1999). The North Atlantic Oscillation. *Science*, **283**, 948–9.

Urbonas, B. R. and Roesner, L. A. (1993). Hydrologic design for urban drainage and flood control. In *Handbook of Hydrology*, ed. D. R. Maidment. New York, NY: McGraw-Hill, Inc., pp. 28.1–28.52.

U.S. Army Corps of Engineers (1956). *Snow Hydrology: Summary Report of the Snow Investigations*. Portland, OR: North Pacific Division.

U.S. Soil Conservation Service (1964). Hydrology. In *National Engineering Handbook*. Section 4. Washington, D.C.: U.S. Soil Conservation Service.

U.S. Soil Conservation Service (1970). *Irrigation Water Requirements*. Engineering Division, Tech. Release No. 21. Washington, D.C.: U.S. Government Printing Office.

U.S. Soil Conservation Service (1986). *Urban Hydrology for Small Watersheds*. Tech. Release 55. Washington, D.C.: U.S. Department of Agriculture.

Verdon, D. C. and Franks, S. W. (2006). Long-term behaviour of ENSO: interactions with the PDO over the past 400 years inferred from paleoclimate records. *Geophys. Res. Lett.*, **33**, L06712, doi:10.1029/2005GL025052.

Vicente, G. A., Scofield, R. A. and Menzel, W. P. (1998). The operational GOES infrared rainfall estimation technique. *Bull. Amer. Meteor. Soc.*, **79**, 1883–98.

Vikulina, Z. A., Gronskaya, T. P., Kashinova, T. D. and Natrus, A. A. (1978). Water balance of lakes and reservoirs. In *World Water Balance and Water Resources of the Earth*, ed. V. I. Korzun. Paris: UNESCO, pp. 533–44.

Villalba, R., Grau, H. R., Boninsegna, J. A., Jacoby, G. C. and Ripalta, A. (1998). Tree-ring evidence for long-term precipitation changes in subtropical South America. *Int. J. Climatol.*, **18**, 1463–78.

Vincent, R. K. (2003). Radar: synthetic aperture radar (land surface applications). In *Encyclopedia of Atmospheric Sciences*, ed. J.R. Holton, J. A. Curry and J. A. Pyle. London: Academic Press, pp. 1851–8.

Vinnikov, K. Y., Robock, A., Qiu, S. *et al.* (1999). Satellite remote sensing of soil moisture in Illinois, United States. *J. Geophys. Res.*, **104**, 4145–68.

Wahl, K. L., Vining, K. C. and Wiche, G. J. (1993). *Precipitation in the Upper Mississippi River Basin, January 1 Through July 31, 1993*. U.S. Geological Survey Circular 1120-B. Washington, D.C.: U.S. Government Printing Office.

Waliser, D. E., Lau, K. M. and Kim, J.-H. (1999). The influence of coupled sea surface temperatures on the Madden-Julian Oscillation: a model perturbation experiment. *J. Atmos. Sci.*, **56**, 333–58.

Walker, G. T. (1923). World weather I. *Mem. Indian Meteor. Depart.*, **24**, 75–131.

Walker, W. R., Hrezo, M. S. and Haley, C. J. (1991). Management of water resources for drought conditions. In *National Water Summary 1988–89 – Hydrologic Events and Floods and Droughts*. U.S. Geological Survey Water-Supply Paper 2375. Washington, D.C.: U.S. Government Printing Office, pp. 147–56.

Wallace, J. M. and Gutzler, D. S. (1981). Teleconnections in the geopotential height field during the Northern Hemisphere winter. *Mon. Wea. Rev.*, **109**, 784–812.

Wang, G. and Schimel, D. (2003). Climate change, climate modes, and climate impacts. *Ann. Rev. Environ. Resour.*, **28**, 1–28.

Ward, A. D. (1995). Surface runoff and subsurface drainage. In *Environmental Hydrology*, ed. A. D. Ward and W. J. Elliot. New York, NY: Lewis Publishers, pp. 133–73.

Ward, A. D. and Elliot, W. J., ed. (1995). *Environmental Hydrology*. New York, NY: Lewis Publishers.

Ward, R. C. and Robinson, M. (2000). *Principles of Hydrology*. 4th edn. London: McGraw-Hill.

Weaver, J. F., Gruntfest, E. and Levy, G. M. (2000). Two floods in Fort Collins, Colorado: learning from a natural disaster. *Bull. Amer. Meteor. Soc.*, **81**, 2359–66.

Wells, N., Goddard, S. and Hayes, M. J. (2004). A self-calibrating Palmer Drought Severity Index. *J. Climate*, **17**, 2335–51.

Westerling, A. L., Cayan, D. R., Brown, T. J., Hall, B.I. and Riddle, L. G. (2004). Climate, Santa Ana winds and autumn wildfires in southern California. *Eos, Trans. Amer. Geophys. Un.*, **85**, 289–300.

Westerling, A. L., Gershunov, A., Brown, T. J., Cayan, D. R. and Dettinger, M. D. (2003). Climate and wildfire in the western United States. *Bull. Amer. Meteor. Soc.*, **84**, 595–604.

Westerling, A. L., Hidalgo, H. G., Cayan, D. R. and Swetnam, T. W. (2006). Warming and earlier spring increase western U.S. forest wildfire activity. *Science*, **313**, 940–3.

Wetzel, M., Meyers, M., Borys, R. *et al.* (2004). Mesoscale snowfall prediction and verification in mountainous terrain. *Wea. Forecast.*, **19**, 806–28.

Wilheit, T., Kummerow, C. D. and Ferraro, R. (2003). Rainfall algorithms for AMSR-E. *IEEE Trans. Geosci. Remote Sens.*, **41**, 204–14.

Wilhite, D. A. and Glantz, M. H. (1987). Understanding the drought phenomenon: the role of definitions. In *Planning for Drought: Toward a Reduction of Societal Vulnerability*, ed. D. A. Wilhite, W. E. Easterling and D. A. Wood. Boulder, CO: Westview Press, pp. 11–27.

Willmott, C. J. (1996). Evaporation. In *Encyclopedia of Climate and Weather*, ed. S. H. Schneider. New York, NY: Oxford University Press, Vol. 1, pp. 303–5.

Wilson, K. B., Baldocchi, D. D., Aubinet, M. *et al.* (2002) Energy partitioning between latent and sensible heat flux during the warm season at FLUXNET sites. *Water Resour. Res.*, **38**, 1294, doi:10.1029/2001WR000989.

Wilson, R. J. S., Luckman, B. H. and Esper, J. (2005). A 500 year dendroclimatic reconstruction of spring-summer precipitation from the lower Bavarian Forest region, Germany. *Int. J. Climatol.*, **25**, 611–30.

Winter, T. C. (2000). The vulnerability of wetlands to climate change: a hydrologic landscape perspective. *J. Amer. Water Resour. Assoc.*, **36**, 305–11.

Winter, T. C. and LaBaugh, J. W. (2003). Hydrologic considerations in defining isolated wetlands. *Wetlands*, **23**, 532–40.

Wohl, E. E. (2000). Inland flood hazards. In *Inland Flood Hazards: Human, Riparian, and Aquatic Communities*, ed. E. E. Wohl. New York, NY: Cambridge University Press, pp. 3–36.

Wolter, K. and Timlin, M. S. (1998). Measuring the strength of ENSO-how does 1997/98 rank? *Weather*, **53**, 315–24.

Wood, E. F. (1995). Heterogeneity and scaling land-atmospheric water and energy fluxes in climate systems. In *Space and Time Scale Variability and Interdependencies in Hydrological Processes*, ed. R. A. Feddes. Cambridge: Cambridge University Press, pp. 3–19.

Woodhouse, C. A. and Overpeck, J. T. (1998). 2000 years of drought variability in the central United States. *Bull. Amer. Meteor. Soc.*, **79**, 2693–714.

Woodhouse, C. A., Gray, S. T. and Meko, D. M. (2006). Updated streamflow reconstructions for the Upper Colorado River basin. *Water Resour. Res.*, **42**, W05415, doi:10.1029/2005WR004455.

Woodhouse, C. A., Lukas, J. J. and Brown, P. M. (2002). Drought in the western Great Plains, 1845–56: impacts and implications. *Bull. Amer. Meteor. Soc.*, **83**, 1485–93.

Woodside, C. A. (2001). A tree-ring reconstruction of streamflow for the Colorado Front Range. *J. Amer. Water Resour. Assoc.*, **37**, 561–9.

World Meteorological Organization (WMO) (1986a). *Intercomparison of Models of Snowmelt Runoff.* Operational Hydrology Report 23. WMO-No. 646. Geneva: World Meteorological Organization.

World Meteorological Organization (WMO) (1986b). *Manual for Estimation of Probable Maximum Precipitation.* Operational Hydrology Report 1. WMO-No. 332, 2nd edn. Geneva: World Meteorological Organization.

World Meteorological Organization (WMO) (1996). *Guide to Meteorological Instruments and Methods of Observation.* WMO-No. 8, 6th edn. Geneva: World Meteorological Organization.

Wulfmeyer, V. and Henning-Müller, I. (2006). The climate station of the University of Hohenheim: analyses of air temperature and precipitation time series since 1878. *Int. J. Climatol.*, **26**, 113–38.

Xie, P. and Arkin, P. A. (1997). Global precipitation: a 17-year monthly analysis based on gauge observations, satellite estimates, and numerical model outputs. *Bull. Amer. Meteor. Soc.*, **78**, 2539–58.

Xie, P. and Arkin, P. A. (1998). Global monthly precipitation estimates from satellite-observed outgoing longwave radiation. *J. Climate*, **11**, 137–63.

Xie, P., Janowiak, J. E., Arkin, P. A. *et al.* (2003). GPCP pentad precipitation analyses: an experimental dataset based on gauge observations and satellite estimates. *J. Climate*, **16**, 2197–214.

Yamaguchi, Y. and Shinoda, M. (2002). Soil moisture modeling based on multiyear observations in the Sahel. *J. Appl. Meteor.*, **41**, 1140–6.

Yang, F., Kumar, A., Schlesinger, M. E. and Wang, W. (2003). Intensity of hydrological cycles in warmer climates. *J. Climate*, **16**, 2419–23.

Yuter, S. E. (2003). Radar: precipitation radar. In *Encyclopedia of Atmospheric Sciences*, ed. J. R. Holton, J. A. Curry and J. A. Pyle. London: Academic Press, pp. 1833–51.

Zaitchik, B. F., Macalady, A. K., Bonneau, L. R. and Smith, R. B. (2006). Europe's 2003 heat wave: a satellite view of impacts and land-atmosphere feedbacks. *Int. J. Climatol.*, **26**, 743–69.

Zeng, N. (2003). Drought in the Sahel. *Science*, **302**, 999–1000.

Zhang, C. and Dong, M. (2004). Seasonality in the Madden-Julian oscillation. *J. Climate*, **17**, 3169–80.

Zhang, Y. J., Wallace, M. and Battisti, D. S. (1997). ENSO-like interdecadal variability: 1900–93. *J. Climate*, **10**, 1004–20.

Zierl, B. and Lischke, H. (2002). Trends in drought in Swiss forested ecosystems. In *Climatic Change: Implications for the Hydrological Cycle and for Water Management*, ed. M. Beniston. Dordrecht: Kluwer Academic Publishers, pp. 329–47.

Index

Printed in the United States
By Bookmasters